METHODS IN VIROLOGY

VOLUME VIII

METHODS IN VIROLOGY

Advisory Board

METHODS IN VIROLOGY

EDITED BY

KARL MARAMOROSCH

WAKSMAN INSTITUTE OF MICROBIOLOGY
RUTGERS UNIVERSITY
NEW BRUNSWICK, NEW JERSEY

AND

HILARY KOPROWSKI

THE WISTAR INSTITUTE OF ANATOMY AND BIOLOGY
PHILADELPHIA, PENNSYLVANIA

Volume VIII

1984

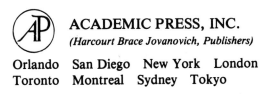

ACADEMIC PRESS, INC.
(Harcourt Brace Jovanovich, Publishers)

Orlando   San Diego   New York   London
Toronto   Montreal   Sydney   Tokyo

ACADEMIC PRESS, INC.
Orlando, Florida 32887

United Kingdom Edition published by
ACADEMIC PRESS, INC. (LONDON) LTD.
24/28 Oval Road, London NW1 7DX

LIBRARY OF CONGRESS CATALOG CARD NUMBER: 66-30091

ISBN 0-12-470208-2

PRINTED IN THE UNITED STATES OF AMERICA

84 85 86 87   9 8 7 6 5 4 3 2 1

Contents

Chapter 1. **Hybridization of Viral Nucleic Acids: Newer Methods on Solid Media and in Solution**
NANCY RAAB-TRAUB AND JOSEPH S. PAGANO

Chapter 2. **Applications of Oligonucleotide Fingerprinting to the Identification of Viruses**
OLEN M. KEW, BALDEV K. NOTTAY, AND JOHN F. OBIJESKI

Chapter 3. **Immunosorbent Electron Microscopy in Plant Virus Studies**
R. G. MILNE AND D.-E. LESEMANN

Chapter 4. Quantitative Transmission Electron Microscopy for the Determination of Mass-Molecular Weight of Viruses
H. M. MAZZONE, W. F. ENGLER, AND G. F. BAHR

Chapter 5. Use of Thin Sectioning for Visualization and Identification of Plant Viruses
GIOVANNI P. MARTELLI AND MARCELLO RUSSO

Chapter 6. Purification and Immunological Analyses of Plant Viral Inclusion Bodies
ERNEST HIEBERT, DAN E. PURCIFULL, AND RICHARD G. CHRISTIE

Chapter 7. Use of Mosquitoes to Detect and Propagate Viruses
LEON ROSEN

Chapter 8. **Prions: Methods for Assay, Purification, and Characterization**
STANLEY B. PRUSINER, MICHAEL P. MCKINLEY, DAVID C. BOLTON,
KAREN A. BOWMAN, DARLENE F. GROTH, S. PATRICIA COCHRAN,
ELIZABETH M. HENNESSEY, MICHAEL B. BRAUNFELD,
J. RICHARD BARINGER, AND MARK A. CHATIGNY

Chapter 9. **Detection of Genome-Linked Proteins of Plant and Animal Viruses**
STEPHEN D. DAUBERT AND GEORGE BRUENING

Contributors

Numbers in parentheses indicate the pages on which the authors' contributions begin.

G. F. BAHR, Department of Cellular Pathology, Armed Forces Institute of Pathology, Washington, DC 20306 (103)

J. RICHARD BARINGER, Department of Neurology, University of Utah School of Medicine, Salt Lake City, Utah 84132 (293)

DAVID C. BOLTON, Department of Neurology, University of California, San Francisco, San Francisco, California 94143 (293)

KAREN A. BOWMAN, Department of Neurology, University of California, San Francisco, San Francisco, California 94143 (293)

MICHAEL B. BRAUNFELD, Department of Neurology, University of California, San Francisco, San Francisco, California 94143 (293)

GEORGE BRUENING, Department of Biochemistry and Biophysics, University of California, Davis, Davis, California 95616 (347)

MARK A. CHATIGNY, School of Public Health, University of California, Berkeley, Naval Biosciences Laboratory, Oakland, California 94625 (293)

RICHARD G. CHRISTIE, Department of Agronomy, University of Florida, Gainesville, Florida 32611 (225)

S. PATRICIA COCHRAN, Department of Neurology, University of California, San Francisco, San Francisco, California 94143 (293)

STEPHEN D. DAUBERT, Department of Plant Pathology, University of California, Davis, Davis, California 95616 (347)

W. F. ENGLER, Department of Cellular Pathology, Armed Forces Institute of Pathology, Washington, DC 20306 (103)

DARLENE F. GROTH, Department of Neurology, University of California, San Francisco, San Francisco, California 94143 (293)

ELIZABETH M. HENNESSEY, Department of Neurology, University of California, San Francisco, San Francisco, California 94143 (293)

ERNEST HIEBERT, Department of Plant Pathology, University of Florida, Gainesville, Florida 32611 (225)

OLEN M. KEW, Division of Viral Diseases, Center for Infectious Diseases, Centers for Disease Control, Atlanta, Georgia 30333 (41)

D.-E. LESEMANN, Institut für Viruskrankheiten der Pflanzen, Biologische Bundesanstalt für Land- und Forstwirtschaft, 3300 Braunschweig, Federal Republic of Germany (85)

GIOVANNI P. MARTELLI, Dipartimento di Patologia Vegetale, University of Bari, and Centro di Studio del CNR sui Virus e le Virosi delle Colture Mediterranee, I-70126 Bari, Italy (143)

H. M. MAZZONE, United States Department of Agriculture, Forest Service, Hamden, Connecticut 06514 (103)

MICHAEL P. MCKINLEY, Departments of Neurology and of Anatomy, University of California, San Francisco, San Francisco, California 94143 (293)

R. G. MILNE, Istituto di Fitovirologia Applicata del CNR, 10135 Turin, Italy (85)

BALDEV K. NOTTAY, Division of Viral Diseases, Center for Infectious Diseases, Centers for Disease Control, Atlanta, Georgia 30333 (41)

JOHN F. OBIJESKI, Vaccine Development, Genentech Inc., South San Francisco, California 94080 (41)

JOSEPH S. PAGANO, Departments of Medicine and of Microbiology and Immunology, and Lineberger Cancer Research Center, University of North Carolina, Chapel Hill, North Carolina 27514 (1)

STANLEY B. PRUSINER, Departments of Neurology and of Biochemistry and Biophysics, University of California, San Francisco, San Francisco, California 94143 (293)

DAN E. PURCIFULL, Department of Plant Pathology, University of Florida, Gainesville, Florida 32611 (225)

NANCY RAAB-TRAUB, Department of Microbiology and Immunology, and Lineberger Cancer Research Center, University of North Carolina, Chapel Hill, North Carolina 27514 (1)

LEON ROSEN, Arbovirus Program, Pacific Biomedical Research Center, University of Hawaii at Manoa, Leahi Hospital, Honolulu, Hawaii 96816 (281)

MARCELLO RUSSO, Dipartimento di Patologia Vegetale, University of Bari, and Centro di Studio del CNR sui Virus e le Virosi delle Colture Mediterranee, I-70126 Bari, Italy (143)

Preface

In a letter to Dr. John L. Dorsey dated May 23, 1804, Benjamin Rush said that "A wide field opens for medical investigations in the United States. The walls of the Old School are daily falling about the ears of its masters and scholars. Come, and assist your Uncle and his friends in erecting a new fabric upon its ruins." In the seventeen years since the publication of the first volume of *Methods in Virology,* many "walls of the Old School" have fallen down to be replaced by "new fabric." The editors, being cognizant of these events, have tried in these two volumes to present the reader with up-to-date, modern revisions of techniques applied to animal, plant, and insect virology.

The early volumes in this series were published at the dawn of the era of molecular virology. Today, techniques applied to the study of molecular virology are becoming standard household techniques, and the contents of Volumes VII and VIII reflect the existence of the "new world" of virology. A series of books in existence for seventeen years may be regarded as a "venerable" one. We feel, however, that none of the preceding volumes can be classified as obsolete. New techniques and methods must be considered as improvements of previously described techniques, but not necessarily as their replacements. Moreover, one never knows what surprises the future holds for students of virology.

Volumes VII and VIII will be of considerable usefulness to all who are engaged in virus research, including graduate students interested in becoming familiar with modern techniques. Infectious disease specialists, bacteriologists, immunologists, vertebrate, invertebrate, and plant pathologists, parasitologists, biochemists, veterinarians, geneticists, and biotechnicians will find these volumes of interest. These two books are an important addition not only to the series but to the rapidly growing list of works dealing with viruses as well.

We express our thanks and appreciation to those who have contributed chapters to Volumes VII and VIII. The authors were chosen on the basis of their outstanding knowledge of given methods, as recognized authorities in their specialized fields, or as creators of new techniques. We also wish to express our appreciation to the staff of Academic Press for continuous encouragement and advice throughout the planning and completion of these volumes.

KARL MARAMOROSCH
HILARY KOPROWSKI

Contents of Previous Volumes

METHODS IN VIROLOGY

VOLUME VIII

1 Hybridization of Viral Nucleic Acids: Newer Methods on Solid Media and in Solution

Nancy Raab-Traub and
Joseph S. Pagano

I. Introduction

The technology of nucleic acid hybridization was the forerunner of recombinant DNA biotechnology, which in turn has spread the use of hy-

1

bridization techniques into many areas of molecular biology. Now developments in recombinant DNA research have had an impact on hybridization technology, primarily by providing pure, well-defined probes of reduced complexity. Although the basic techniques of molecular hybridization have changed little since the previous review of the methodology in Chap. 13, Vol. VI of this series (Huang and Pagano, 1977), the emphasis on different techniques has changed. Also, the application of the techniques to pathobiology has widened with the increasing appreciation of the contributions that molecular biological technology can make toward understanding mechanisms of disease on the cellular, molecular, and epidemiological level.

The concept of nucleic acid hybridization is based on the homology between the two hydrogen-bonded strands of nucleic acid either in double-stranded DNA or in hybrid molecules that consist of strands of RNA and DNA. In all cases a single-stranded nucleic acid probe that has been labeled in some fashion is allowed to form a hybrid with homologous sequences. The double-stranded labeled material is then either quantitated, visualized, or further analyzed. The principal uses of nucleic acid hybridization techniques are for detection of specific genomes or portions of them, determination of homology between genomes, characterization of DNA structure, and detection and analysis of RNA transcripts.

The purposes of this chapter are to provide in one place a compendium of the most useful techniques of molecular hybridization in a practical rather than a theoretical form. We present here a brief recapitulation of some of the basic techniques that were presented in detail in Vol. VI of this series with some examples of areas in which they have continued utility, and then we expand with greater detail on methods largely based on restriction enzyme technology. As most scientific questions can be approached in several ways, the choice of a particular method is largely governed by the availability of material in a particular system. However, the emergence of recombinant DNA and restriction enzyme technology dictates the choice of methods that rely on abundant DNA and provide an analysis of DNA fine structure in accurate detail.

In this chapter, we use Epstein–Barr virus (EBV) systems to provide examples and illustrations. EBV is a herpesvirus containing double-stranded DNA with a molecular weight of approximately 115×10^6, or 170,000 base pairs. The genome of the virus has a complex structure with several regions of repetitive DNA interspersed with unique DNA. Much of the current knowledge of the biochemistry of EBV and its identification as a pathologic agent in several malignant and benign disease states have been dependent on hybridization technology. However, the same techniques as applied to detection and characterization of EBV genomes and transcripts can be utilized in most other viral and nonviral systems.

II. Purification of Nucleic Acid

A. PURIFICATION OF VIRAL DNA

In those types of hybridization analyses in which a test DNA is analyzed for the presence of viral sequences, purity of the nucleic acid used as a radiolabeled probe is essential for specific hybridization. However, the availability of specific viral sequences in recombinant DNA, which is propagated as a plasmid in bacteria, has largely circumvented purification difficulties. When nucleic acid purified from virus must be used for radiolabeling, a rigorous purification scheme is called for. After purification of virions, the nucleic acid extracted from them can be separated from contaminating DNA on the basis of physical properties of the viral DNA such as sedimentation coefficient, buoyant density, or physical configuration.

For example, in the EBV system virus is harvested from the extracellular fluid from producer lymphocyte lines. The supernatant fluids are clarified by centrifugation at 4000 g for 15 min. The virus can then be pelleted from the clarified fluids by centrifugation at 15,000 g for 90 min followed by banding in a 10–30% dextran sulfate gradient. The light-scattering viral band is aspirated from the gradient with a syringe fitted with an 18-gauge needle, and this virus is centrifuged through 0.5 mM sodium phosphate, pH 6.5. The pellet is treated with 2% sodium dodecyl sulfate (SDS) for 5 min at 60°C followed by phenol/chloroform extraction. The nucleic acid is then purified to equilibrium through a cesium chloride gradient, and the fractions corresponding to the buoyant density of EBV, 1.718 g/cm^3, are pooled, dialyzed, and concentrated (Dolyniuk et al., 1976).

An alternative method, essentially that of Adams and Lindahl (1975), is to pellet the virus from the clarified supernatant fluid by treatment with polyethylene glycol (PEG). To each liter of supernatant fluid 20 g of sodium chloride is added and stirred until dissolved, followed by 200 ml of PEG solution [polyethylene glycol 6000, 50% (w/v) in 0.5 M NaCl]. This suspension is placed at 4°C overnight. The PEG complexes are centrifuged at 8000 g for 15 min and then suspended in 2% RPMI. The virus is pelleted at 25,000 rpm for 90 min in an SW27 rotor. The virus pellet is suspended in phosphate-buffered saline solution (PBS) with 5 mM MgCl$_2$ and incubated with pancreatic DNase, 50 μg/ml, for 2 hr at room temperature. The virus suspension is then purified on a 10–50% sucrose gradient; the viral band is collected and repelleted. The virions are lysed with Sarkosyl, treated with Pronase, made into a CsCl solution with a density of 1.718 g/cm³, and centrifuged to equilibrium. The fractions with the appropriate density are pooled and dialyzed. EBV DNA of high molecular weight can be obtained by velocity sedimentation centrifugation in a 5–20% sucrose gradient.

B. Purification of DNA from Cultured Cells

In selected instances the nucleic acid to be analyzed is obtained from cells in culture. The preparation of DNA from such samples is relatively straightforward. The cells are washed in cold PBS and lysed with 1% Sarkosyl or SDS in the presence of 10 mM EDTA, and 100 μg/ml proteinase K. The suspension of lysed cells is incubated at 50°C for 2–3 hr with occasional rolling. If the sample is to be analyzed by digestion with restriction endonucleases, any treatment that may shear the DNA should be avoided. The nucleic acid is then extracted several times with equal volumes of Tris-equilibrated phenol and chloroform containing 2% isoamyl alcohol and then dialyzed into 50 mM Tris (pH 8.0), 1 mM EDTA, 10 mM NaCl. The sample can then be treated with DNase-free RNase (100 μg/ml) for 3 hr at 37°C, extracted with phenol/chloroform, and dialyzed into 100 mM Tris, 1 mM EDTA.

If further purification or enrichment for viral sequences is necessary and the virus has not integrated into cellular DNA, the DNA can be centrifuged to equilibrium in CsCl. The tube is punctured from the bottom with a large heated metal rod approximately 1/8 in. in diameter, and fractions are slowly collected. The fractions with the density of the viral DNA are pooled and dialyzed.

The use of supernatant fluids from high-salt precipitation of nucleic acid is an effective way to enrich viral DNA (Hirt, 1967). Cells are lysed in a solution of 10 mM Tris (pH 8.0), 10 mM EDTA, 0.6% SDS, for 15 min at room temperature. NaCl is added to a concentration of 1 M and kept at 4°C. Under these conditions high-molecular-weight cellular DNA readily precipitates whereas small viral DNAs will remain in solution overnight. The larger herpesvirus DNAs should be collected at 4–10 hr (Pater et al., 1976). In modification of this type of enrichment, the cells are lysed with 0.25% Triton X-100 in 10 mM Tris (pH 7.9), 10 mM EDTA for 10 min at room temperature. NaCl is added to 0.2 M, and the lysate is gently inverted several times and immediately centrifuged at 1000 g for 10 min at 4°C. This method works well for purifying viral DNA from productively infected cells late in infection but is not useful for isolating EBV episomal DNA located in the nucleus from nonproducer cells. The nuclear membrane remains intact under these conditions (Pignatti *et al.,* 1979).

C. PURIFICATION OF CYTOPLASMIC RNA

To obtain an RNA fraction consisting of ribosomal RNAs and mRNAs that have been processed and transported to the ribosomes, the cells are separated into nuclear and cytoplasmic fractions. The cellular membrane can be lysed with a nonionic detergent such as 0.25% Triton X-100 or 0.5% Nonidet P-40 (NP-40) and kept on ice for 10 min. The cell suspension should be slightly vortexed or homogenized by 10–20 strokes with a tight-fitting Dounce homogenizer. An RNase inhibitor such as RNAsin, 100 units/ml (Scheel and Blackburn, 1979), or 10 mM vanadyl ribonucleoside complexes (Berger and Birkenmeier, 1979) should be included. All glassware used in any procedure involving RNA should be autoclaved or sterilized by dry heat. The nuclei are pelleted at 5000 rpm for 5 min in an SS34 rotor. The cytoplasmic supernatant fluid can be treated in one of several ways to deproteinize the nucleic acid–protein complexes. The complexes can be disrupted with 2% SDS or Sarkosyl and then extracted with phenol/ chloroform–2% isoamyl alcohol and precipitated with ethanol. If 10 mM vanadyl ribonucleoside has been used as the RNase inhibitor, the phenol should contain 1% 8-hydroxyquinoline to extract the vanadyl complexes efficiently. All phenol extractions should be a mixture of equal volumes of phenol and chloroform–2% isoamyl alcohol; without chloroform, RNA with long tracts of polyadenylic acid will be extracted into the organic phase. Alternatively, the cytoplasmic supernatant fluid can be made into a 4 M guanidine hydrochloride or guanidine thiocyanate solution (Chirgwin *et al.,* 1979), CsCl is added to 0.5 g/ml, and this is layered over a 1.57 M CsCl cushion. The RNA is purified from proteins, membrane debris, and contaminating DNA and forms a clear gelatinous pellet after centrifugation at 44,000 g for 18 hr (Seeburg *et al.,* 1977).

D. PURIFICATION OF NUCLEIC ACID FROM TISSUE SPECIMENS

To obtain a reasonable yield of intact DNA and RNA from solid tissues, the tissue should be quickly frozen in dry ice and maintained at −70°C without thawing. The tissue can then be homogenized with a blender or a Tissumizer (Tekmar, Inc., Cincinnati, OH) in a strong denaturing solution, such as 4 M guanidine thiocyanate. Frozen tissue can be pulverized with a mortar and pestle filled with liquid nitrogen or with an instrument such as the Braun Mikrodismembrator (Braun, B., Melsungen, Federal Republic of Germany). The pulverized tissue is solubilized with 4 M guanidine thiocyanate. This solution can then be treated as described in Section II,C and separated into DNA and RNA fractions by pelleting the RNA through a CsCl cushion. The RNA pellets through the cushion, and the DNA bands at the interface of the cushion with the diluted CsCl solution.

A recently described method rapidly dissolves the pulverized frozen tissue in a mixture of equal volumes of 0.3 M sodium acetate (pH 7.5), 0.5% SDS, 5 mM EDTA, and phenol equilibrated overnight with 0.3 M sodium acetate, 1% hydroxyquinoline (Krieg *et al.*, 1983). The mixture is gently extracted for an additional 5 min. An equal volume of chloroform is added, and the mixture is extracted for 5 min. The aqueous and organic phases are separated by centrifugation at 4000 rpm for 5 min, and the interphase and aqueous phase are extracted with an equal volume of chloroform–2% isoamyl alcohol. The total nucleic acids are precipitated with two volumes of 95% ethanol. The nucleic acid is then dissolved and the RNA selectively precipitated with an equal volume of 4 M lithium chloride at 4°C for at least 4 hr. The precipitate is pelleted in an SS34 rotor at 10,000 rpm for 20 min. The DNA can then be precipitated with two volumes of ethanol at −20°C for at least 1 hr.

An advantage of either of these two methods is that both RNA and DNA can be obtained from a single tissue sample.

E. Purification of Polyadenylated RNA

RNA can be separated into polyadenylated and nonpolyadenylated fractions by chromatography on oligodeoxythymidine [oligo(dT)] cellulose (Aviv and Leder, 1972). This material, which is commercially obtained, can bind 100 A_{260} units of polyadenylated RNA per gram of cellulose with retention of 99% of the mRNA. The cellulose should be suspended and poured in elution buffer consisting of 10 mM Tris (pH 7.5), 0.1% SDS, 1 mM EDTA and then washed with 5–10 bed volumes of binding buffer to equilibrate the cellulose. Binding buffer consists of 10 mM Tris (pH 7.5), 0.5 M NaCl, 1 mM EDTA, 0.1% SDS. The precipitated RNA obtained by any of the previously described methods should be dissolved in elution buffer. Direct addition of a NaCl stock solution to bring the RNA solution to the binding salt concentration may result in precipitation of the RNA; therefore an equal volume of 10 mM Tris (pH 7.5), 1 M NaCl should be added while swirling the solution.

The RNA solution should be passed through the column two to three times to assure complete binding. After loading the sample, the column should be washed with 10 bed volumes of loading buffer followed by 2 volumes of 10 mM Tris (pH 7.15), 0.1 M NaCl, SDS 0.1%, 1 mM EDTA. The polyadenylated RNA can then be eluted with 2–3 volumes of sterile elution buffer. After the addition of the appropriate salt, the RNA can be precipitated with 2.5 volumes of ethanol. RNA should be stored at $-70°C$ in 70% ethanol.

III. Preparation of Labeled Nucleic Acid Probes

A. DNA LABELED in Vivo

In general, nucleic acid labeled in vivo is rarely used as a viral probe in hybridization studies. Although the label is uniformly incorporated into DNA, the specific activity is not sufficient for most hybridization assays.

Efficient incorporation of a label into viral nucleic acid is dependent on the ability to infect cells uniformly and productively in a small amount of medium containing the radioactive label. In the EBV system the most efficient system is based on the ability of the EBV (HR1) strain to superinfect the latently EBV-infected Raji cell line. With a high multiplicity of infection, at 10 hr postinfection ^{32}P-labeled organic phosphate is selectively incorporated into viral DNA. All of the restriction enzyme fragments of the input viral genome EBV (HR1) are labeled approximately equally (Shaw et al., 1977). However, the EBV (HR1) strain contains a series of defective molecules that contain certain EBV sequences in an unusual linkage. Thus, when DNA labeled in this fashion is hybridized to Southern blots of DNA of other EBV strains that do not contain these defective molecules, the intensity of hybridization to restriction enzyme fragments is variable, not equimolar. In other words, some sequences of EBV are overrepresented in the DNA synthesized in HR1 superinfected Raji cells.

B. PREPARATION OF cRNA

In those instances in which the amount of viral nucleic acid that can be obtained and labeled directly in vitro is limited, the use of Escherichia coli DNA-dependent RNA polymerase to synthesize in vitro a labeled cRNA copy of a DNA template has the advantage that an amplified amount of probe is synthesized with a limited amount of template. However, the specific activity of such probes can only be estimated, as the actual amount of cRNA synthesized is unknown. The procedure for the preparation of EBV cRNA is that of Nonoyama and Pagano (1971) with modifications by Carolyn Smith. The reaction mixture contains 0.04 M Tris (pH 7.9), 0.01 M $MgCl_2$, 0.1 mM dithiothreitol, 0.15 M KCl, 0.5 mg/ml bovine serum albumin, 0.15 mM GTP, ATP, and CTP, 500 μCi of [^{3}H]UTP (35–50 Ci/mmol), 20 units of enzyme (380 units/mg protein), and 2 μg of high-molecular-weight EBV DNA in 1.0 ml. The mixture is incubated at 37°C. Two-microliter aliquots are removed from the reaction mixture and precipitated with trichloroacetic acid at various intervals to monitor cRNA synthesis. When incorporation plateaus (2–5 hr), the DNA is digested with DNase (RNase-free, Sigma Chemical Co., St. Louis, MO) at a concentration of 50 μg/ml for 3 min at 37°C. The reaction is stopped by the addition of Sarkosyl to 1%, and the labeled material is separated from unincorporated [^{3}H]UTP by chromatography through Sephadex G-50 (Pharmacia). The excluded peak is extracted with phenol and chloroform and precipitated with ethanol. Generally 0.5–1 × 10^8 cpm are synthesized.

C. Direct Labeling of DNA and RNA with Iodine

The utility of ^{125}I as a radioactive tag of nucleic acid is largely based on the ability to label RNA and single-stranded DNA directly to high specific activity. ^{125}I has an emission energy severalfold higher than that of ^{3}H, yet considerably lower than that of ^{14}C, ^{35}S, or ^{32}P. This intermediate emission energy is particularly useful for *in situ* hybridization, as the hybridized probe has an intense signal that is not obliterated by a high background. In other types of hybridization, the 60-day half-life may be advantageous compared with the short half-life of 14.3 days for ^{32}P.

Although it is still frequently used to label protein components, ^{125}I is highly volatile and difficult to work with; therefore the use of ^{125}I has declined. The availability of efficient methods for direct labeling or synthesis of a labeled copy of both DNA and RNA enzymatically has eliminated the need to iodinate nucleic acid directly. Although several researchers have synthesized ^{125}I-labeled nucleotides for use in enzymatic procedures, the high emission energy of ^{32}P makes it the isotope of choice for most hybridizations to Southern, Northern, or dot blots. The development of biotin-labeled nucleotides and the fluorescein-conjugated avidin system has revolutionized *in situ* hybridization and eliminated the advantages of ^{125}I (Singer and Ward, 1982).

The details of direct iodination methods for DNA, RNA, and nucleotides were covered in some detail in Vol. VI of this series (Huang and Pagano, 1977). In general, the technique of Commerford (1971) has been modified to label RNA, denatured DNA, and nucleotides. In EBV systems, directly iodinated RNA has been used to map regions of transcription (Rymo, 1979) or indirectly label DNA (Rymo, 1979; Shaw *et al.*, 1975) with iodinated nucleotides.

D. Direct Labeling of DNA *in Vitro*

Most virus-specific probes are labeled enzymatically *in vitro* with the use of *E. coli* polymerase I (Rigby *et al.*, 1977). This enzyme has three separate enzymatic activities: a 5′→3′ polymerase activity that adds nucleotide residues to a 3′-hydroxyl group, a 5′→3′ exonuclease that removes nucleotides from a free 5′-end, and a 3′→5′ exonuclease that degrades DNA from a free 3'-hydroxyl end. When one strand of duplex DNA is nicked, DNA polymerase I will processively remove nucleotides 5'→3' and will add nucleotides to the 3'-hydroxyl group utilizing the single-stranded template provided by the exonuclease. The 3'→5' exonuclease activity is blocked by the polymerase activity. The result is movement of the nick 5'→3' on the DNA (nick translation) and replacement of the nucleotides with labeled nucleotides. Since the nicks are randomly located, labeling is generally uniform.

All types of double-stranded DNA can be labeled efficiently with a high percentage of incorporation of labeled nucleotides, the resultant specific activity of the DNA probe reflecting the concentration of labeled nucleotides in the reaction mixture. Perhaps the most useful aspect of this type of reaction is that the percentage of incorporation is independent of the type of label used in the reaction. Therefore, one can label DNA with ^{3}H, ^{35}S, ^{32}P, ^{125}I, and biotin with equal efficiency.

The size of the single-stranded pieces of DNA after denaturation is dependent on the number of nicks introduced into the double-stranded DNA with DNase I. Ideally the DNA should be about 400 bases long. Commercial preparations of DNase I are prepared as a stock solution of 1 mg/ml in 0.01 N HCl and stored at $-20°$C in small aliquots. The aliquots are diluted in 10 mM Tris (pH 7.5), 5 mM MgCl$_2$, 1 mg/ml bovine serum albumin and stored at 4°C for 2 hr before use. Immediately before use the DNase is diluted to attain the desired concentration in the final reaction mixture. The specific activity of the resultant probe is dependent on this concentration, which usually ranges between 1 and 10 pg/μl of the reaction mixture. This concentration should be determined for the type of DNA that is to be labeled, such as viral, cloned, or intracellular DNA.

The reaction mixture usually includes 1 μg of DNA template in 50–120 μl of solution, buffered with 50 mM Tris (pH 7.2–7.5) or 50 mM potassium phosphate (pH 7.6) and contains 5–10 mM MgCl$_2$. The DNA can be treated with the predetermined amount of DNase I for 5 min at 37°C followed by inactivation of the enzyme for 10 min at 80°C. Alternatively, a lower concentration of DNase can be maintained in the reaction mixture. The final reaction mixture contains 10–20 μM cold deoxynucleotides, 1 μg bovine serum albumin, 1 mM dithiothreitol, and 120 μCi labeled deoxynucleotide, or approximately 200 pmol. When high specific activity is required all four deoxynucleotides can be labeled. The reaction is held on ice before addition of 5–15 units of the polymerase and then incubated at 14°C for 2 hr. Synthesis at higher temperatures results in selective loss of the 5'→3'-exonuclease activity. When this occurs the DNA polymerase may reverse and copy newly synthesized DNA. This aberrant reaction produces "snap-back" DNA, which rapidly renatures, is resistant to digestion by Sl nuclease, and is useless for hybridization.

E. Synthesis of Labeled cDNA

Different RNAs are copied into DNA with varying efficiencies such that the conditions for synthesis of long stretches of cDNA must be determined for each RNA. However, a general set of conditions has been established that works well in providing the greatest yield of cDNA from a total RNA population. The RNA may be representative of that obtained from the nucleus or cytoplasm, nonpolyadenylated or polyadenylated (Taylor et al., 1976).

Total RNA populations can be copied by priming the synthesis with random deoxynucleotides generated by digesting calf thymus DNA for 4 hr at 37°C with 70 μg/ml DNase I. The DNA is then treated with 0.5 M NaOH. High concentrations of this DNA are included during cDNA synthesis. The sequences represented in polyadenylated RNA are best identified by synthesizing a cDNA of the polyadenylated RNA selected by oligodeoxythymidine chromatography and priming the synthesis with oligodeoxythymidine.

Excellent incorporation can be obtained with a 100-μl reaction mixture containing 50 mM Tris (pH 8.3), 10 mM MgCl$_2$, 20 nmol of each unlabeled deoxynucleotide triphosphate, 200 pmol of labeled deoxynucleotide triphosphate, 30 mM β-mercaptoethanol, and 20 units of avian myeloblastosis virus reverse transcriptase. The reaction mixture should contain 25 μg of random deoxynucleotide primer or 0.1 A_{260} unit of oligodeoxythymidine.

After incubation for 3 hr at 37°C the reaction can be stopped by adding SDS. Unincorporated nucleotides are separated by chromatography through Sephadex G-50.

IV. Basic Methods of Nucleic Acid Hybridization

Once the probes have been procured, hybridization techniques consist of different methods of obtaining and exposing the test nucleic acid to annealing with the probe. The test DNA may be extracted from cells and fixed to membrane filters after denaturation; it may be solubilized and exposed to the probe in liquid media; or the DNA may be left unextracted and tested in cells *in situ*. We present here some of the features of the basic techniques that were presented in detail in Vol. VI of this series.

A. Hybridization to DNA on Nitrocellulose Filters

This classic method of hybridization anneals denatured DNA immobilized on nitrocellulose filters with radioactive virus-specific nucleic acid. It was first used in herpesvirus systems by zur Hausen and Schulte-Holthausen (1970), who used an EBV DNA probe radiolabeled in cell culture with

[^{3}H]thymidine. The specificity and sensitivity of hybridization, although limited by the low specific activity and partial purity of the probe, were sufficient to permit detection of EBV DNA sequences in a non-virus-producing cell line suspected of harboring EBV genomes.

Alternatively the nitrocellulose filters can be hybridized to a labeled cRNA probe (Nonoyama and Pagano, 1971). The advantage of a cRNA probe in this type of analysis is that unhybridized cRNA can readily be eliminated by RNase digestion followed by quantitation of undigested or hybridized cRNA. The concentration of DNA remaining affixed to the filter is determined by diphenylamine tests after hybridization.

This method continues to be the most convenient for estimations of number of genome copies in EBV systems. The calculation of viral genome equivalents is based on reconstruction curves established by hybridization of the cRNA probe to known amounts of purified viral DNA. However, such calculations are based on the assumption that the test DNA contains the entire genome. This methodology cannot distinguish between hybridization due to multiple copies of defective genomes or to the entire genome or detect the preferential hybridization of reiterated sequences. Therefore in test situations in which any of these variables is possible, filter hybridization cannot provide a reliable copy number. However, cRNA:DNA hybridization is a simple and reliable method for determining relative genome numbers within a standard test situation. For example in the EBV system the entire genome is present in relatively constant copy number in latently infected nonproducer cell lines. In contrast, in virus-producer cell lines the copy number is much higher and quite variable and represents the average of the small percentage of cells producing virus and the majority of latently infected cells. Defective molecules are not generally produced, and all strains are basically identical in sequence complexity. Therefore, one can rapidly assess, for example, the effect of an antiviral drug on viral replication by determining a relative copy number before and after treatment.

Duplicate samples of total cellular DNA (75 μg of DNA based on A$_{260}$) are denatured by heating at 80°C for 10 min in the presence of 0.25 M NaOH containing 0.5 mM neutralized EDTA. Several concentrations of Raji DNA (60 EBV genome copies/cell) are used to standardize the results; calf thymus DNA in addition to blank filters is used to establish nonspecific background hybridization. The DNA samples are immediately chilled in ice to prevent renaturation. Neutralization buffer [1 M Tris (pH 7.4):H$_2$0:concentrated HCl, 60:140:3] is used to neutralize the samples to pH 7.5, and 20× SSC is added to a final concentration of 6× SSC. The DNA samples are then immobilized by slowly filtering the samples through nitrocellulose filters (24 mm, 0.45 μm; (Schleicher & Schuell, Inc., Keene, NH), which are rinsed with 6× SSC before and after the DNA solutions is filtered. After air drying, the samples are heated at 80°C for 4 hr under vacuum.

Hybridization is carried out with batches of filters in a solution containing 6× SSC, 0.1% Sarkosyl, 100 μg yeast RNA/filter, and 1 × 10⁵ cpm cRNA/filter (1 ml of solution/filter) in a 66°C water bath for 22–24 hr with gentle shaking. At the end of hybridization, the filters are washed four times with 2× SSC at room temperature, once at 56°C for 30 min, quickly washed twice with 56°C 2× SSC, and allowed to cool. The filters are then treated with 40 μg RNase/ml per filter at 37°C for 1 hr to remove unhybridized cRNA; cRNA that is bound to the DNA on the filters will not be digested. After RNase treatment, the filters are washed four additional times with 2× SSC. The filters are dried and counted. After counting, the filters are rinsed with toluene to remove scintillators, and the amount of DNA on the filters is determined by the diphenylamine reaction (Burton, 1956).

The counts per minute per microgram of DNA are calculated for each filter after correcting for background. The DNA counts per microgram obtained for the calf thymus DNA filters are subtracted from the other values to account for nonspecific sticking of the cRNA. Based on the Raji standard of 60 EBV genome copies per cell, the genome copies per cell can then be calculated for each of the other DNA samples.

The results of such an experiment are presented in Table I, in which the relative number of EBV genomes in the producer HR1 cell line are assessed after treatment with several nucleoside analogs (Lin et al., 1983).

B. DOT BLOTS

Another method of hybridization to DNA bound to nitrocellulose filters was developed by Kafatos et al. (1979) and involves the binding of multiple samples of identical amounts of cloned DNA to uniformly sized spots of nitrocellulose. The filter is hybridized to a radioactive DNA mixture to test for the presence of sequences homologous to the cloned DNA. By comparison with dilutions of single-stranded radioactive DNA, the relative concentrations of homologous sequences in an unknown test DNA can be determined. Although spot or dot hybridization is not as accurate as reassociation kinetic analyses, the technique is useful for analyzing DNA or RNA populations for sequence homology from a large number of samples.

The conditions established by Kafatos for quantitative binding of DNA and preparation of the dot blots are particularly useful for screening gradient fractions for viral DNA in one simple hybridization procedure. A Plexiglas apparatus is commercially available for preparing dot blots. The device consists of multiple tapered wells that apply the solution to the filter in a circle 1 mm in diameter and have an air inlet such that a very weak vacuum can be used to draw the solution slowly through the filter. All the samples can be applied to the wells and simultaneously pulled through the filters.

TABLE I

Effects of Nucleoside Analogs on Epstein–Barr Virus Genome
Replication in P³HR-1 Cells[a]

Drug[b]	Molarity (μM)	cpm $\times 10^3$ hybridized per 50 μg of DNA[c]	EBV genome copies per cell
FMAU	5	3.6	69
BVCU	10	4.4	85
FIAC	5	6.7	129
ACV	100	3.9	75
None	—	43.3	838

[a] Data from Lin et al. (1983).

[b] FMAU, 1-(2-deoxy-2-fluoro-β-D-arabinofuranosyl)-5-methyluracil; BVDU, (E)-5-(2-bromovinyl)-2'-deoxyuridine; FIAC, 1-(2-deoxy-2-fluoro-β-D-arabinofurano-syl)-5-iodocytosine; ACV, acyclovir.

[c] Average hybridization values of duplicate DNA filters each standardized to 50 μg of DNA per filter. Counts per minute bound by calf thymus DNA (200 cpm) was subtracted as background. Counts per minute by 50 μg of Raji DNA was 3126 cpm and was taken as 60 genome equivalents per cell.

In Fig. 1 the purification of crude EBV obtained by pelleting virus from the supernatant fluids of an EBV producer cell line through a 5–35% Nycodenz (Accurate Chemical Co, Westbury, NJ) gradient has been analyzed by several criteria (Fowler et al., 1984). The triangular symbols with dotted line represent proteins extrinsically labeled with ^{125}I in the crude virus preparation. The squares with solid line represent [³H]thymidine incorporation in cell culture. The triangles with solid line at the bottom of the curve represent the percentage of positive cells in an immunofluorescence assay that tests for binding of virus to the EBV-specific receptor with human EBV antiserum (Simmons et al., 1983). Each fraction was analyzed for EBV DNA by dot hybridization, which is shown at the bottom of Fig. 1. The fractions that contain EBV DNA correspond with the [³H]thymidine peak; empty EBV capsids that bind to the EBV receptor but do not contain EBV DNA are located in fractions 27–35. Thus the viral DNA content across a gradient can be rapidly analyzed in one hybridization. The dot blots are prepared by digesting an aliquot of each fraction with proteinase K (2.5 mg/ml). The sample is made 0.5 M NaOH to denature any DNA and is neutralized by adding an equal volume of 2 M ammonium acetate. The nitrocellulose sheets are soaked with distilled H_2O and washed with 1 M ammonium acetate and then placed in the dot blot apparatus (BRL, Bethesda, MD). Each well is rinsed with 1 M ammonium acetate, and the fraction aliquots in amounts of 100–200 μl are slowly drawn through the filter. The wells are rinsed with 1 M ammonium acetate, and the filter is removed, washed in 4× SSC (1× = 0.15 M NaCl, 0.015 M sodium citrate), and dried under vacuum at 80°C

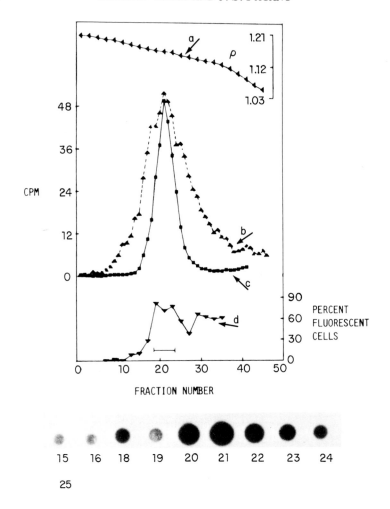

FIG. 1. Dot hybridization for EBV DNA across a Nycodenz gradient. Upper panel: Crude EBV (0.5 ml) was layered on a 7.5-ml 15–35% Nycodenz gradient and centrifuged for 2 hr at 102,000 *g* in a Ty65 rotor. Fractions (0.2 ml) were collected, and the TCA-precipitable counts were measured by scintillation counting. Aliquots were taken for dot hybridization and assayed for the ability to bind to Raji cells. (a) Densities calculated from the refractive indices; (b) EBV extrinsically labeled with ^{125}I; (c) EBV intrinsically labeled with [^{3}H]thymidine; (d) EBV detected in immunofluorescent binding assay. Bar indicates dot blot-positive fractions. Lower panel: Dot hybridization of gradient fractions. DNA from a 10-µl aliquot of each fraction was applied to a nitrocellulose sheet and hybridized to a ^{32}P-labeled *Bam*HI V probe. The bound radioactivity was detected by autoradiography.

for 2 hr. Each filter contains salmon sperm DNA as a negative control and known amounts of plasmid DNA containing an EBV restriction enzyme fragment, *Bam*HI V. The filters are pretreated for 6 hr with 6× SSC, 10× Denhardt's reagents (Denhardt, 1966) [1× = 2% Ficoll, 0.2% poly-(vinylpyrrolidone), 2% bovine serum albumin], 2% SDS, 100 μg/ml polyadenylic acid, and 100 μg/ml single-stranded calf thymus DNA. The filters are hybridized overnight to 10^7 cpm nick-translated *Bam*HI V EBV DNA at 72°C in this same solution. The filter is washed in decreasing concentrations of SSC from 2× to 0.1× with 0.2% SDS at 72°C, dried, and exposed to X-Omat AR film.

Dot blots are useful also for rapid screening of numerous samples of cells for viral content (Brandsma and Miller, 1980). Replicate samples of 10^5–10^6 cells in 10 μl are spotted onto nitrocellulose, where they are lysed and the DNA is simultaneously denatured by soaking the filter in 0.5 M NaOH. The procedure for lysing the cells and fixing the DNA to the filter is that developed by Grunstein and Hogness (1975) for screening bacterial clones for DNA sequences. The nitrocellulose filter is transferred through a series of shallow dishes with Whatman 3MM chromatography paper moistened with several solutions. The filter is consecutively treated with 0.5 M NaOH for 7 min, twice in 0.6 M NaCl, 1 M Tris (pH 6.8) for 1 min, and once in 1.5 M NaCl, 0.5 M Tris (pH 7.4) for 5 min. The filter is then dried on a Büchner funnel, floated on 95% ethanol, air dried, and baked at 80°C under vacuum for 3 hr. The genome content is estimated by comparison with a known standard, such as the Raji cell line for EBV or a reconstruction with a mixture of viral DNA (purified or recombinant) and uninfected cells.

C. Renaturation Kinetics

Renaturation kinetic analyses are based on the ability of single-stranded pieces of DNA to reassociate with homologous DNA. If such a reaction is allowed to take place under constant conditions, the rate of the formation of double-stranded DNA is a function of the concentration of the reactants such that rate = constant[reactant]^n. The constant must be determined experimentally. If n were equal to zero the velocity would be independent of the concentration of the reactant. If n were equal to 1, it would be a first-order reaction. However, renaturation is dependent on the concentrations of both strands such that A + B = C + D, therefore $n = 2$, and the reaction exhibits second-order kinetics.

The reaction is described by the equation

$$D_t/D_0 = 1/(1 + kD_0t) \tag{1}$$

D_t equals the amount of DNA remaining single stranded in moles of DNA per liter at time t, in seconds. The constant k is dependent on temperature of hybridization, salt concentration, base composition, size of DNA fragments, and viscosity. It is expressed as liter sec^{-1} mol^{-1} DNA. It is necessary to convert the amount of DNA from moles of DNA to moles of nucleotides. C_0 in moles of nucleotide per liter equals D_0 (moles of DNA per liter) times M (grams of DNA per mole of DNA) divided by 320 g of DNA per mole of nucleotide. M equals the molecular weight of the unique sequences in the DNA. This is the sequence complexity of the DNA. DNA containing a single copy of every sequence per molecule has a sequence complexity equal to the molecular weight.

Solving the equation for D_0

$$D_0 = C_0(320/M) \tag{2}$$

This value for D_0 can be substituted into the reaction equation, Eq. (1), and a new constant can be defined: $K = k(320/M)$. This reaction rate constant is inversely proportional to the complexity of the DNA.

There are two ways to plot this reaction. The C_0t plot is the percentage of single-stranded nucleic acid at time t plotted versus the C_0t on a log scale. The DNAs that are more complex plot farther to the right on this type of plot. The Wetmur–Davidson plot converts the equation into $D_0/D_t = 1 + KC_0t$ (Wetmur and Davidson, 1968). D_0/D_t is calculated as $1/(1 - \%\text{hybridized})$. For a second-order reaction this plot is a straight line that intercepts the ordinate at 1 with a slope equal to K.

To test DNA for the presence of viral sequences, one compares the rate of hybridization of a small amount of labeled single-stranded viral DNA with itself in the presence of control DNA to the rate of hybridization of the same concentration of viral DNA in the presence of test DNA. The concentration of viral sequences in the test DNA can be determined by calculating C_0 for the test DNA or by comparing the differences in slope to a reconstruction hybridization. The ratio of slopes is roughly equivalent to the ratio of genome equivalents.

There are two relatively simple and generally acceptable methods for determining the amount of hybridization that has occurred. The first method employs the differential binding capacity of hydroxyapatite (HAP) for single-stranded DNA compared to double-stranded DNA. The second method employs the greater substrate affinity of S1 nuclease for single-stranded rather than double-stranded DNA. Hydroxyapatite tends to give an overestimate of the amount of double-stranded DNA because molecules that are partially double stranded with long single-stranded tails will be retained when the single-stranded DNA is eluted.

Hydroxyapatite has high binding capacity for DNA, approximately 1.2 g of DNA per gram of HAP, and can be regenerated for reuse. One disadvantage is that the columns are ideally run at 60°C; this necessitates the use of water-jacketed columns in conjunction with a circulating water bath. A typical use of HAP in an experimental situation will illustrate the utility of this procedure. Henry *et al.* (1979) characterized the sequence arrangements of the genomes of several defective interfering virus populations of equine herpesvirus type 1 (EHV-1) generated by high-multiplicity passage of the virus compared with the genome of standard wild-type EHV-1 (see Fig. 2). The reaction mixtures consisted of 0.1 μg of ^{3}H-labeled standard virus DNA probe (8–10 $\times$ 10 cpm/μg) and 0.9 μg of unlabeled test DNA, sonicated to provide pieces of approximately 6 S, in 100 μl of 0.06 M NaPO$_4$, pH 6.8. Equivalent amounts of calf thymus DNA and unlabeled standard viral DNA were used as negative and positive controls, respectively. Aliquots of the hybridization solution were sealed in 100-μl capillary pipets,

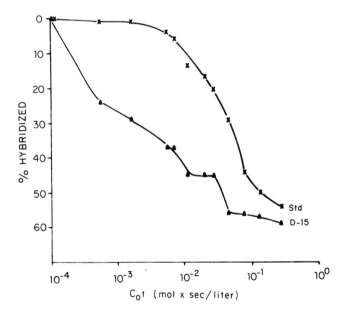

FIG. 2. Comparison of the reassociation kinetics of standard EHV-1 DNA and DNA isolated from particles of defective passage D-15. Hybridization mixtures consisted of ^{3}H-labeled standard DNA reannealed to unlabeled standard DNA and ^{3}H-labeled D-15 DNA reannealed to unlabeled D-15 DNA; all DNAs were isolated from purified virions. Reaction aliquots (100 μl) were sealed in capillary pipets, heated to 100°C for 15 min to denature the DNA mixtures, and allowed to reanneal at 63°C. Samples were taken at various times and stored at -70°C. Quantitation of single-stranded and double-stranded DNA was accomplished by hydroxyapatite chromatography. ($\times$) ^{3}H-labeled EHV-1 DNA reannealed to standard EHV-1 DNA; ($\blacktriangle$)^{3}H-labeled D-15 DNA reannealed to unlabeled D-15 DNA. Reprinted from Henry *et al.* (1979), *Virology* **92**.

denatured at 110°C for 15 min in an ethylene glycol bath, and allowed to reanneal at 63°C (25° below the T_m under the conditions employed). Samples were taken at appropriate times, quickly chilled in ice, and frozen until assayed.

Reassociation of labeled viral DNA was monitored by HAP (BioGel, DNA grade, Bio-Rad Laboratories, Richmond, CA) chromatography at 60°C. Water-jacketed columns (1 × 10 cm) containing 0.5 g of HAP were used. The DNA was bound to the column in the presence of 0.06 M NaPO$_4$ (pH 6.8), followed by several volumes of 0.14 M NaPO$_4$, which will elute the single-stranded DNA while allowing the double-stranded DNA to remain attached to the matrix. NaPO$_4$ (0.4 M) is then passed through the column to elute the remaining double-stranded DNA molecules. Subsequently, the DNA was precipitated with trichloroacetic acid (TCA), collected on nitrocellulose filters, and counted by liquid scintillation spectroscopy. A total of 3% of single-stranded DNA adsorbed to the column in 0.14 M NaPO$_4$ buffer; the fraction of single-stranded and double-stranded DNA observed during reassociation was corrected accordingly. The results obtained by this procedure were calculated by the relative amounts of the two DNA species present at a given time and plotted as shown in Fig. 2. The multiphase C_0t curve for this defective DNA indicates that there are different families of sequences that are individually less complex than viral DNA and are repeated a different number of times. A Wetmur–Davidson plot would not generate a straight line for this reaction.

The second standard procedure for the quantitation of single-stranded vs double-stranded DNA involves the use of S1 nuclease. This enzyme exhibits a differential specificity for single-stranded DNA over double-stranded DNA and functions specifically by splitting the phosphodiester band of denatured DNA. S1 nuclease digestion eliminates single-stranded DNA so that it is necessary to retain an untreated aliquot(s) from each hybridization sample in order to provide a reference point for subsequent calculations. This is normally accomplished by diluting the sample obtained from each time point into a given volume of digestion buffer (see below); this in turn is divided into three equal aliquots. Two of the aliquots are treated with S1 nuclease, and the remaining portion of the sample is left untreated in order to provide the total number of counts per time point.

Each lot of S1 nuclease must be titered against known quantities of single-stranded and double-stranded DNA so that the desired amount of enzyme can be determined and to provide sufficient S1 nuclease to result in complete degradation of the denatured DNA in each sample. At low substrate concentrations (single-stranded DNA) of less than 5 μg/ml, S1 exhibits altered kinetics with a substantial portion of the single-stranded DNA remaining undigested, so that exogenous unlabeled single-stranded DNA must be added to the reaction. Excess unlabeled double-stranded DNA must be included in the digestion mixture due to a small but measurable activity of S1 toward double-stranded DNA. S1 nuclease displays optimal activity at a salt concentration of 0.2–0.3 M, so that the digestion buffer must have a salt concentration within this range. In addition, this relatively high salt concentration significantly reduces the double-strand activity of the enzyme. If EDTA has been included in the hybridization buffer, then the zinc ion concentration must be of sufficient quantity to overcome the chelating effect of EDTA.

We have utilized DNA:DNA hybridization coupled with S1 nuclease digestion to determine the resident number of Epstein–Barr virus (EBV) genomes present in EBV-transformed Raji cell clones (B. Henry, II, unpublished results). The hybridization solution contained ^{32}P-labeled EBV DNA incubated with total cell DNA from each clone at a concentration of 200 μg/ml in 6× SSC. The hybridization solution was aliquoted into micropipets, sealed, and incubated at 70°C. Samples were frozen at each time point. Each time point was diluted into 1.3 ml of buffer containing 30 mM sodium acetate (pH 4.5), 0.2 M NaCl, 2 mM ZnCl$_2$, 25 μg/ml single-stranded DNA, and 50 μg/ml native DNA. The time-point dilutions were mixed with a vortex instrument and divided into three 0.4-ml aliquots. The results of these hybridization experiments are shown in Fig. 3. The hybridization with DNA from each clone generated lines with different slopes indicating a different copy number. The results were compared with several reconstruction hybridizations using purified EBV DNA. The hybridization reaction with 100 ng of purified EBV DNA in 200 μg of control DNA had a slope equal to 0.0061. The ratio of the slopes for the test DNAs to that for the known copy number gives the relative copy number in each clone.

D. *In Situ* HYBRIDIZATION

One great advantage of *in situ* hybridization is the ability to localize virus-specific sequences to a particular cell type. It is also possible to detect viral nucleic acid in a rare cell when biochemical extraction of the nucleic acid from the total sample would have diluted viral sequences beyond detection. The availability of cloned fragments for providing virus-specific probes of reduced complexity and the development of new ways of labeling DNA have greatly increased the sensitivity of these techniques since the work reported in Vol. VI of this series (Huang and Pagano, 1977). Although the methods for fixation and the details of hybridization are presented in Vol. VII, Chap. 7 (Haase, Brahic, and Stowring), we wish to present some comparisons of *in situ* hybridization with different labeled probes and show their utility for analyzing cell samples.

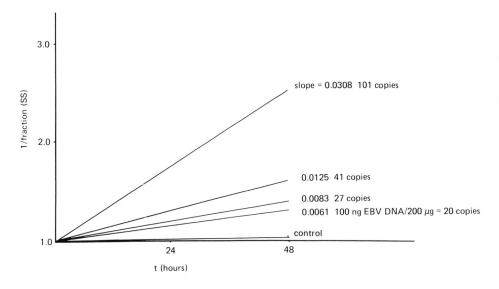

FIG. 3. Hybridization of EBV DNA to intracellular DNA prepared from cellular clones of the Raji cell line. Hybridization mixtures consisted of ^{32}P-labeled EBV DNA and 200 μg test DNA. A reconstruction hybridization consisting of 100 ng purified EBV DNA with 200 μg calf thymus DNA is equivalent to 20 copies of EBV DNA per cell. The number of copies of EBV in each of the clones was obtained by determining the ratio of the slope from each hybridization to the slope of the reconstruction hybridization.

With [^{3}H]cRNA synthesized from the total viral genome, it was formerly not possible to detect reliably EBV DNA *in situ* in the nonproducer Raji cell line, which contains 60 genome equivalents per cell, but only in the few percent of cells that produce virus in the HRl cell line. In Fig. 4 the EBV cloned fragment, *Bam*HI V, which contains the internally repeated sequence present approximately 10 times per EBV genome, is labeled and used as a probe. The cytohybridization to an EBV-negative cell line, BJAB, (Fig. 4a and d), Raji cells (Fig. 4b and e), and HR1 cells (Fig. 4c and f) are presented. Figure 4a–c are cytohybridizations with a tritiated cRNA probe synthesized from *Bam*HI V, and Fig. 4d–f use biotinylated nick-translated *Bam*HI V. With both probes, hybridization to the producer cells of HR1 is easily detected. However, the biotinylated probe can also detect viral sequences in the Raji cells and nonproducing cells of HR1. Hybridization is visualized as small punctate regions of fluorescence. No hybridization is detected in BJAB cells, and there is no background in the biotinylated preparation.

This technique has been applied to identify cells that are producing virus in patients with infectious mononucleosis (Sixbey *et al.,* 1984). Sloughed

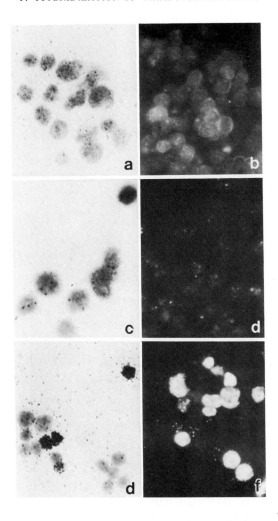

FIG. 4. Cytohybridization for EBV DNA by two techniques. (a,b) EBV-negative lymphoid cell line BJAB fails to hybridize with ³H-labeled cRNA probe or with biotin-labeled DNA as detected by autoradiography and FITC–avidin, respectively. Latently infected Raji cells (c,d) and productively infected P3HR-1 (e,f) reflect their varying EBV genome copy number by the different degrees of binding of these probes. Reprinted from Sixbey *et al.* (1984), *N. Eng. J. Med.* **310.**

cells in throat washings were hybridized to ³H-labeled and biotinylated *Bam*HI V. Figure 5 shows the cytohybridization to epithelial cells from an EBV-seronegative donor (Fig. 5a and b) and to cells from patients with infectious mononucleosis (Fig. 5c–h). The morphological deterioration of the cells in Fig. 5g and h suggests viral cytolysis.

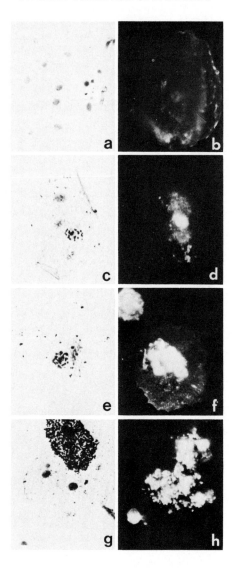

Fig. 5. EBV DNA in oropharyngeal epithelial cells by cytohybridization.   (a,b) Epithelial cells from the throat washings of an EBV-seronegative donor do not hybridize to either probe.   (c-h) The clear nuclear location of hybridization in the epithelial cells of patients with infectious mononucleosis suggests an intracellular location of the virus. Note the apparent morphological deterioration of cells reflecting a lytic phase as virus extends from nucleus to cytoplasm (g,h). Reprinted from Sixbey *et al.* (1984), *N. Eng. J. Med.* **310.**

By varying the conditions of hybridization, it has now become possible to detect EBV RNA *in situ*. In hybridizations for RNA the cells are not treated to denature DNA, so that presumably only RNA is available for hybrid formation. Moreover, the T_m of RNA:DNA hybrids in aqueous solutions containing 70% formamide is approximately 15°C higher than the corresponding DNA:DNA hybrid (Casey and Davidson, 1977). Therefore, by hybridizing in 70% formamide at an elevated temperature, approximately 5–10°C lower than the T_m of the DNA:DNA hybrid, the formation of RNA:DNA hybrids will proceed at the maximal rate, and the hybrids will tend to be more stable.

In Fig. 6 EBV RNA is detected with biotinylated EBV *Bam*HI H fragment. Two abundant early replicative RNAs are transcribed from this fragment (Hummel and Kieff, 1982). Figure 6a shows the diffuse fluorescence over approximately 30% of superinfected Raji cells. This hybridization is not detected if the cells are pretreated with RNase (Fig. 6b). Epithelial cells from EBV-seronegative donors did not contain EBV RNA (Fig. 6c), which was readily detected in the oropharyngeal epithelial cells from patients with infectious mononucleosis (Fig. 6d–f) (Sixbey *et al.*, 1984).

V. Southern Blot Hybridization

A major change in emphasis in hybridization methodology has resulted from the emergence of restriction enzyme and DNA-transfer technology. The ability to divide a large DNA genome into discrete identifiable segments that can individually be probed for homology after transfer to a nitrocellulose membrane has provided the means to analyze DNA structure and complexity at a much more detailed level. With this approach, unknown DNA can be analyzed for homology to viral DNA, and the region of the viral genome that contains the homologous sequences will be identified. The careful combined choice of which DNA to affix to the filter and which population to label as a probe enables identification of regions of homology, deletions, and regions of transcription. Much of the information obtainable by renaturation kinetic analyses can be obtained by blot hybridization in a much less laborious procedure.

A. RESTRICTION ENDONUCLEASE DIGESTION AND DNA TRANSFER

The type II restriction endonucleases, which are primarily used for cleavage of large DNA genomes into smaller segments, recognize a particular DNA sequence, usually four to six nucleotides, that contain a 2-fold axis of symmetry (Roberts, 1982). However, the enzymes generally cleave sev-

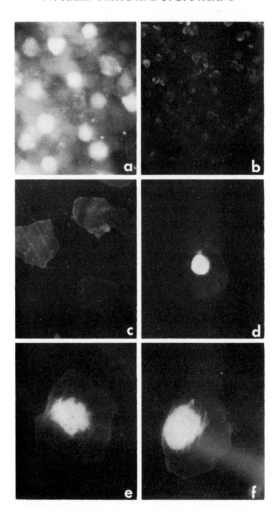

Fig. 6. EBV RNA detected with biotinylated EBV DNA probes. (a) DNA:RNA cytohybridization produced diffuse fluorescence over 30% of superinfected Raji cells (SIRC); the remaining cell population is totally negative, a pattern distinct from DNA:DNA hybridization (Fig. 4d) in Raji cells. (b) RNase-pretreated SIRC and latently infected infectious mononucleosis lymphoblastoid lines as well as (c) epithelial cells from EBV-seronegative donors did not hybridize for RNA. (d–f) Note the nuclear location of the label for RNA in oropharyngeal epithelial cells of patients with infectious mononucleosis and its diffusion (e, f) beyond a visibly well-demarcated nuclear membrane. Reprinted from Sixbey *et al.* (1984), *N. Eng. J. Med.* **310**.

eral nucleotides away from the axis on each strand. This staggered cut yields DNA fragments with cohesive termini that can readily recombine with other such termini to form recombinant molecules.

The optimal reaction conditions for each particular enzyme are indicated on the manufacturer's specification sheet. To avoid possible errors in buffer conditions it is generally advisable to prepare a $10 \times$ stock solution of the appropriate buffer that can be stored at $-20°C$ indefinitely. Reaction volumes should be kept to a minimum for maximum enzyme efficiency. However, the volume of the added restriction enzyme should be less than one-tenth the final volume of the reaction mixture to avoid inhibition of the enzyme by excess glycerol. After digestion at the recommended temperature for 1 hr or more, the reaction can be stopped by the addition of EDTA or SDS.

The DNA fragments generated by the restriction endonuclease digestion are then separated by electrophoresis through an agarose gel. The distance through which a DNA fragment has migrated is inversely proportional to the log of the molecular weight (Helling *et al.,* 1974). The range of molecular weights for which this is a linear relation is dependent on the agarose gel concentration. Therefore the optimal gel concentration for the separation of the fragments of interest should be determined with DNA of known molecular weights. Usually agarose gels are prepared with 0.5 $\mu g/ml$ of ethidium bromide incorporated into the gel and the running buffer, which provides a convenient means of visualizing the DNA in the gel and after transfer to nitrocellulose.

The transfer of DNA fragments from the agarose gel to the nitrocellulose is generally accomplished as described by Southern (1975) with various modifications. The gel is trimmed to the desired size, and the DNA is denatured within the gel matrix by soaking in 0.5 M NaOH, 1.5 M NaCl for 1 hr and then neutralized for 1 hr in 0.5 M Tris (pH 6.8), 1.5 M NaCl. The gel is then placed on some supporting matrix in a dish of $20 \times$ SSC. A well-soaked sponge sheet approximately 1/2-in. thick overlaid with one layer of Whatman filter paper provides a continuous and even source of $20 \times$ SSC. The gel is then masked on all sides with sheets of Parafilm. A sheet of nitrocellulose, cut to fit the gel and moistened in $2 \times$ SSC, is carefully placed on the gel, eliminating any air bubbles that may form between the gel and the nitrocellulose. A sheet of Whatman filter paper cut to fit is then placed on top and allowed to become moist. This is followed by a stack of absorbent material (e.g., paper towels) and an evenly distributed weight (e.g., a sheet of Plexiglas). Assembled in this fashion the Parafilm allows only the salt solution to pass through the gel and carry the single-stranded DNA which will stick to the nitrocellulose as the solution is drawn into the towels. The transfer is generally allowed to proceed overnight followed by disassembly of the transfer apparatus and fixing of the DNA to the nitrocellulose by baking at 80°C for 2 hr under vacuum.

B. DETERMINATION OF A RESTRICTION ENZYME CLEAVAGE MAP

Many of the questions that can be answered through the use of Southern blots are dependent on a detailed knowledge of the DNA structure of the entire viral genome or a segment of particular interest. Therefore the first step in most instances is to locate the sites at which any particular restriction enzyme cleaves the DNA. A DNA linkage map represents the linear array of these sites. The two principal methods to determine fragment order are analysis of different enzyme digestions after hybridization with particular labeled fragments to blots, and analysis of partial digestion products.

In order to determine the linkage of fragments, one must determine the size of the fragments cleaved with single enzymes and obtain purified fragments generated by one enzyme for hybridization to blots of fragments generated by the other enzymes. It is usually necessary to work with three enzymes or more to determine the linkage and to cross-check this linkage by constructing probes from the fragments generated by each enzyme.

Perhaps the simplest method to obtain the fragments generated by each enzyme is to slice the fragment from the gel and place it in a dialysis bag with sterile electrophoretic buffer. The bag is then placed perpendicular to the current in the electrophoresis chamber. The DNA is electroeluted from the gel and onto the inner wall of the dialysis tubing. The DNA is released from the bag by reversing the current for several minutes. The gel slice can be restained with ethidium bromide, 0.5 μg/ml, to determine whether all the DNA has been eluted. The DNA in the buffer can then be concentrated for labeling or further purified. In general, DNA that is not further purified does not label to high specific activity, but usually a sufficient number of counts can be incorporated for mapping purposes. To eliminate the various contaminants that elute from the gel and inhibit most subsequent enzymatic procedures, the DNA is loaded in the electrophoresis buffer onto a DEAE-Sephacel column and eluted with 10 mM Tris (pH 7.6), 1 mM EDTA, 0.6 M NaCl and then precipitated with ethanol.

A simple method of obtaining the restriction enzyme fragment order for only one enzyme is to analyze the partial digestion products (Smith and Bernstiel, 1976). This is particularly easy with a cloned recombinant DNA fragment, for which the identification of the terminal fragment is straightforward. For example, to determine the order of a fragment generated by one enzyme within an *Eco*RI fragment, one digests the cloned fragment with the enzyme of interest alone or also with *Eco*RI. The internal fragments will comigrate in either digestion, whereas the terminal fragments will be larger in the single-enzyme digestion, as they are still attached to a piece of vector DNA. Either of the terminal fragments can be excised from the gel from the double-enzyme digestion and used as probe on a blot of the partial digestion with the enzyme to be mapped. Hybridization with a terminal fragment to a partial digestion will identify the partial digestion products as they increase in size from that terminal. The differences in the molecular weights of these fragments identify the sequential order of the internal fragments by molecular weight from that end.

C. ANALYSES OF DNA COMPLEXITY

The Southern blot of viral DNA, restricted with an enzyme that cuts the genome into relatively small pieces, can be used to analyze DNA complexity in a manner similar to that of dot blots of a series of clones. One can also use a series of Southern blots of recombinant DNA fragments that span the viral genome, each cut with additional restriction enzymes to provide a finer analysis. Total intracellular DNA from the test sample, which can be enriched for viral sequences by any of the methods mentioned, is labeled *in vitro* by nick translation with ^{32}P-labeled deoxynucleotides and hybridized to the blot(s). Comparison to a replicate blot that has been hybridized to viral DNA or the appropriate recombinant DNA, which will hybridize to all the fragments, will identify those fragments containing sequences not represented in the test DNA and therefore not detected with the labeled intracellular DNA. Small deletions of sequences within a fragment will not be detected; however, weak hybridization to any fragment may indicate a fragment from which sequences are deleted in the test DNA. This procedure provides a quick indication of viral DNA content.

For all hybridizations to nitrocellulose blots with greater than 10^8 cpm, the filter should be pretreated with hybridization solution containing $10 \times$ Denhardt's reagents as described in Section IV, B. The hybridization solution is usually $4 \times$ to $6 \times$ SSC and contains several competitors to inhibit nonspecific binding to nitrocellulose, such as single-stranded calf thymus or salmon sperm DNA.

The nitrocellulose sheet can be exposed to the hybridization solution in a flat, sealed plastic bag that is submerged in a slowly shaking water bath overnight at the desired temperature. Hybridization should generally be performed at 25°C lower than the T_m of the DNA. The temperature can be lowered 0.6°C for every 1% formamide included in the hybridization solution. After hybridization the blot is washed by the method used for the dot blots.

This approach was used to identify the EBV sequences present in three nonproducer cell lines (Raab-Traub *et al.,* 1980). Reassociation kinetic analyses had previously determined that the virus from the B95-8 cell line, the only available producer cell line established with virus of infectious mononucleosis origin, lacked approximately 15% of the sequences in the HRl strain established from a Burkitt's lymphoma (Pritchett *et al.*, 1975). The residual single-stranded labeled HRl DNA after extensive absorptive hybridizations with B95-8 DNA had localized this sequence to a large *Eco*RI fragment (Raab-Traub *et al.*, 1978). Further mapping determined that these sequences were located in three *Bam*HI fragments, B1, I1, and W1.

Duplicate blots of virus purified from two cell lines of Burkitt's lymphoma origin, AG876 and W91, are shown in Fig. 7. The DNAs have been digested with *Bam*HI, which generates approximately 32 fragments. There is some variation from strain to strain. A blot of each strain was hybridized to labeled AG876 DNA (designated T. P. for total probe), and to ^{32}P-labeled intracellular DNA from the nonproducer cell lines, C, F, and M, established from patients with infectious mononucleosis. The intracellular DNA was enriched for viral sequences by buoyant density centrifugation. The labeled DNA intracellular DNA hybridized to all of the viral fragments with approximately the same intensity as the total probe.

To provide stronger evidence that the sequences deleted from B95-8 were present in the C, F, and M cell lines, recombinant DNA clones of the B1, I1, and W1 fragments were digested with an enzyme that generated fragment(s) consisting entirely of sequences deleted from B95-8 and to which B95-8 viral probe would not hybridize. In Fig. 8 digestion of the fragment B1 with *Tha*I (A) generated two fragments of 1.05 and 0.66 $\times$ 10^6 daltons. Although the B95 probe does not hybridize to them, the sequences are clearly contained in the C, F, and M cell lines. The *Bam*HI W1 fragment is located between B1 and I1 in the genome and is entirely deleted from B95-8. Digestion with *Taq*I or *Tha*I (B) generated several fragments to which C, F, and M hybridize. Similarly, digestion of *Bam*HI I1 with *Sal*I (C) generated a 0.9 $\times$ 10^6 fragment that does not hybridize to B95 but does hybridize to the labeled intracellular DNAs. By determining the linkage of such fragments, this method can be used to map deletions and to screen test DNAs for the presence of particular sequences.

D. Detection of Viral DNA

Southern blot hybridization is useful also for the unambiguous detection of viral sequences. For this purpose the blot is prepared from purified test DNA digested with an appropriate restriction enzyme. As all cells may not contain the viral genome, it may be necessary to enrich for viral sequences prior to digestion. The blot is then hybridized to a virus-specific probe. For large viruses, such as the herpesviruses, it is best to use a cloned fragment that has reduced sequence complexity and will hybridize more efficiently.

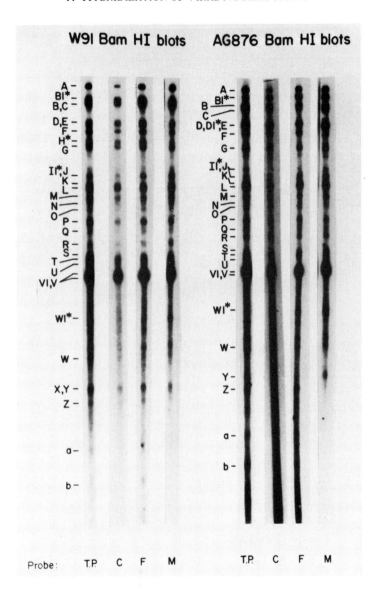

FIG. 7. Sequence complexity of three cell lines, C, F, and M, of infectious mononucleosis origin. Intracellular DNA from the cell lines C, F, and M was enriched for viral sequences by isopycnic centrifugation, labeled with ^{32}P-dCTP, and hybridized to blots of *Bam*HI fragments, of AG876 and W91 DNA. T.P. refers to blot hybridized with ^{32}P-labeled W91 or AG876 DNA. Reprinted from Raab-Traub *et al.* (1980), *Cell.*© M.I.T.

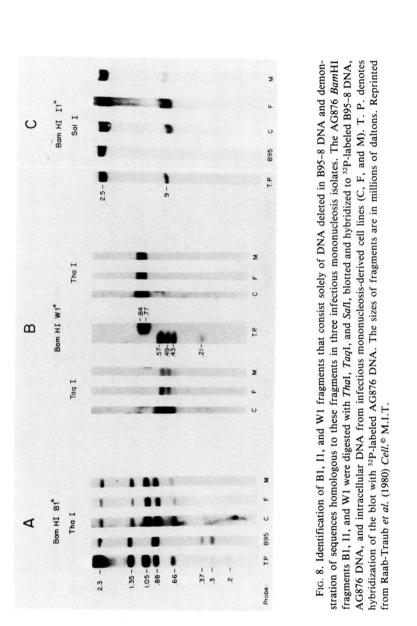

FIG. 8. Identification of B1, I1, and W1 fragments that consist solely of DNA deleted in B95–8 DNA and demonstration of sequences homologous to these fragments in three infectious mononucleosis isolates. The AG876 BamHI fragments B1, I1, and W1 were digested with ThaI, TaqI, and SalI, blotted and hybridized to ^{32}P-labeled B95–8 DNA, AG876 DNA, and intracellular DNA from infectious mononucleosis-derived cell lines (C, F, and M). T. P. denotes hybridization of the blot with ^{32}P-labeled AG876 DNA. The sizes of fragments are in millions of daltons. Reprinted from Raab-Traub et al. (1980) Cell.© M.I.T.

The Epstein–Barr virus contains a large internally repeated sequence of 2×10^6 daltons that is invariably present in all virus strains in approximately 5–10 copies. The enzyme *Bam*HI makes one cut in this sequence generating a 2×10^6-dalton fragment, *Bam*HI V, which is 5–10 *M* (Dambaugh *et al.*, 1980). Therefore one can restrict test DNAs with *Bam*HI and hybridize to ^{32}P-labeled *Bam*HI V cloned DNA. The fact that the 2×10^6-dalton fragment is overrepresented on the blot and that the size of the fragment is constant provides a most sensitive unambiguous detection of viral DNA. In Fig. 9 viral DNA has been detected in this fashion in several tissue samples from patients with nasopharyngeal carcinoma (NPC).

This same approach can be used to analyze test DNA structure. A large amount of test DNA is restricted and run on a wide well. The paper can

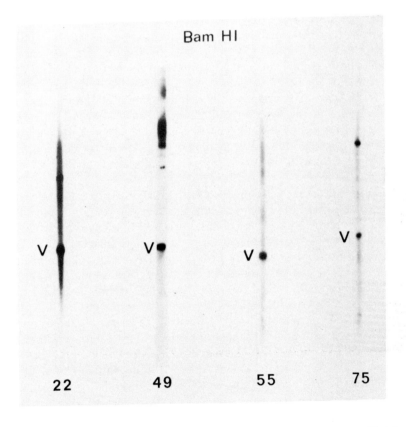

FIG. 9. Detection of EBV DNA in tissue samples. Total intracellular DNA purified from four nasopharyngeal carcinoma tissues obtained from biopsy were digested with the restriction endonuclease, *Bam*HI, subjected to electrophoresis through an 0.5% agarose gel, and transferred to nitrocellulose. The filter was hybridized to 5×10^7 cpm of a ^{32}P-labeled EBV fragment, *Bam*HI V. In each case, the 2×10^6-dalton *Bam*HI V fragment can be identified.

be cut into multiple strips, each of which can be hybridized to a cloned fragment. The detection of the corresponding viral fragment in the test DNA will identify any fragments that have different molecular weights. Such differences may represent altered restriction enzyme cut sites, DNA rearrangements, insertions, deletions, or integrations. This procedure is somewhat laborious, and, depending on viral DNA copy number, all fragments may not be detectable. Moreover, the tissue sample may not provide sufficient DNA for this technique.

In those cases in which it is necessary to determine detailed DNA structure it may be necessary to clone the viral sequences in the test DNA. Cloning will provide unlimited DNA for further mapping studies. For example, in NPC, an epithelial cell malignancy associated with the Epstein–Barr virus, cells cannot be grown *in vitro,* and tissue is available only from biopsy. In order to compare the EBV DNA contained in these tissues with viral DNA from cell lines established from Burkitt's lymphoma, intracellular DNA purified from an NPC biopsy was enriched for viral sequences by buoyant density equilibrium centrifugation and cloned into a cosmid vector (Meyerowitz *et al.,* 1980). Cosmids were designed for cloning large inserted DNA fragments. They are generally derived from a plasmid containing a drug-resistance factor, such as pBR322, are small in size, and contain the λ bacteriophage cohesive ends (cos) site and one or more unique restriction sites. These features enable cloning of large DNA fragments, up to 45 kilobases in size, with the λ *in vitro* packaging system for high-efficiency cloning.

Ligations are set up with high concentrations of vector and target DNA to favor the formation of long concatamers that will contain the test DNA flanked by vector DNA. These concatamers will be cleaved at the cos site by the λ packaging system and packaged into bacteriophage particles. After infection of *Escherichia coli,* the recombinant DNA will replicate as a plasmid, the bacteria can be selected for the drug-resistance marker, and bacterial clones can be screened for viral DNA. The recombinant viral clones can then be compared to viral DNA.

In Fig. 10 the recombinant *Eco*RI B and C fragments obtained from the NPC DNA are compared with those of the EBV (W91) strain. Within *Eco*RI B the *Bam*HI E fragment, the terminal *Bam*HI fragment within *Eco*RI B, is larger in the NPC fragment. This is the result of a shifted *Eco*RI cut site. The *Bam*HI K fragment within *Eco*RI B is 200,000 daltons larger in the NPC fragment. This fragment contains a simple repeated sequence that is present in a varying number of copies, and as a result the fragment varies in size from strain to strain. Within *Eco*RI C, the *Bam*HI fragments of both W91 and the NPC fragment are identical and include B1, I1, and W1. Therefore the deletion found in the EBV (B95) strain is not present in the NPC isolate.

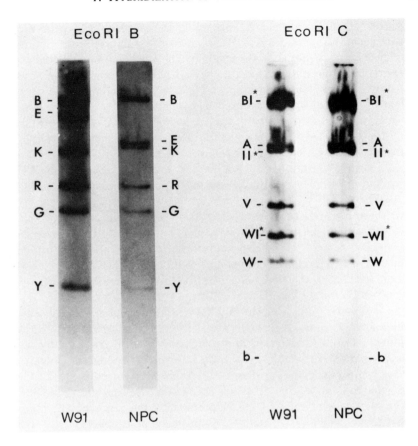

FIG. 10. Recombinant DNA clones of the *Eco*RI B and C fragments obtained from DNA purified from a nasopharyngeal (NPC) biopsy compared to EBV (W91) virion DNA. Recombinant *Eco*RI B and C fragments from NPC material and EBV virion DNA were digested with *Bam*HI, transferred to nitrocellulose after electrophoresis, and hybridized to ³²P-labeled EBV DNA. The *Bam*HI fragments within each fragment are indicated.

E. IDENTIFICATION OF SEQUENCES THAT ENCODE RNA

Southern blots are also used to identify the viral sequences that encode RNA either at different times after infection, in different types of infection, after treatment which chemical inducers, or in different disease states. The procedure is particularly useful for initial RNA mapping studies of complex viral genomes. After identification of the subset of viral sequences that are transcribed, the size and relative abundance of the RNAs encoded by these sequences can be determined by preparing RNA blots.

The approach is similar to that used for analyzing intracellular viral DNA sequences. However, instead of labeling intracellular DNA, a labeled probe prepared from the intracellular RNA is hybridized to a series of Southern blots of fragments that span the genome. The RNA can be prepared from nuclear or cytoplasmic fractions, or polyadenylated sequences can be separated by oligo(dT) chromatography. The RNA population can be directly iodinated (Rymo, 1979) or be used as a template to construct a labeled cDNA probe (King et al., 1980). Alternatively, the RNA can be hybridized in solution to labeled viral DNA and the DNA:RNA hybrids separated from double- or single-stranded DNA by buoyant density gradient centrifugation through cesium sulfate. The RNA is then hydrolyzed by treatment with alkali, and the labeled DNA is hybridized to Southern blots (Dambaugh et al., 1979).

All these methods have been successfully utilized in the EBV system. However, with the availability of high-quality pure reverse transcriptase and high-specific-activity ^{32}P-labeled nucleotides, cDNA synthesis is probably the most sensitive and efficient method. With this methodology it has been possible to identify the EBV sequences that are transcribed in NPC tissue (Raab-Traub et al., 1983).

The tissues are homogenized in guanidine thiocyanate, and total intracellular RNA is purified through a cesium chloride cushion. The RNA is separated into polyadenylated and nonpolyadenylated fractions by oligodeoxythymidine chromatography. For a further selection for polyadenylated sequences the cDNA synthesis with reverse transcriptase is primed with oligodeoxythymidine. The ^{32}P-labeled cDNAs are then hybridized to Southern blots of recombinant EBV DNAs spanning the genome. The blots of each fragment are hybridized to ^{32}P-labeled nick-translated EBV virion DNA. This step determines the position of the EBV fragments on the blot and provides a comparison of the relative strength of each blot.

The results of this type of experiment are shown in Fig. 11. The strip at the left of each fragment is the total probe hybridization. The EcoRI A and B fragments have been digested with EcoRI and BamHI. The BamHI fragments within each EcoRI fragment are indicated at the left. Polyadenylated RNA is apparently transcribed from the three fragments, BamHI V, K, and EcoRI DIJhet. These fragments encode the most abundant RNAs in a latently infected cell line (King et al., 1980).

F. Nuclease Mapping of RNA

The DNA sequences represented in processed mRNA molecules can be determined by electron microscopy of RNA:DNA hybrids or by identifying the splice points by S1 nuclease digestion of RNA:DNA hybrids followed by electrophoresis of the nuclease-resistant DNA fragments (Berk and

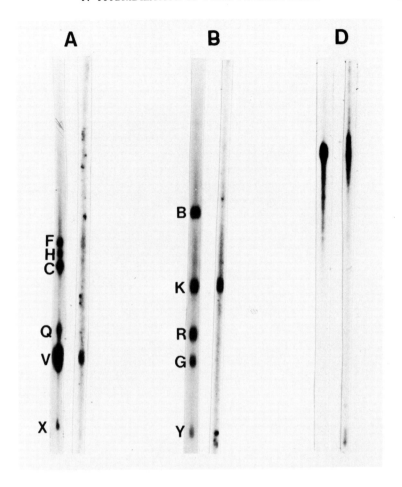

FIG. 11. Identification of the EBV sequences that encode polyadenylated RNA in a naso-pharyngeal carcinoma (NPC) biopsy. ^{32}P-labeled cDNA synthesized to the polyadenylated RNA from NPC 49 was hybridized to duplicate strips of blots of restriction enzyme fragments of EBV. The EcoRI fragments A and B were digested with EcoRI and BamHI. The BamHI fragments within each of the EcoRI fragments are indicated at the left. A duplicate strip of each blot that was hybridized to ^{32}P-labeled HR1 viral DNA is shown to the right of the cDNA hybridization. Data from Raab-Traub et al. (1983).

Sharp, 1977). The viral DNA is frequently labeled with P in vivo, and the DNA bands are detected by autoradiography of the gel. Alternatively, the gel can be transferred to nitrocellulose and hybridized to a ^{32}P-labeled probe. A fine-structure restriction enzyme map of the DNA fragments will identify those fragments that can be used as probes to map exactly the exon regions of the DNA. In addition, the use of the Southern transfer technique permits identification of the template strand by hybridization with a strand-specific probe.

In either approach an excess of purified DNA, generally from a recombinant DNA clone that spans the region of interest, is hybridized to a small amount of the RNA in 80% formamide. As RNA:DNA hybrids are more stable in high concentrations of formamide, the nucleic acids are annealed at a temperature above the T_m of the DNA:DNA hybrid but below the T_m of the RNA:DNA hybrid. A schematic of the resultant hybrid is shown in Fig. 12. The hybrid has single-stranded DNA tails at the 5′ and 3′ ends of the RNA and an intervening sequence on the DNA strand, which forms an

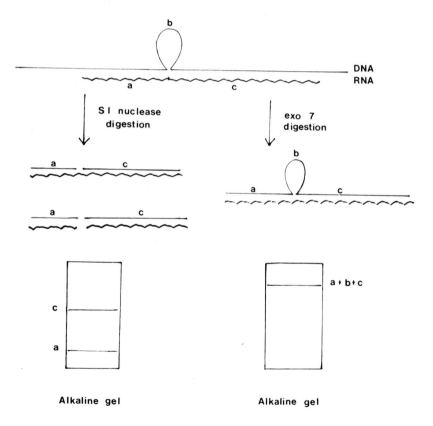

FIG. 12. Schematic of nuclease digestion of a spliced RNA:DNA hybrid. S1 nuclease digestion degrades all single-stranded DNA at the termini and the single-stranded DNA intron loop (b). Electrophoresis through an alkaline agarose gel will identify the size of the continuous nonspliced regions a and c. Exonuclease VII (exo 7) degrades single-stranded DNA at the termini and will generate a DNA strand spanning the sequences encoding the entire RNA.

internal loop, designated b. Digestion with S1 nuclease will degrade the single-stranded tails and the internal loop and generate DNA segments a and c. S1 cleavage of a ribophosphodiester bond opposite a nick is inefficient; therefore electrophoresis on a neutral agarose gel will yield three species equal in size to a, c, and a + c. Electrophoresis in an alkaline gel will disclose only the individual DNA species a and c. Exonuclease VII will digest only the single-stranded tails, not any internal loops, and will identify the length of the DNA fragment between the 5′ and 3′ ends of the RNA.

The reaction is set up by precipitating the appropriate amount of DNA and RNA with ethanol. To detect RNAs of very low abundance, the concentration of RNA can be increased up to 500 μg in a 100-μl hybridization reaction mixture. The nucleic acid pellet is suspended in hybridization buffer containing 80% formamide, 0.4 M NaCl, 1 mM EDTA, buffered with 40 mM PIPES (pH 6.5). The solution is incubated at the temperature necessary to separate the DNA strands, usually 15–20°C higher than the T_m. The T_m of the DNA in formamide, T_m^*, is equal to the $T_m - [0.5(\%GC) + 0.75(\%AT)]\%$ formamide. The tube is then transferred to a bath at the desired temperature, approximately 10°C above the T_m of the DNA, which is 5°C lower than the T_m of the RNA:DNA hybrid. The solution is never allowed to cool below this temperature. After hybridization for 3 hr, the solution is diluted with 0.3 ml of cold S1 nuclease buffer with a variable amount of S1. The concentration of S1, NaCl, and the incubation temperature will have to be optimized for each type of RNA:DNA hybrid. After digestion the reaction is stopped by the addition of EDTA, extracted with phenol/chloroform, and precipitated with isopropanol. The precipitate is dissolved in 10 mM Tris, 1 mM EDTA, adjusted to 0.5 M NaOH, and run on an alkaline agarose gel. After electrophoresis, the DNA is transferred to nitrocellulose.

VI. Northern Blots

There are two denaturing gel methods that are frequently used to separate RNAs by molecular weight prior to transfer to nitrocellulose. The RNA is either denatured by treatment with glyoxal and dimethyl sulfoxide (DMSO) and subjected to electrophoresis through an agarose gel, or denatured with formamide and formaldehyde and separated through an agarose gel containing formaldehyde.

To denature RNA with glyoxal and DMSO, ethanol-precipitated RNA is dissolved in a solution of 50% DMSO, 0.01 M NaH$_2$PO$_4$ (pH 7.0), and 1 M glyoxal and heated to 50°C for 60 min. The RNA is rapidly cooled to 20°C and loaded immediately onto a 1–1.5% agarose gel that does not contain ethidium bromide. The gel buffer is 0.01 M Na_2HPO_4 (pH 7.0). The pH of the running buffer is somewhat critical, as glyoxal will separate from RNA at pH greater than 8.0. Therefore the buffer should be recirculated between the two electrophoretic chambers with a peristaltic pump. The RNAs can be transferred with 20× SSC exactly as described for Southern blots without any pretreatment of the agarose slab.

To denature RNA with formaldehyde and formamide, ethanol-precipitated RNA is dissolved in a solution containing 50% formamide, 6.5% formaldehyde in 1× running buffer. Formaldehyde is usually obtained as a 37% solution, which should make up 17.5% of the denaturing solution. The running buffer is 0.02 M morpholinopropanesulfonic acid (MOPS) (pH 7.0), 5 mM sodium acetate, 1 mM EDTA. DNA restriction fragments can be used as molecular weight markers. They can be denatured and run exactly as the RNA samples are. Formaldehyde gels can be run and transferred in the presence of ethidium bromide. However, the RNA or DNA cannot be visualized until after transfer to nitrocellulose and can then be marked on the nitrocellulose sheet.

Prior to transfer, the gel should be soaked for 5 min each in several changes of water, for 45 min in 50 mM NaOH, 10 mM NaCl, to hydrolyze partially high-molecular-weight RNAs, neutralized for 45 min in 0.1 M Tris (pH 7.5), and soaked in 20× SSC for 1 hr. The gel is then transferred to nitrocellulose with 20× SSC. The rate of migration of RNA through a denaturing gel is linearly related to the log of the molecular weight. Therefore by using DNA standards or the 18 S and 28 S ribosomal RNAs that still contaminate most polyadenylated RNA preparations and can be visualized on the nitrocellulose, it is possible to determine the molecular weights of the RNAs. The number of copies of each RNA can be roughly estimated with reconstructions using DNA fragments.

References

Adams, A., and Lindahl, T. (1975). *Proc. Natl. Acad. Sci. U.S.A.* **72**, 1477–1481.
Aviv, H., and Leder, P. (1972). *Proc. Natl. Acad. Sci. U.S.A.* **69**, 1408–1412.
Berger, S. L., and Birkenmeier, C. S. (1979). *Biochemistry* **18**, 5143–5149.
Berk, A. J., and Sharp, P. (1977). *Cell* **12**, 721–732.
Brandsma, J., and Miller, G. (1980). *Proc. Natl. Acad. Sci. U.S.A.* **77**, 6851–6855.
Burton, K. (1956). *Biochem. J.* **62**, 315–323.
Casey, J., and Davidson, N. (1977). *Nucleic Acids Res.* **4**, 1539–1546.
Chirgwin, J. M., Przbyla, A. E., MacDonald, R. J., and Rutter, W. J. (1979). *Biochemistry* **18**, 5294–5299.
Commerford, S. L. (1971). *Biochemistry* **10**, 1993–2000.
Dambaugh, T., Nkrumah, F. K., Biggar, R. J., and Kieff, E. (1979). *Cell* **16**, 313–322.
Dambaugh, T., Beisel, C., Hummel, M., King, W., and Fennewald, S. (1980). *Proc. Natl. Acad. Sci. U.S.A.* **77**, 2999–3003.

Denhardt, D. T. (1966). *Biochem. Biophys. Res. Commun.* **23**, 641–646.

Dolyniuk, M., Wolff, W., and Kieff, E. (1976). *J. Virol.* **18**, 289–297.

Fowler, E., Raab-Traub, N., and Hester, S. (1984). Submitted.

Grunstein, M., and Hogness, D. S. (1975). *Proc. Natl. Acad. Sci. U.S.A.* **72**, 3961–3965.

Helling, R. B., Goodman, H. M., and Boyer, H. W. (1974). *J. Virol.* **14**, 1235–1244.

Henry, B. E., Newcomb, W. W., and O'Callaghan, D. J. (1979). *Virology* **92**, 495–506.

Hirt, B. (1967). *J. Mol. Biol.* **26**, 365–369.

Hummel, M., and Kieff, E. (1982). *J. Virol.* **43**, 262–272.

Kafatos, F. C., Jones, C. W., and Efstratiadis, A. (1979). *Nucleic Acids Res.* **7**, 1541–1552.

King, W., Thomas-Powell, A. L., Raab-Traub, N., Hawke, M., and Kieff, E. (1980). *J. Virol.* **36**, 506–518.

Krieg, P., Amtmann, E. and Sauer, G. (1983). *J. Anal. Biochem.* **134**, 288–294.

Lin, J.-C., Smith, M. C., Cheng, Y. C., and Pagano, J. S. (1983). *Science* **221**, 578–579.

Meyerowitz, E. M., Guild, G. M., Prestidge, L. S., and Hogness, D. S. (1980). *Gene* **11**, 271–276.

Nonoyama, M., and Pagano, J. S. (1971). *Nature (London) New Biol.* **233**, 103–106.

Pater, M., Hyman, R. W., and Rapp, F. (1976). *Virology* **75**, 481–483.

Pignatti, P. F., Cassai, E., Meneguzzi, G., Chenciner, N., and Milanesi, G. (1979). *Virology* **93**, 260–264.

Pritchett, R. F., Hayward, S. D., and Kieff, E. (1975). *J. Virol.* **15**, 556–569.

Raab-Traub, N., Pritchett, R., and Kieff, E. (1978). *J. Virol.* **27**, 388–398.

Raab-Traub, N., Dambaugh, T., and Kieff, E. (1980). *Cell* **22**, 257–269.

Raab-Traub, N., Hood, R., Yang, C.-S., Henry, B., II, and Pagano, J. S. (1983). *J. Virol.* **48**, 165–175.

Rigby, P. W. J., Dieckmann, M., Rhodes, C., and Berg, P. (1977). *J. Mol. Biol.* **113**, 237–251.

Roberts, R. (1982). *Nucleic Acids Res.* **10**, 117–129.

Rymo, L. (1979). *J. Virol.* **32**, 8–18.

Scheel, G., and Blackburn, P. (1979). *Proc. Natl. Acad. Sci. U.S.A.* **769**, 4898–4904.

Seeburg, P. H., Shing, J., Martial, J. A., Ullrich, A., Baxter, J. D., and Goodman, H. M. (1977). *Cell* **12**, 157–165.

Shaw, J. E., Huang, E.-S., and Pagano, J. S. (1975). *J. Virol.* **16**, 132–140.

Shaw, J. E., Seebeck, T., Li, J.-L., and Pagano, J. S. (1977). *Virology* **77**, 762–771.

Simmons, J. G., Hutt-Fletcher. L., Fowler, E., and Feighny, R. J. (1983). *J. Immunol.* **130**, 1303–1308.

Singer, R. H., and Ward, D. C. (1982). *Proc. Natl. Acad. Sci. U.S.A.* **79**, 7331–7335.

Sixbey, J. W., Nedrud, J. G., Raab-Traub, N., Hanes, R. A., and Pagano, J. S. (1984). *N. Eng. J. Med.* **310**, 1225–1230.

Smith, H. O., and Bernstiel, M. L. (1976). *Nucleic Acids Res.* **3**, 2387–2395.

Southern, E. (1975). *J. Mol. Biol.* **98**, 502–517.

Taylor, J. M., Illmensee, R., and Summer, J. (1976). *Biochem. Biophys. Acta* **442**, 324–330.

Wetmur, J. G., and Davidson, N. (1968). *J. Mol. Biol.* **31**, 349–370.

zur Hausen, H., and Schulte-Holthausen, H. (1970). *Nature (London)* **227**, 245–248.

2 Applications of Oligonucleotide Fingerprinting to the Identification of Viruses

Olen M. Kew, Baldev K. Nottay, and
John F. Obijeski

I. Introduction

Fingerprinting is a technique by which oligonucleotides, produced by cleavage of RNA molecules with specific ribonucleases, are separated in two dimensions. Originally developed as a preparatory method, fingerprinting

41

is currently most frequently applied as an analytical method for comparing the genomes of RNA viruses. Comparisons are based on the principle that the large, structurally unique oligonucleotides separate into patterns, or "fingerprints," that are highly characteristic of the RNA sequence from which they derive. Accordingly, the sequence, and thus genetic, relationships can be readily assessed. Since the characteristic oligonucleotides originate from all regions of the RNA molecule, the distributions of sequence similarities and differences may be surveyed over the entire viral genome. Although not providing specific base sequence information, fingerprinting does offer a simple and comparatively rapid means of sequence comparison, with the results obtained in a pictorial form.

The applications of fingerprint analysis in virology are very broad. A few examples of the applications include analysis of RNA genome size and complexities, identification of subgenomic RNA molecules, analysis of the messenger RNA species produced by both DNA and RNA viruses, characterization of defective-interfering viral genomes, biochemical description of genetic reassortants and recombinants, and the estimation of the degree of sequence divergence between closely related virus strains.

Oligonucleotide fingerprinting has emerged as an important tool for the identification of RNA virus strains (Brown, 1982; Clewley and Bishop, 1982). Fingerprinting provides a reliable means of following the natural distribution and transmission of a specific viral genotype over time. This capability is of considerable value in human and veterinary epidemiology, in which genetically distinct but antigenically similar epidemic and zoonotic strains in coetaneous circulation can be distinguished, and their transmission separately followed. Because viral antigens tend to be more conserved than the overall genomic RNA sequence, fingerprinting is generally much more sensitive than serology in detecting divergence between strains. On the other hand, fingerprint analysis is not complicated by antigenic drift. Although the availability of monoclonal antibodies promises to greatly improve the precision and discrimination of viral serodiagnosis (Yewdell and Gerhard, 1981), fingerprinting is, in a sense, a more open technique, not requiring the production of new reagents in order to monitor the circulation of a given virus genotype. Further, unlike serology, fingerprint comparisons are not restricted to the small set of genomic sequences specifying the amino acid residues on a polypeptide surface.

In this chapter, we describe the principles and methodologies of oligonucleotide fingerprinting and briefly review the applications of this technique to the identification of RNA viruses. Particular emphasis will be placed on the use of this method to follow the distribution and transmission of viral agents of infectious disease. Several excellent reviews on oligonucleotide fingerprinting have appeared. Those of Beemon (1978) and Wang (1978) concentrate on the retrovirus group. Clewley and Bishop (1982) also describe methodology and the broad applications of fingerprinting in virology. Pedersen and Haseltine (1980b) and De Wachter and Fiers (1982) describe some current fingerprinting methods.

II. Principles and Mechanisms of Fingerprint Analysis

A. OVERVIEW OF FINGERPRINT ANALYSIS

Oligonucleotides for fingerprinting studies are produced by complete digestion of viral RNA with the specific ribonucleases (RNases) T1 and A. RNase T1, from the fungus *Aspergillus,* specifically cleaves phosphodiester bonds adjacent to guanosine (G) residues, producing guanosine 3′-phosphate (Gp) and a set of oligonucleotides of varying length terminating with Gp. Because of its high specificity under moderate digestion conditions, RNase T1 is the enzyme of choice for routine fingerprint studies. The other known single base-specific RNases also cleave at G (Takahashi and Moore, 1982) and thus confer no significant advantage over RNase T1. Pancreatic RNase A, which preferentially cleaves at pyrimidine residues, is infrequently used in routine comparative studies because of the reduced complexity of the fingerprint patterns of its cleavage products. RNase A was used in conjunction with RNase T1 in classical studies of RNA structure (Barrell, 1971), and has been applied more recently in detailed fingerprint comparisons between strains (Nomoto *et al.,* 1981) as well as in studies on total genome sequences (Kitamura and Wimmer, 1980; Kitamura *et al.,* 1981). Many other RNases with different base specificities are known. Some are used in RNA sequence analysis to produce specific partial fragments of end-labeled RNA molecules (Simoncsits *et al.,* 1977; Donis-Keller, 1980; Donis-Keller *et al.,* 1977). However, fingerprinting requires complete digestion of the RNA, conditions for which the specificities of the other known RNases are too broad for them to be applicable.

The number of RNase T1 oligonucleotides in a complete digest is one more than the number of G residues in the RNA. For a random polynucleotide, the size distribution of oligonucleotide fragments is predicted from its composition by the relationship (Beemon, 1978)

$$F_n = (G)^2(1-G)^{n-1}$$

where F_n is the fraction of oligonucleotides of chain length $n,$ and G is the mole fraction of guanosine residues in the RNA. For an RNA of 8000 bases containing 25% G, this relationship predicts approximately 500 Gp monomers ($n=1$), 160 5-mers, 38 10-mers, 9 15-mers, 2 20-mers, and 6 oligonucleotides for which $n>20$, to occur among the 2001 oligonucleotides produced by complete digestion. Oligonucleotides with chain lengths $n \geq 10$

have a high statistical probability of containing unique sequences that occur only once in a haploid viral genome. For the hypothetical 8000-base viral genome described in the example above, oligonucleotides containing 10 or more base residues constitute approximately 25% of the total sequences. Although the structurally unique oligonucleotides of a typical RNA molecule are too numerous to be resolved effectively by a single mechanism of separation, the two-dimensional fingerprinting technique, in which a different separation mechanism is employed for each dimension, allows good resolution to be achieved in practice. The proportion of the genome that can be resolved experimentally into separate oligonucleotide spots is primarily dependent upon the size and complexity of the RNA. This is because, for polynucleotides of a given guanosine content, the number of unique oligonucleotides increases with chain length, but the mean oligonucleotide chain length remains constant (Aaronson *et al.*, 1982). For large RNA molecules, such as viral genomes, an important fraction of the potentially informational oligonucleotides is incompletely resolved, and is thus effectively unavailable for analysis. For nonsegmented genomes, from 5 to 15% of the unique oligonucleotides can be resolved and individually analyzed. With viruses having segmented genomes it is possible to examine higher percentages of the total genome by separate analysis of each segment.

B. Mechanisms of Two-Dimensional Separation of Oligonucleotides

Two different fingerprinting methods have been developed. The earlier method, developed by Brownlee and Sanger (1969), utilized high-voltage electrophoresis in the first dimension and chromatography in the second. In the second method, that of De Wachter and Fiers (1972), separation is achieved in both dimensions by electrophoresis through polyacrylamide gels. In both methods, oligonucleotides are separated in the first dimension according to their net charges and in the second dimension primarily according to chain length.

First-dimension electrophoresis for both is generally performed near pH 3.5 in the presence of 6–7 M urea. At this pH, the greatest charge differences among the bases occur (Markham and Smith, 1952), such that the net negative charges, and thus electrophoretic mobilities, of the nucleoside monophosphates increase in the order $Cp < Ap < Gp < Up$. In the absence of significant frictional forces, oligonucleotides migrate according to the total charge contributions of their constituent base residues. The combination of acidic pH and high urea concentrations disrupts the secondary structure of the oligonucleotides and reduces interchain aggregation, largely excluding effects of conformation on mobilities. Further, for the poly-acrylamide gel system, the use of large-pore gels eliminates molecular siev-ing effects on all but the largest oligonucleotides (Frisby *et al.*, 1976), thus maximizing the contributions of charge to electrophoretic mobilities.

Second-dimension separation in the two fingerprinting systems occurs by fundamentally different mechanisms. The method of Brownlee and Sanger (1969) utilizes homochromatography, in which oligonucleotides are com-petitively displaced from a cationic surface during ascending chromatog-raphy. The displacement rate, and thus migration rate, is approximately inversely related to chain length, but base composition and secondary struc-ture significantly influence mobilities. De Wachter and Fiers (1972) em-ployed a highly cross-linked 20% polyacrylamide gel buffered at pH 8 ("neutral gel"). In the neutral pH range, the four bases are largely un-charged. All significant charges on the oligonucleotides therefore derive from ionization at the phosphodiester bonds (and at the terminal phos-phate), and the charge:mass ratios for all long oligonucleotides are nearly constant. Thus the polyacrylamide gel acts strictly as a molecular sieve per-mitting fractionation according to size and shape. To a first approximation, electrophoretic mobilities in the second dimension decrease with the loga-rithm of the oligonucleotide chain length (De Wachter and Fiers, 1982). Deviations from this relationship, arising from contributions of secondary structure and base composition, are usually small (De Wachter and Fiers, 1982; Lee and Wimmer, 1976; Pedersen and Haseltine, 1980a).

A typical two-dimensional polyacrylamide gel fingerprint of a virus RNA is shown in Fig. 1. In this example, the smaller oligonucleotides have been run off the second-dimension gel in order to improve resolution. The oli-gonucleotides are arranged in a graticulated pattern formed by a series of bands curving diagonally away from the quadrant nearest the origin of first-dimension electrophoresis (lower left, Fig. 1). This pattern is most evident in the region of the more numerous shorter-chain oligonucleotides. The bands form according to uridine (U) content. Oligonucleotides of the band closest to the origin contain no U residues, those of the next band contain one U, those of the next, two, and so on. Within each band, the cytidine (C)-rich oligonucleotides are diagonally opposite (above and behind in the orientation of the Fig. 1 pattern) the adenosine (A)-rich oligonucleotides of equivalent chain length (Pedersen and Haseltine, 1980a; De Wachter and Fiers, 1982). The lower part of each fingerprint contains the large, RNase T1-resistant oligonucleotides, whose mobilities are diagnostic for a virus genotype. Fingerprints of related and unrelated polioviruses are shown in Fig. 2. The diagnostic oligonucleotides are distributed in a roughly trian-gular region, a pattern that occurs because with increasing chain length oli-gonucleotides generally approach the mean base composition, and hence first-dimension mobility, of the entire RNA molecule.

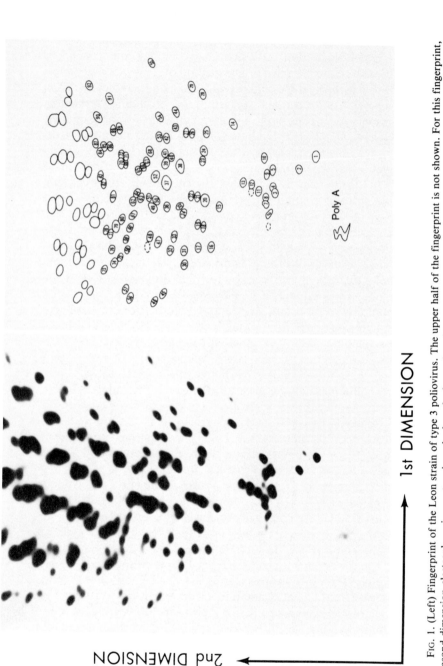

1st DIMENSION

2nd DIMENSION

FIG. 1. (Left) Fingerprint of the Leon strain of type 3 poliovirus. The upper half of the fingerprint is not shown. For this fingerprint, second-dimension electrophoresis was terminated when the bromphenol blue dye migrated 30 cm. (Right) Tracing of the fingerprint. Oligonucleotide spots are numbered arbitrarily. Identification of the less well resolved spots is based upon replicate fingerprints and oligonucleotide mapping studies (Fig. 5). Some spots, for example 37 and 52, may contain more than one oligonucleotide. Oligonucleotides that could not be accurately mapped because their spots did not give reproducible intensities are indicated by dashed lines.

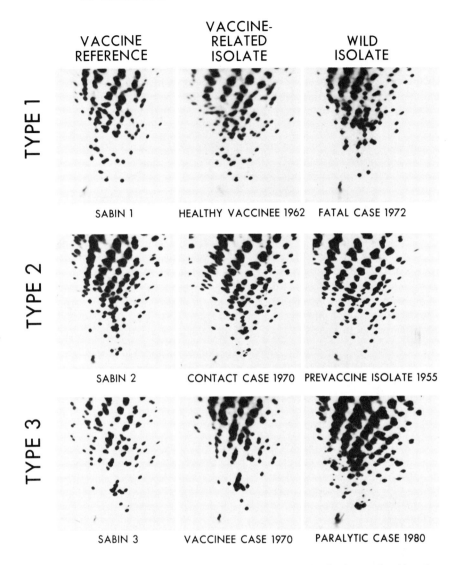

FIG. 2. Fingerprint comparisons of closely and distantly related poliovirus strains. Note that the vaccine-related isolates (center column) have fingerprints that are very similar to those of the respective oral vaccine reference strains (left column). Wild isolates from each of the three serotypes have fingerprints (right column) that are entirely distinct from those of the vaccine-related group, indicating a lack of close genetic relatedness. Note also the strong similarities in the fingerprints of Sabin type 3 (Leon 12 a_1b) and its nonattenuated parent, Leon (Fig. 1).

C. SENSITIVITY OF FINGERPRINTING IN DETECTING GENETIC RELATEDNESS AMONG RNA VIRUSES

Oligonucleotide fingerprinting is a tool of extraordinary power and precision for demonstrating close genetic relationships among viruses and, therefore, for determining the natural distribution and transmission of any given viral genotype. The typical fingerprint contains 30–60 diagnostic spots, each having an independent mobility and, as a result, they represent many independent points of reference for comparison. Because the potential number of patterns formed by the spots of the diagnostic oligonucleotides is very large, equivalent migration of a majority of the spots of two viral RNA samples implies a close genetic relationship between the viruses. Frequently fingerprints of different isolates are compared by simple visual inspection, and similarities in patterns can be perceived when as few as 50% of the diagnostic oligonucleotides spots are shared. The simplest method for confirming genetic relatedness is codigestion and co-electrophoresis of equivalent isotopic amounts [as measured in disintegrations per minute (dpm)] from the two viral RNAs. This approach appears to be sufficient when 50% or more of the oligonucleotide spots comigrate. When fewer spots are shared, additional evidence, such as composition or sequence analysis of corresponding spots, should be obtained to establish genetic relatedness clearly. Moreover, when less than about 25% of the oligonucleotide spots of two different strains comigrate, the remaining pattern similarities are usually difficult to detect, and further comparisons may not be performed unless genetic relatedness is suspected on other grounds, such as epidemiological data or the existence of strains with intermediate fingerprint patterns.

The number of oligonucleotide spots common to two RNA molecules decreases rapidly with differences in their overall sequence homology. For example, Aaronson et al. (1982) used computer-simulated mutation of the genome segment encoding the influenza hemagglutinin to examine the relationship between the number of large oligonucleotides shared and the extent of overall base sequence homology. According to their model, two RNA molecules that diverge at 1, 5, or 10% of their total sequences share 85, 50, and 25%, respectively, of their diagnostic oligonucleotides. These estimates are in good agreement with experimental comparisons of foot-and-mouth disease virus genomes by fingerprinting and competition hybridization (Robson et al., 1979). From these studies it is clear that the lower threshold for recognition of relatedness by fingerprinting is near the level of 90% base sequence homology. Thus, the evolutionary range of fingerprinting is quite short.

The basis for the pronounced sensitivity of oligonucleotide fingerprinting to genetic divergence is evident in view of the mechanisms of fractionation.

Consider a typical diagnostic oligonucleotide of 15 nucleotide residues. A change in only one base (7% of the total) involving C, A, or U residues will result in an altered first-dimension mobility, and the spot will appear to have shifted laterally in the fingerprint of the mutant RNA. The greatest mobility shifts are expected from C→U transitions, the smallest from C→A transversions. Changes involving G residues will either eliminate or create a cleavage site for RNase T1, and spots may appear to move both laterally and vertically within the diagnostic region, or disappear from it altogether. Similarly, new diagnostic spots may arise from other regions of the genome upon mutation of a G residue.

D. Fingerprinting Compared with Other Methods of Virus Identification

With the exception of genome sequencing, other approaches for measuring genetic relatedness, both molecular and serological, are less sensitive than fingerprinting in detecting small mutational changes, but, instead are able to detect more distant relationships among viruses. For example, competition nucleic acid hybridization can detect homologies between strains that share only 25% of their overall sequences (Minson and Darby, 1982). Restriction endonuclease analysis of DNA genomes may be used for quantitative estimates of sequence divergence as great as 20–30% (Brown *et al.,* 1979). Serological methods, especially those involving polyclonal antibodies directed against internal or nonstructural viral protein antigens, are often able to detect conserved homologies in polypeptide structure that are beyond the limits of sensitivity of standard nucleic acid hybridization methods. Genome sequencing has the ultimate sensitivity to detecting mutational differences among viruses, and routine comparative sequence analysis of selected genome regions is possible. However, unlike fingerprinting, current sequencing methods are not well suited for providing a survey of the total sequence divergence among many strains.

III. Experimental Procedures

A. Preparation of Radiolabeled RNA

1. *In Vivo Labeling*

The genomic RNAs of most animal RNA viruses can be labeled to activities suitable for production of high-quality fingerprints by biosynthetic incorporation of ortho[^{32}P]phosphate. The general strategies for isotopic labeling of viral macromolecules have been reviewed by Henry (1967). Conditions for optimal incorporation of radiophosphate vary for different virus and cell culture systems. Important factors to be considered first when developing a labeling strategy are duration of the replicative cycle and the multiplicity of infection of input virus.

For viruses having short infectious cycles and for which high-titer inocula are available, infection occurs as a one-step cycle. Labeling periods may precede or coincide with infection and extend throughout the cycle, or be adjusted to correspond to an optimal interval within the cycle. Efficiencies of incorporation may be improved by maintenance of the cells in phosphate-deficient medium for several hours prior to infection, thus depleting intracellular pools of inorganic phosphate. Because exogenous phosphate equilibrates slowly with intracellular nucleotide pools (Henry, 1967), label should be added well before the expected period of maximal RNA synthesis. Chilling (0–4°C, 60–120 min) infected suspension cultures immediately prior to addition of ^{32}P has been used to improve incorporation of label into poliovirus RNA (Wimmer, 1972). For high-specific-activity labeling, ortho[^{32}P]phosphate is added without carrier to a synthetic medium lacking phosphate. If serum is to be included in the medium, it should be extensively dialyzed against 0.15 M NaCl. The volumes of the labeling medium should be kept to a minimum, and frequent agitation of the medium is recommended for uniform distribution of isotope. The increased cell densities (1–5 × 10^6 cells/ml medium) generally used during radiolabeling may necessitate inclusion of supplemental buffers, such as 25 mM N-2-hydroxyethylpiperazine-N'-2-ethanesulfonic acid (HEPES) or 0.3% sodium citrate, pH 7.2 (Ogra et al., 1968). These buffers are occasionally inhibitory, and their effects on total yield should be determined before routine use.

For systems in which virus production occurs over an extended period or that require multiple rounds of infections, optimal labeling conditions should balance total virus yield against the specific activity of the final RNA product. Cell viability is prolonged by inclusion of some carrier phosphate in the culture medium, and multiple harvests may be required to assure recovery of a high proportion of intact virions. In all systems, virus should be harvested and purified as early as possible to minimize radiation damage from unincorporated isotope.

Typical inputs of ortho[^{32}P]phosphate are 100–1000 μCi/ml culture. The total isotope input should be adjusted according to expected yield of label incorporated into intact viral RNA, which may vary even among closely related virus strains, from nearly 1% to less than 0.01%. In some systems, inclusion of actinomycin D (1–5 μg/ml) in the labeling medium may increase incorporation of ^{32}P into virus RNA by selectively inhibiting host transcription. However, the effects of this antibiotic on labeling efficiencies should be examined carefully before routine use, especially when many different agents are to be compared. For example, some strains of poliovirus, a group generally regarded as actinomycin D resistant, incorporate ^{32}P poorly in the presence of this antibiotic (Nomoto *et al.*, 1979a; Nottay *et al.*, 1981).

The highest quality fingerprints are obtained with RNA labeled *in vivo* and extracted from purified virions. Some rapid diagnostic fingerprinting procedures utilize labeled intracellular viral RNA (La Torre *et al.*, 1982). RNA labeled *in vivo* is frequently described as "uniformly labeled," implying that each phosphodiester bond is labeled to an equivalent specific activity. Uniformity of label is demonstrated by equivalence of base composition whether the RNA is degraded to 3'-nucleotides (by RNase T2 digestion) or to 5'-nucleotides (by digestion with nuclease P1 or snake venom exonuclease). Although demonstrated in some systems (Billeter *et al.*, 1974), deviations from uniformity have been observed in others, complicating determinations of oligonucleotide base composition (Lee *et al.*, 1979). With long-chain oligonucleotides, the specific activity differences among residues tend to average out, so that the radioactivity in a large oligonucleotide labeled *in vivo* is in close proportion to its chain length.

2. In Vitro Labeling

In vitro labeling of viral RNA presents several advantages over labeling *in vivo*. For some viruses, *in vitro* methods offer the only effective means of labeling genomic RNA to high specific activities (Frisby, 1977). Moreover, *in vitro* isotopic labeling can be performed rapidly and requires substantially lower inputs of radioisotope. Only small quantities (1 μg or less) of chemically pure viral RNA are needed for analysis. Repeated *in vivo* labelings are avoided, as it is generally necessary to prepare a viral RNA sample only once, with small amounts being withdrawn for labeling whenever needed. Finally, 5'- and 3'-radiophosphate end-labeled oligonucleotides can be recovered from the fingerprints for sequence analysis.

Three basic approaches are applied for *in vitro* labeling of RNA for fingerprinting.

  a. 5'-End Labeling. The most widely used *in vitro* labeling procedure utilizes T4 polynucleotide kinase (PNK) to transfer radiophosphate from [γ-^{32}P]ATP to the free 5'-hydroxyl groups of RNase-generated oligonucleotides (Frisby, 1977). Alkaline phosphatase is added to the RNase T1 digestion mixture to remove all 3'-terminal phosphates from the oligonucleotides. The combined activities of alkaline phosphatase and PNK catalyze a net transfer of phosphate from the 3'- to the 5'-ends, a process that does not significantly alter the electrophoretic mobilities of the treated oligonucleotides (Brownlee and Sanger, 1967). Also, because some PNK preparations contain a contaminating 3'-phosphatase activity (Szekely and

Sanger, 1969), prior removal of the 3'-terminal phosphates eliminates potential charge and migration heterogeneity brought about by partial dephosphorylation of an oligonucleotide. Several approaches have been used to eliminate alkaline phosphatase activity from the subsequent PNK reaction. These include removal by phenol extraction (Frisby, 1977), inactivation with 0.1 N HNO$_3$ (Lee and Fowlks, 1982), or inhibition by inclusion of inorganic phosphate in the PNK reaction (Chaconas and van de Sande, 1980). The last approach, included in the labeling method of Pedersen and Haseltine (1980b), is operationally very simple. We have used their method, described below, to obtain high-quality fingerprints of viral RNA.

Dried RNA (200–500 ng) in a 0.5-ml polypropylene tube (Eppendorf) is resuspended in 1 μl of sterile water to which is added 1 μl of 40 mM Tris-HCl (pH 8.0) containing for each microgram of RNA 0.2 unit of RNase T1 (Sankyo, distributed by Calbiochem) and 2 × 10^{-4} units of *Escherichia coli* alkaline phosphatase (BAP) (Worthington BAPF; P/L Biochemicals). It is important to omit EDTA, normally present during RNase T1 digestion of prelabeled RNA, because it is a strong inhibitor of BAP. For less than 100 ng of RNA, addition of 100 ng of polyguanylic acid (Miles Biochemicals) to the digests facilitates cleavage by RNase T1 (Pedersen and Haseltine, 1980b). Incubation is for 30 min at 37°C. The PNK reaction immediately follows by addition of 50 μl of PNK reaction mixture and 5 units of PNK (P/L Biochemicals; New England Nuclear). The PNK reaction mixture is prepared by first lyophilizing 50–200 μCi [γ-^{32}P]ATP in a 0.5-ml Eppendorf tube, then dissolving the labeled ATP (>2000 Ci/mmol) in 50 μl of 10 mM K$_2$HPO$_4$–K$_3$PO$_4$ (pH 9.5), 10 mM Mg(OAc)$_2$, 5 mM dithiothreitol. Incubation is at 37°C for 1–16 hr. Because the rates of phosphorylation by PNK vary widely for different oligonucleotides (Frisby, 1977), prolonged incubation favors more uniform incorporation of label into the oligonucleotides. However, with increased incubation, nuclease contamination of the PNK preparations may become evident, and the optimal conditions may involve a compromise between maximizing uniform incorporation and minimizing the appearance of secondary spots. The reaction is terminated by addition of 50 μl of 0.6 M NH$_4$OAc, and 100 μg of carrier yeast tRNA. The oligonucleotides and carrier are precipitated by mixing with 300 μl of cold (-20°C) 95% ethanol and chilling in a dry ice-ethanol bath for 20 min. The precipitate is collected by centrifugation in a microcentrifuge (15 min at 4°C), washed with 400 μl of cold 95% ethanol, and centrifuged again (5 min at 4°C). After lyophilization, the labeled digest is ready to be resuspended in first dimension electrophoresis sample buffer.

b. 3'-End Labeling. Use of the base-specific chemical cleavage method of Peattie (1979) for sequence determination of oligonucleotides requires radiolabeling of the 3'-ends. Currently, this is most frequently accomplished by use of the T4 RNA ligase-mediated transfer of cytidine 3'-[5'-^{32}P]diphosphate (pCp) to the 3'-terminal hydroxyls of dephosphorylated oligonucleotides (England and Uhlenbeck, 1978). Lee and Fowlks (1982) have combined this method with fingerprinting to produce 3'-labeled oligonucleotides by a simple procedure. Only an outline of the procedures is given here, as detailed descriptions are presented in the original articles. Digestion with RNase T1 and BAP is performed essentially as described above in the subsection on 5'-end labeling. Afterward, BAP is inactivated by making the digest mixture 100 mM in HNO$_3$ (pH 2; 10 min at room temperature), followed by addition of ZnSO$_4$ to a final concentration of 1 mM (2 min at room temperature) to inactivate RNase T1. The digest is precipitated with ethanol, washed with 95% ethanol, and lyophilized. The precipitate is resuspended in the ligase reaction system (30 μl) containing 50 mM HEPES (pH 8.3), 5 mM ATP, 10 mM MgCl$_2$, 3.3 mM dithiothreitol, 10% (v/v) dimethyl sulfoxide, 15% (v/v) glycerol, 200 μCi [5'-^{32}P]pCp, and 6 units of T4 RNA ligase (P/L Biochemicals). Incubation (4°C, 16 hr) is terminated by ethanol precipitation of the labeled oligonucleotides as described for 5'-end labeling.

   c. Radioiodination. Commerford (1971, 1980) described a method for the radioiodination of cytosine residues of RNA in the presence of thallium ions that is applicable to oligonucleotide fingerprinting (Robertson *et al.*, 1980; Clewley and Bishop, 1982; Lee and Fowlks, 1982). Radiolabeling with ^{125}I has the advantages of being very rapid, producing labeled RNA with high specific activities and having a longer half-life (60 days) than ^{32}P-labeled RNA (14.3 days). The factors governing the efficiencies of iodine labeling are discussed in detail elsewhere (Huang and Pagano, 1977; Commerford, 1980). We have used the method of Lee and Fowlks (1982), described below, to produce fingerprints of high quality.

   From 1 to 20 μg of viral RNA in water (5 μl in a 1.5-ml Eppendorf tube) is combined with 12.5 μl of 0.3 mM thallic nitrate buffer: 30 μl of 0.1 M thallic nitrate [Tl(NO)$_3$; K and K; Alfa] in 1 N HNO$_3$ mixed with 9.97 ml of 0.2 M NaOAc (pH 4.4). Carrier-free ^{125}I (0.3–1.5 mCi, pH 7–11) is added, and the total volume is brought to 25 μl with water. After 4 min of incubation at 70°C, 475 μl of sodium dodecyl sulfate (SDS) buffer [10 mM Tris-HCl (pH 7.5), 0.12 M NaCl, 0.1% SDS, 1 mM EDTA] is added, and the mixture is incubated for an additional 10 min at 70°C. Then 1 ml of cold 95% ethanol is added, and the mixture is chilled in dry ice–ethanol for 20 min, then centrifuged (30 min, 4°C). The precipitate is washed four or five times by gently filling the tube with cold 80% ethanol and decanting. The final supernatants are monitored with a Geiger counter; after confirmation that no soluble label remains, the pellet is lyophilized. The labeled RNA is ready for digestion with RNase T1.

   d. Comments. Fingerprints produced from oligonucleotides labeled *in vitro* generally correspond well with fingerprints of the same RNA labeled *in vivo*. However, depending on the method of labeling, small differences may exist. As mentioned, spots of 5'- or 3'-end-labeled oligonucleotides may vary in intensity as a result of preferential activities of PNK and RNA ligase (Frisby, 1977; Lee and Fowlks, 1982). Moreover, the total number of spots may differ among fingerprints labeled *in vivo* and *in vitro*. Spots of 5'-labeled material, for example, may be missing if PNK phosphorylation is very inefficient or blocked by the presence of a 5'-terminal cap structure (Banerjee, 1980) or a protein (Wimmer, 1982). Spots may also be missing from fingerprints of 3'-labeled material (Lee and Fowlks, 1982), possibly as a result of incomplete dephosphorylation. Secondary oligonucleotide fragments, produced by trace nucleases, may also be labeled, thus giving rise to additional spots (Szekely and Sanger, 1969).

Fingerprints of iodinated RNA may also be deficient in certain spots. Because radioiodination of RNA occurs primarily by the production of 5-iodocytosine, 5-iodouracil being a minor product (Commerford, 1971), only cytosine-containing oligonucleotides are effectively labeled. Although the larger diagnostic oligonucleotides generally contain some cytosine, the spots of cytosine-deficient oligonucleotides may be light. The poly(A) streak observed in the fingerprints of *in vivo*-labeled RNA of many positive stranded viruses is absent from fingerprints of iodinated RNA.

Fingerprints produced with *in vitro*-labeled material are generally satisfactory for comparisons of the genomic RNAs of different viruses. Only when the fingerprints of the most closely related strains are being compared does the potential existence of artifactual spots present uncertainties in estimating the extent of sequence divergence.

B. Digestion of RNA

Digestion of *in vivo*-labeled RNA or iodinated RNA is performed in 1.5-ml Eppendorf tubes. Viral RNA plus 100 μg of carrier yeast tRNA are dissolved in 1 ml of 20 mM Tris-HCl (pH 7.8), 150 mM NaCl, transferred to a 1.5-ml Eppendorf tube, and mixed with 200 μl of 1% (w/v) cetyltrimethylammonium bromide (CTAB; obtained from Sigma or BDH). The CTAB precipitation step purifies the RNA of uncharged polymeric contaminants and nucleases that are precipitated by ethanol (Ralph and Bergquist, 1967). After 20 min on ice, the insoluble cetyltrimethylammonium ribonucleate salts are collected by centrifugation in a microcentrifuge (5 min at 4°C). The flocculent pellet is washed twice with cold 70% ethanol/30% 150 mM NaCl in water, to convert the RNA back to the sodium salt, then once with 70% ethanol–water, and finally with 95% ethanol. Each wash step is followed by 2 min centrifugation at 4°C. The RNA is dried under vacuum, and the pellet is resuspended in 20 μl of 10 mM Tris-HCl (pH 7.5), 1 mM EDTA, containing 0.2 unit of RNase T1 per μg RNA (enzyme: RNA mass ratio of 1:25). Samples of double-stranded RNA are denatured by boiling for 2 min and quick chilling (Walker *et al.*, 1980) or freezing (Sugiyama *et al.*, 1982) before addition of enzyme. Digestion is at 37°C for 30 min. To 20 μl of digest is added 40 μl of urea–dye mix (9 M urea/50% glycerol saturated with bromphenol blue/0.3% trypan red).

Digestion of RNA with RNase A (Calbiochem) is at 37°C for 30 min in 10 mM Tris-HCl (pH 7.5), 1 mM EDTA, 0.3 M NaCl, at an enzyme:RNA mass ratio of 1:10. Increased ionic strength suppresses secondary cleavage by RNase A at adenosine residues. However, the salt will impair resolution in the first dimension and should be removed by ethanol precipitation prior to electrophoresis (Nomoto *et al.*, 1981).

Stocks of the RNase T1 and RNase A are stored as small aliquots in their respective digestion buffers at −20°C at concentrations of 1 mg/ml.

C. Fingerprinting Methods

1. *Two-Dimensional Gel Electrophoresis*

Many variations of the basic technique of De Wachter and Fiers (1972) have been described (Frisby *et al.*, 1976; Lee and Wimmer, 1976; Kennedy, 1976; Clewley *et al.*, 1977b; Pedersen and Haseltine, 1980a,b; Stewart and Crouch, 1981). Our fingerprinting procedures essentially follow the modifications described by Lee and Wimmer (1976) and Lee *et al.* (1979).

a. Reagents. All stock solutions (Tables I and II) should be prepared with deionized water using chemical reagents of the highest grades available. Especially critical to the production of high quality fingerprints are the chemical purities of the acrylamide N,N'-methylene bisacrylamide (Bis) and urea components.

Highly purified, electrophoresis-grade acrylamide and Bis from commercial sources (e.g., Bio-Rad Laboratories; Eastman Organic Chemicals) can be used without further purification. Reagent-grade acrylamide can be purified by recrystallization in chloroform, and Bis by recrystallization in acetone, as described by Adesnik (1971). Alternatively, monomers can be mixed in the specified ratios (Tables I and II) and purified by stirring (12 hr, room temperature) with mixed-bed resin (20 g/liter; Bio-Rad AG 501-X8), followed by filtration through Whatman No. 1 paper. Purified solutions are passed through a Whatman GF/C filter and stored in the dark at 4°C.

Solutions of ultrapure urea (Research Plus; BDH AnalAR) require only filtration (0.22 μm) before use. Reagent grade urea is purified by stirring

TABLE I

Stock Solutions for First-Dimension Gel System

Component	Volume required for one gel
Acrylamide: Bis (40:0.5)[a]	18.0 ml
Urea, 9M	60.0 ml
Deionized H_2O	8.5 ml
Saturated citric acid	0.7–1.0 ml
$FeSO_4 \cdot 7H_2O$ [0.16% (w/v)]	0.5 ml
Ascorbic acid [1% (w/v)]	2.0 ml
H_2O_2 (30%)	75.0 μl

Electrode buffer: 25 mM citric acid adjusted to pH 3.3 with 10 N NaOH

Sample buffer: 20 μl of RNA digest mixed with 40 μl of 9 M urea–50% glycerol, saturated with bromphenol blue–0.3% trypan red

[a] Acrylamide (400 g/liter) and N,N'-methylene bisacrylamide (5 g/liter).

(12 hr) at 9 M solution with mixed-bed resin (20 g/liter). The urea solution is further purified by filtration through a bed of silicic acid (Lee and Wimmer, 1976), prepared by suspending about 100 g of powdered silicic acid (Sigma SIL-R) in 1 liter of deionized H_2O, and filtering the suspension under suction through a 15-cm filter of Whatman No. 4 paper. Solutions of purified urea may be stored for up to 3 months at 4°C.

Saturated citric acid is stored at room temperature.

TABLE II

Stock Solutions for Second-Dimension Gel System

Component	Volume required for one gel	
	Connecting gel	Separating gel
Acrylamide:Bis (23.2:1.55)[a]	—	209.0 ml
Acrylamide:Bis (40:0.5)[b]	2.5 ml	—
Tris–borate, 1 M, adjusted to pH 8.2[c]	1.0 ml	11.0 ml
Deionized H_2O	6.4 ml	—
TEMED	25 μl	0.1 ml
Ammonium persulfate (10%)	125 μl	0.4 ml

Electrode buffer: 50 mM Tris–borate (pH 8.2)

[a] Acrylamide (232) g/liter and N,N'-methylene bisacrylamide (15.5 g/liter)
[b] From Table I.
[c] Tris base (121 g/liter) and boric acid (63 g/liter).

Ferrous sulfate [0.16% (w/v), in 1-ml aliquots] and ascorbic acid [1% (w/v), in 2-ml aliquots] are filtered (0.22 μm) and stored at −20°C.

Hydrogen peroxide (30%) is dispensed into 1-ml volumes and stored at 4°C.

Electrophoresis grade N,N,N',N'-tetramethylethylenediamine (TEMED; Bio-Rad Laboratories) is stored at 4°C.

Solutions of ammonium persulfate (10%) are prepared fresh before use.

Solutions of Tris-borate (1.0 M; 121 g/liter Tris base and 63 g/liter boric acid adjusted to pH 8.2) are autoclaved and stored at room temperature.

   b. Apparatus. First-dimension plates (12 × 45 cm) are cut from 1/8-in. glass. Second-dimension plates are made from 3/16 in. glass cut to 36 × 43 cm. Spacers are 1/16 in. (0.16 cm) × 1 cm strips of polyvinyl chloride, cut to lengths of 45 cm for the first-dimension assembly and 43 cm for the second-dimension plates. The first-dimension well former is 1/16 in. Teflon with two or three teeth, 1.25 cm wide × 4 cm long, separated by 1.25 cm. Plates should be cleaned immediately after use with a mild nonabrasive detergent, rinsed thoroughly with deionized water, and dried. Just before use, plates are washed with 95% ethanol and air dried. Cleaning acid should be avoided, because the glass surface may be altered, causing the first-dimension gels to adhere so tightly to the glass that removal and transfer are prevented.

   The plates and spacers are assembled to form a mold for the gel, and the assembly is clamped with medium steel binder clips. Precautions must be taken to prevent leakage of the unpolymerized gel from the mold. We prefer to seal the sides and bottom with 1-in. Teflon tape (3M catalog No. 5490) taking special care to reinforce the corners, where leaks are most prone to occur. Alternatively, spacers may be coated with a thin film of silicone vacuum grease before assembly. Chamber bottoms may also be sealed by use of a temporary spacer (Clewley *et al.,* 1977b) or with plasticine (De Wachter and Fiers, 1972). Problems of leakage of the first-dimension gels are reduced by setting the chamber at a nearly horizontal angle during polymerization. To assure that the first-dimension gel strip is in firm contact with the glass plates during polymerization of the second-dimension gel, the bottom of the assembly is clamped with several steel binder clips, and air pockets are removed by hydrostatic pressure. Good contact can be achieved also by using slightly thinner spacers for the second-dimension chamber than those used for the first dimension (Pedersen and Haseltine, 1980a), or by using spacers of compressible silicone rubber (Clewley *et al.,* 1977b).

   The electrophoresis units used in our laboratory, purchased from Lee's Unique Instruments (De Soto, TX), are shown in Fig. 3. The slab gel chambers are placed in the lower buffer reservoirs and connected to the upper reservoirs with filter paper wicks of Whatman 3 MM or equivalent, precut

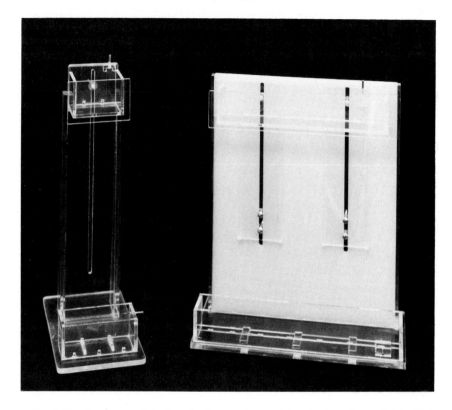

Fig. 3. The electrophoresis buffer chambers used in our laboratory for fingerprint analysis: first-dimension apparatus (left), and second-dimension apparatus (right). The units are constructed of Plexiglas and can be purchased from Lee's Unique Instruments (De Soto, Texas).

to 9 × 9 cm (first dimension) and 6.5 × 34 cm (second dimension). The wicks are saturated with electrode buffer, and the exposed parts are wrapped in Saran wrap or between Mylar sheets to prevent evaporation. Upper and lower reservoir capacities are 375 ml each for the first-dimension unit and 1000 ml each for the second-dimension apparatus. Although the second-dimension unit can accommodate up to three gels, we use only one gel per unit and do not recirculate reservoir buffers.

c. First-Dimension Electrophoresis. The first-dimension gel is 8% polyacrylamide (acrylamide:Bis ratio, 80:1), 6 M urea, pH 3.3. The electrode buffer is 25 mM citric acid adjusted to pH 3.3 with 10 N NaOH. Stock solutions and mixing ratios are given in Table I, and the conditions for electrophoresis are shown schematically in Fig. 4.

The acrylamide–urea solution is titrated to pH 3.3 with saturated citric

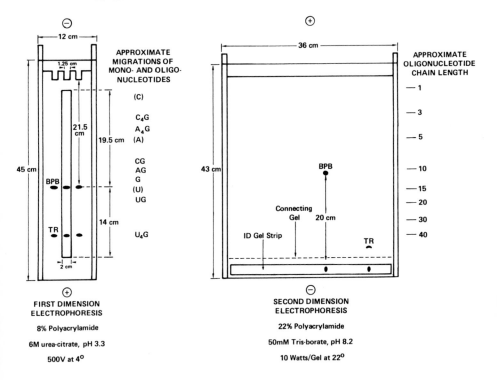

FIG. 4. Schematic representation of the two-dimensional polyacrylamide gel electrophoresis fingerprinting method. Separation in the first dimension is according to oligonucleotide charge, which follows from composition. The approximate positions relative to the dye bands of the 3′-mononucleotides and some small oligonucleotides are shown. The 3′-mononucleotides of cytosine (C), adenosine (A), and guanosine (G) are not produced by RNase T1 digestion, as indicated by the enclosure of their symbols in parentheses. Second-dimension mobilities decrease with the logarithm of the oligonucleotide chain lengths.

acid (usually less than 1 ml is required; substantially greater amounts indicate the presence of impurities in the urea). The solution is warmed to 37°C, then mixed with the catalysts of Jordan and Raymond (1969) in the order $FeSO_4 \cdot 7H_2O$, ascorbic acid, and hydrogen peroxide, and poured to within about 1.5 cm from the top of the prewarmed (37°C) mold. A 9 × 10 cm sheet of Parafilm, inserted about 1 cm into the mold and supported and curved with one hand, serves as a convenient pouring guide. Bubbles are removed by briefly holding the mold upright. The mold is then set at an angle of about 15°, and the well former is inserted about 2 cm into the gel.

After polymerization (30 min), the bottom tape seal and the well former are removed, and the top of the gel is washed with and then overlayed with electrode buffer. The sample (up to 60 μl and 100 μg of RNA digest), in loading buffer, is briefly heated (60°C for 3 min) to disrupt aggregates, and layered in the sample well with a microsyringe (Hamilton). The loaded gel is transferred to the electrophoresis unit in a cold room, and the wick is inserted, saturated with electrode buffer, and wrapped. Electrophoresis is performed at 4°C at a constant potential of 450 V (about 7 mA). Total power (watts = volts × amperes), a measure of heat production, should not exceed 3.2 W. The run is terminated when the bromphenol blue dye marker (green at pH 3.3) migrates 21.5 cm from the bottom of the sample well (about 16 hr). The trypan red marker migrates ahead of the bromphenol blue (Fig. 4), but behind the leading oligonucleotide fragments.

   d. *Second-Dimension Electrophoresis.* After first-dimension electrophoresis, a glass plate is removed, and, using a Plexiglas template as a guide, the gel is cut with a rotary pizza cutter into 2.5 cm-wide strips containing the migration lanes of oligonucleotides. The strips are cut at their ends 19.5 cm above and 14 cm below the bromphenol blue dye mark (Fig. 2). A piece of Parafilm, about 4 × 45 cm, is centered and then pressed gently on top of a gel strip, with care taken to remove air pockets. One end of the Parafilm and gel strip is gently lifted with forceps and pulled directly back to transfer the gel, without stretching, from the glass plate to the Parafilm strip. At this point the gel strips may be frozen at −70°C if desired. Otherwise, each strip is soaked for 5 min in 500 ml of second-dimension electrode buffer (50 mM Tris–borate, pH 8.2) to remove urea, the presence of which will otherwise inhibit polymerization at the junction between the first- and second-dimension gels. The lower reservoir of the second-dimension electrophoresis apparatus is a convenient vessel for this operation, after which the buffer is discarded. The strip is drained of buffer and placed, gel side up, along the edge of a laboratory bench. A second-dimension glass plate is placed over the gel, and contact is made when the strip is centered, parallel to, and about 1–2 mm from, the short (36 cm) edge of the plate. The ends of the Parafilm are folded over the plate to hold the gel in place, and then the plate is turned over, and the Parafilm is removed. After assembling, sealing, and clamping the chamber, about 10 ml of second-dimension electrode buffer is dispensed into the chamber, and the chamber is placed upright in a tank of cold tap water (40 cm deep) to allow the hydrostatic pressure to express bubbles between the gel and the glass plate. The chamber is removed from the tank, and the buffer is allowed to overlay the gel for 10 min, then poured out.

   The second-dimension gel is 22% polyacrylamide (acrylamide:Bis ratio, 15:1) in 50 mM Tris-borate, pH 8.2. Components and mixing volumes for the second-dimension gel are given in Table II. A 10% low-cross-linked polyacrylamide connecting gel is cast first to form a sealant and junction between the first- and second-dimension gels. After addition of catalysts, about 5 ml of connecting gel solution is pipetted down a spacer; the chamber is quickly tilted from side to side to assure good coverage of the gel strip and then set on a level surface for polymerization (1–2 min). The second-dimension separating gel solution is transferred (via tubing attached to a glass reservoir) until the chamber is about one-third full. The chamber is tilted as before to mix any remaining unpolymerized connecting gel, placed upright in the water tank, filled to about 2 cm from the top, and then overlaid with water. Immersion at this stage serves two purposes. The cold water acts as a heat sink to reduce convection during polymerization, and the balance in hydrostatic pressure retards leakage of material from the chamber (Lee and Wimmer, 1976). Polymerization is complete in about 30–40 min.

Electrophoresis is performed at room temperature at a constant voltage of 600 V (about 16 mA/gel), or at a constant power of 10–12 W/gel. Regulation at constant power is preferred because as the resistance of the second-dimension gel system increases during the run, the voltage may be allowed to rise (up to 1000 V) under conditions of constant Joule heating. For recovery of all size classes of oligonucleotides, electrophoresis is terminated when the bromphenol blue dye migrates 20 cm (about 16 hr). Although separation in the first dimension is limited by the size of the gel strip that can be accommodated by the second-dimension gel mold, resolution of the diagnostic RNase T1-resistant oligonucleotides is improved in the second dimension by continued electrophoresis. Migration of the bromphenol blue dye marker to 30 cm or more is frequently advantageous.

e. Comments. Several conditions must be carefully controlled to assure production of high-quality fingerprints. It is particularly important to avoid excessive heating of the gels during electrophoresis. Overheating in the first-dimension separation may cause extensive streaking of the oligonucleotides. Poor contact between the first- and second-dimension gels, caused, for example, by incomplete removal of urea, may also result in localized heating and thus produce distorted fingerprints. Bubbles in the gels or nonuniform polymerization of the second-dimension gel can also cause distorted patterns.

The formulations described here incorporate several important modifications of the original De Wachter and Fiers (1972) fingerprinting system. The Bis cross-linker content is lower in the first-dimension and higher in the second-dimension gels. Urea is omitted from the first-dimension electrode buffer. First-dimension electrophoresis is run at pH 3.3 because it was found empirically that better separation of the oligonucleotides was obtained (Lee and Wimmer, 1976). The original Tris–citrate second-dimension buffer system is replaced by a Tris–borate system, of lower conductivity (Frisby *et al.*, 1976), thus allowing for higher voltage gradients. Additional changes have been introduced by other investigators. Polymerization of the first-dimension gel may be catalyzed with TEMED and ammonium persulfate (Holland *et al.*, 1979). Pedersen and Haseltine (1980b) omitted urea in the first-dimension gels and included 7 M urea in the second-dimension gels. Shorter oligonucleotides are better resolved, and the urea-containing gels may be frozen without breaking at $-70°C$, allowing for higher fluorescent yields with intensifying screens (Laskey, 1980). Second-dimension gels may also be frozen if they contain 0.5% linear polyacrylamide (Freeman *et al.*, 1979; Stern and Kennedy, 1980).

2. *Homochromatography*

The original systems for fingerprinting complete viral genomes were based on the homochromatographic technique of Brownlee and Sanger (1969). Homochromatography was developed to overcome problems encountered earlier in resolving long-chain oligonucleotides (Sanger *et al.*, 1965), and its availability opened the way for a series of important structural and comparative studies of RNA viral genomes (Horst *et al.*, 1972; Duesberg and Vogt, 1973; Billeter *et al.*, 1974; Wang *et al.*, 1975). Homochromatography is also used in sequence analysis of end-labeled RNA (Rensing and Schoenmakers, 1973). A brief description of the elements of this method is given here, and further details are given by Brownlee and Sanger (1969) and Barrell (1971). A mini-fingerprinting technique, in which second-dimension homochromatographic separation is performed on small polyethyleneimine (PEI)-cellulose thin-layer plates, has been described by Volckaert *et al.* (1976).

Separation in the first dimension is by high-voltage electrophoresis on cellulose acetate strips in 5% acetic acid–0.5% pyridine (pH 3.5), 7 M urea. Oligonucleotides are transferred by blotting from the cellulose acetate strip to a thin-layer plate of DEAE-cellulose. After washing the plate with ethanol to remove urea, second-dimension fractionation is performed by ascending chromatography at 60°C. The chromatographic solvent, called a homomixture, contains 7 M urea and 5% yeast tRNA (previously partially base hydrolyzed, then dialyzed against water). Other homomixtures, differing in the pretreatment of the tRNA, are used to resolve shorter chain oligonucleotides. Separation in the second dimension is based upon the relative efficiencies at which the labeled oligonucleotides are displaced by the polynucleotides of the homomixture. Short-chain low-valence oligonucleotides are more readily displaced along the DEAE stationary phase by the polynucleotides of the mobile phase, and thus migrate more rapidly than oligonucleotides with longer chains and higher valences. As a result, oligonucleotides are resolved in the second dimension primarily according to their chain lengths. However, the contributions of base composition are important, and pronounced deviations from an inverse relationship between mobility and molecular weight are frequently observed.

3. One-Dimensional Fingerprints

A rapid one-dimensional fingerprinting procedure has been used to distinguish isolates of foot-and-mouth disease virus (La Torre *et al.,* 1982). RNase T1 digests were resolved according to chain length on sequencing gels of 20% polyacrylamide, 8.3 M urea, 1× 90 mM Tris–90 mM boric acid (pH 8.3)–2 mM EDTA (1× TBE) (Sanger and Coulson, 1978). Patterns were distinguished by the migrations of the larger oligonucleotides as well as by the relative migrations of the tracts of polycytidylic acid [poly(C)], a genomic feature peculiar to the aphthovirus and cardiovirus groups. Band patterns of closely related viruses were very similar, but isolates with distinct two-dimensional fingerprints also had different one-dimensional band patterns. Although much less sensitive to small genetic differences than two-dimensional procedures, this approach may be generally useful as a preliminary screen for genetic relatedness.

D. Autoradiography

After electrophoresis, one plate is removed and the gel is covered with polyethylene wrap. A screen-type X-ray film, 35 × 43 cm (e.g., DuPont Cronex 4, Kodak XRP, Kodak XAR, or Fuji RX), is placed over the gel. The film may be preflashed to improve sensitivity (Laskey, 1980). A calcium tungstate-type intensifying screen (e.g., DuPont Lightning Plus or Fuji Mach 2) is placed over the film, and a Plexiglas plate (36 cm × 43 cm × 1/4 in.) is set on top to assure good contact between the gel and the film. The assembly is securely taped to prevent accidental displacement of the film along the gel, and sealed in a black plastic bag (Picker International) for exposure at 4°C. Good fingerprints are obtained with picornavirus RNA (approximately 7500 bases) after 16 hr of exposure when 5 × 10^6 dpm are applied to the gel. Because diffusion of the larger oligonucleotides in 22% polyacrylamide is very slow at 4°C, gels can be exposed for up to 2 weeks without detectable loss of resolution. Well-defined fingerprints can thus be obtained with as little as 10^5 dpm of RNA.

E. Analysis of Isolated Oligonucleotides

1. Extraction from Polyacrylamide Gels

If the oligonucleotides are to be analyzed further, reference spots of ^{32}P-labeled radioactive ink are made on the polyethylene sheet at each corner of the gel and covered with clear tape. Oligonucleotides, located by alignment of reference spots on the developed film and the gel, are excised with a sterile scalpel or cork borer and transferred to a 1.5-ml Eppendorf tube. Oligonucleotides can be recovered with good yields (>80%) by mechanical extraction. The brittle second-dimension gel piece is crushed with a flame-sealed Eppendorf pipet tip (bulb diameter 2–3 mm) attached to a glass rod (Lee and Wimmer, 1976). Excessive grinding should be avoided to prevent substantial release of linear polyacrylamide, which copurifies with nucleic acids. The crushed gel is soaked for 2 hr at 37° C in 750 μl of elution buffer [0.5 M NH$_4$ OAc, 10 mM Mg(OAc)$_2$, 0.1% SDS, 2 mM EDTA] (Maxam and Gilbert, 1980) containing 50 μg of carrier tRNA. The gel particles are removed by centrifugation (500 g, 2 min) through a spin filter (see below), washed with 200 μl of elution buffer, and centrifuged again. To the filtrate is added 2.5 ml of cold 95% ethanol. After 20 min in a dry ice–ethanol bath, the precipitate is pelleted in a swinging-bucket rotor (10,000 g, 4°C, 30 min), washed with 95% ethanol, centrifuged again (10,000 g, 4°C, 5 min), and lyophilized.

The spin filter set consists of a 1-ml Eppendorf pipet tip plugged with silanized glass wool [Alltech Associates, or prepared as described by Maxam and Gilbert (1980)], mounted through an adapter to the top of a 12 × 75 mm polypropylene tube. The reusable adapters are made from 1.5-ml Eppendorf tubes from which the bottoms and the snap caps have been removed.

A different approach is described by Stewart and Crouch (1981). The entire fingerprint is transferred to DEAE paper by uniform suction of elution buffer through the gel. After drying, the oligonucleotides are stably associated with the DEAE paper, and the transferred fingerprints are visualized by autoradiography with intensifying screens at −70°C. Individual oligonucleotides are excised from the DEAE paper and efficiently recovered by elution with triethylamine carbonate, which is removed by lyophilization.

2. Composition Analysis

The base compositions of uniformly labeled oligonucleotides can be determined after complete secondary digestion with RNase A. The double digestion products have the general structure (Ap)$_{0-n}$Xp, where X = C, G, or U. Because Gp is present only at the 3′ terminus of an RNase T1-generated oligonucleotide, only one G-containing product should be observed upon secondary digestion, with all other base residues present in integral molar ratios relative to Gp. Two basic approaches are most commonly used to identify the products of secondary digestion. The first, described in detail by Barrell (1971), utilizes high-voltage electrophoresis on DEAE paper at pH 3.5. The products are identified by their mobilities relative to markers of known composition. Identification of longer products (three or more A residues) can be confirmed by elution from the DEAE, digestion with RNase T2 (non-base specific), and electrophoresis as before at pH 3.5. In the second method, developed by Volckaert and Fiers (1977), two-dimensional chromatographic separation occurs on small (6.7 × 10 cm) PEI-cellulose thin-layer plates (CEL 300, Machery-Nagel, Brinkmann). Samples are spotted 1 cm from each edge and resolved in two steps for each dimension. Elution in the first dimension (along the short axis) is first with water until the front moves 1.5 cm; then the plate is resolved with 20–22% formic acid until the solvent reaches the top. After the plate is air dried, second-dimension separation is at pH 4.3, first in 0.1 M formic acid–pyridine (to 2 cm), then in 1.0 M formic acid–pyridine until the front reaches the top. In the acidic conditions of the first dimension (pH < 2), Ap and Cp have no net negative charges and thus have minimal ionic interaction with the PEI support. Because migration rates increase with decreasing negative charge, the digestion products are resolved primarily according to the identities of their 3′-terminal bases, in the order Cp > Gp > Up. At pH 4.3, migration of the products is reduced with increasing chain length. Thus the base composition of a digestion product can be deduced from its position on the chromatogram.

3. *Oligonucleotide Sequence Determination*

For determination of sequences, either 5′- or 3′-^{32}P end-labeled oligonucleotides can be utilized in the technique of partial digestion with base-specific RNases (Simoncsits *et al.,* 1977; Donis-Keller *et al.,* 1977; Donis-Keller, 1980). A commercial kit for enzymatic sequencing of RNA is available from P/L Biochemicals.

Sequences can also be determined from oligonucleotides labeled at either the 5′- or 3′-ends by fingerprinting the products of partial alkaline hydrolysis (wandering-spot analysis). The techniques of separation in wandering-spot analysis are essentially the same as described for total genome fingerprinting and can utilize either electrophoresis–homochromatography (Rensing and Schoenmakers, 1973) or two-dimensional polyacrylamide gel electrophoresis (Nomoto *et al.,* 1981). By examining partial hydrolyzates, wandering-spot analysis can reconstruct the effects on the mobility of an oligonucleotide upon stepwise removal of each base residue. Because, for first-dimension mobilities, removal of a cytosine residue causes the greatest shift toward the anode and removal of a uridine residue causes the greatest shift toward the cathode, wandering-spot analysis is most efficient at locating pyrimidines. Enzymatic sequencing, in contrast, most readily distinguishes purine residues, because of the base specificities of the RNases T1 (G) and U2 (A + G). The complementary strengths of the two sequencing methods may be combined to sequence oligonucleotides (Darlix *et al.*, 1979; Rommelaere *et al.*, 1979; Nomoto *et al.*, 1981). Adenosine residues are located by partial cleavage with RNase U2 and separation on sequencing gels. The same partial hydrolyzate used to form the "ladder" in the sequencing gel is also fingerprinted, and the positions of C and U residues are determined.

The third direct RNA sequencing technique, involving base-specific chemical cleavages, may also be used, but determinations can be performed only with oligonucleotides labeled at their 3'-ends (Peattie, 1979).

F. Isolation of Individual RNA Molecules for Fingerprinting

1. *Gel Electrophoresis*

Resolution of individual genome segments, subgenomic transcripts, or full-length viral genomes from defective-interfering RNA species for separate fingerprint analysis is usually achieved by gel electrophoresis. The choice of gel system depends upon the size range of the RNA molecules and whether the chains are single or double stranded.

Large, single-stranded RNA molecules over the size range of $0.5–4 \times 10^6$ can be separated on preparative gels of the high-resolution acid agarose–urea system (Lehrach *et al.*, 1977). The gel contains 1.5% low-gelling-temperature agarose (Marine Colloids, distributed by Miles Laboratories), 6 M urea, and 25 mM sodium citrate (pH 3.5). The electrode buffer is 25 mM sodium citrate (pH 3.5). If the gels are run horizontally, 6 M urea is included in the electrode buffer. Better resolution is generally obtained with vertical gels, cast between frosted glass plates and supported by a 5% polyacrylamide plug. Electrophoresis is at 4°C at a potential of 2–5 V/cm. RNA is eluted from gel slices by dilution in 5–10 volumes of 20 mM Tris-HCl (pH 7.8), 1 mM EDTA, and melting the mixture at 65°C for 5 min. RNA is recovered from the mixture by phenol extraction, reextraction of the interphase, and ethanol precipitation of the combined aqueous phases (Weislander, 1979). If the RNA is to be subsequently labeled *in vitro*, oyster glycogen (Calbiochem) can be substituted for tRNA as a carrier. Two micrograms of glycogen is used for each microgram of tRNA.

Single-stranded RNA molecules over the size range $0.3–1.5 \times 10^6$ (e.g., influenza virus genome segments) can be resolved on slab gels containing 2–3% polyacrylamide (acrylamide:Bis 20:1), 6 M urea, with both the gel and electrode buffers consisting of Loening's E buffer [36 mM Tris–30 mM NaH$_2$PO$_4$ (pH 7.8)–1 mM EDTA] (Scheurch *et al.*, 1975; Loening, 1969). RNA can be recovered from the gels by mechanical extraction essentially as described for isolation of oligonucleotides.

Good resolution of double-stranded RNA molecules is obtained on a preparative scale by electrophoresis on gels containing 7.5% acrylamide, 0.2% Bis, 6.3 M urea, 0.1% SDS, in Loening's E buffer. The electrode buffer is Loening's E buffer containing 0.1% SDS (Ito and Joklik, 1972). The discontinuous SDS–polyacrylamide gel system of Laemmli (1970) may also be used for preparative isolation of double-stranded RNA molecules (Clewley and Bishop, 1979a). Double-stranded RNA molecules may be recovered from polyacrylamide gels by mechanical extraction or by electroelution into dialysis membranes (Scheurch et al., 1975; McDonell et al., 1977). Gels are soaked for several hours in deionized water to remove SDS and stained for 30 min with 10 μg/ml ethidium bromide in 0.2× TBE. The RNA bands, visualized under long-wavelength ultraviolet light, are excised from the gel. Each gel slice is placed in a dialysis bag (previously boiled in 5 mM EDTA, then autoclaved for 10 min in deionized water) containing a small volume of 0.2× TBE, and the bag is sealed at both ends without trapping air bubbles. Bags are placed longitudinally between the electrodes of an electroelution tank containing just enough 0.2× TBE to submerge the bags. Electrophoresis is at 100 V (0–4°C) until the ethidium bromide-stained RNA migrates from the gel (usually 2–3 hr) and forms a fluorescent streak along the inside of the bag. The polarity of electrophoresis is reversed for 2 min, the buffer is removed from the bag, and the bag is washed once with 0.2× TBE. The RNA-containing buffer is twice extracted with phenol, and the RNA is recovered by ethanol precipitation.

2. Hybridization Selection

Ortin et al. (1980) isolated influenza virus segments for fingerprinting by hybridization of total viral RNA to plasmid DNA, immobilized on nitrocellulose filters, containing the sequences of specific segments. After hybridization at 37°C for 24–36 hr in 50% formamide, 0.5% SDS, 5× SSC (1× SSC = 150 mM NaCl, 15 mM sodium citrate, pH 7.0), the filters were washed with 2× SSC, and the hybrids were treated with 10 units/ml RNase T1 in 10 mM Tris-HCl (pH 7.5), 0.5 M NaCl, in the presence of 20 μg/ml carrier tRNA. The filters were again washed with 2× SSC, and the resistant RNA was eluted at 50°C in formamide and recovered by ethanol precipitation. Hybridization selection should find wider application as a preparative method with the increasing availability of cloned copies of viral genes.

G. Oligonucleotide Mapping

1. Genomes with Terminal Poly(A) Sequences

Physical genomic maps of RNA viruses can be constructed by determination of the order of the unique RNase T1 oligonucleotides along the RNA molecule. Ordering of oligonucleotides is most easily accomplished with

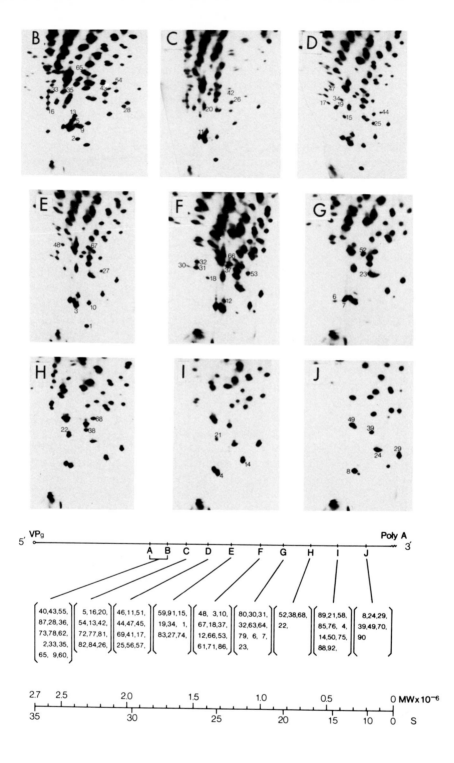

5′ VPg   Poly A
                                                                         ~~~ 3′
                    A B   C   D   E   F   G   H   I   J

{ 40,43,55, } { 5,16,20, } { 46,11,51, } { 59,91,15, } { 48, 3,10, } { 80,30,31, } { 52,38,68, } { 89,21,58, } { 8,24,29, }
{ 87,28,36, } { 54,13,42, } { 44,47,45, } { 19,34, 1, } { 67,18,37, } { 32,63,64, } { 22, } { 85,76, 4, } { 39,49,70, }
{ 73,78,62, } { 72,77,81, } { 69,41,17. } { 83,27,74. } { 12,66,53, } { 79, 6, 7, } { 14,50,75, } { 90 }
{ 2,33,35, } { 82,84,26, } { 25,56,57, } { 61,71,86, } { 23, } { 88,92, }
{ 65, 9,60, }

2.7   2.5   2.0   1.5   1.0   0.5   0 MWx10⁻⁶

35   30   25   20   15   10 0 S

genomes having terminal sequences of polyadenylic acid [poly(A)]. For these, the basic analytical strategy is partial fragmentation of the viral RNA, selection of the poly(A)containing fraction, size fractionation of the co-terminating fragments, and fingerprinting to determine which oligonucleotides disappear with decreasing chain length (Wang *et al.*, 1975). Several conditions have been used to produce partial fragmentation of the viral RNA, including limited base hydrolysis (Wang *et al.*, 1975), sonication (Kennedy, 1976), boiling in Tris–EDTA buffer (Merregaert *et al.*, 1981), partial digestion with RNase III (Stewart *et al.*, 1980), or radioactive decay of ^{32}P-labeled phosphodiester bonds (Faller and Hopkins, 1978). Nested sets of polynucleotides are generally obtained with common 3'-ends by selection of the poly(A)-containing fraction on oligo(dT)-cellulose (Aviv and Leder, 1972) or poly(U)-Sepharose (Lindberg and Persson, 1972). Harris *et al.* (1980) obtained fragments of varying lengths extending from the poly(C) tract near the 5'-end of foot-and-mouth disease virus RNA by selection on oligo(dG)-cellulose. Separation according to chain length is generally performed in denaturing sucrose gradients (Wang *et al.*, 1975; Darlix *et al.*, 1979). Better separation of individual size classes of fragments is achieved by electrophoresis on SDS–polyacrylamide–agarose gels (Stewart *et al.*, 1980), but sedimentation gradients are usually preferred because the fractions are more easily recovered.

Results of a typical mapping experiment, following the procedures of Wang *et al.* (1975), are shown in Fig. 5. The basic steps are given below. Partial base hydrolysis of *in vivo* ^{32}P-labeled virion RNA, incubated at 50°C in low-salt buffer [LSB; 10 mM Tris-HCl (pH 7.8), 10 mM NaCl, 1 mM EDTA], is brought about by addition of 0.05 volume of 1.0 M Na$_2$CO$_3$ to increase the pH to 10.8. At various intervals (e.g., 0, 15 sec, 30 sec, 1 min, 2 min, . . .) small aliquots are neutralized in 0.1 volume of 0.5 M acetic

---

FIG. 5. Mapping of the oligonucleotides of the Leon strain of type 3 poliovirus using the methods of Wang *et al.* (1975) and Nomoto *et al.* (1979b). Virion RNA labeled *in vivo* was partially base hydrolyzed, and the poly(A)-containing RNA fragments were selected by two cycles of affinity chromatography on oligo(dT)-cellulose. The poly(A)-containing fragments were separated according to size by sucrose gradient sedimentation, followed by resedimentation of each of the 10 pooled fractions. Peak fractions from each gradient were fingerprinted (B-J). The fraction A fingerprint is not shown because it was identical to that of untreated virion RNA (Fig. 1). Spots are numbered as in Fig. 1. (Lower panel) A map of oligonucleotide order determined by visual comparisons of relative spot intensities in the fingerprints above. Numbers given in brackets are arranged in sets of ascending values according to the intensities of the spots in the next (lower S value) gradient pool. Spots that are missing in the next fingerprints are listed first; those that are present but visibly lighter than the others are given in succeeding sets depending upon intensity. The overall map is compressed relative to the physical genome, probably because the relationship between genome size and experimentally determined S value deviates from the equation of Spirin (1963) used to calibrate the map.

acid. Optimal incubation and sampling conditions vary with different RNA preparations. The neutralized samples are then pooled, and ethanol precipitated. To select for the poly(A)-containing fraction, the RNA is resuspended in 500 μl of oligo(dT) binding buffer [10 mM Tris-HCl (pH 7.8), 0.5 M NaCl, 1mM EDTA, 0.5% SDS], applied to a column containing 500 mg of oligo(dT)-cellulose, and the column is washed with 20 volumes of binding buffer. Poly(A)-containing RNA is eluted from the oligo(dT) by addition of elution buffer [10 mM Tris-HCl (pH 7.8), 1 mM EDTA, 0.05% SDS], and the peak fractions are pooled. Two cycles of oligo(dT) selection are generally required to eliminate non-poly(A)-containing RNA contaminants. After ethanol precipitation, the poly(A)-containing RNA fragments are resuspended in 0.5 ml of LSB/0.1% SDS, heated at 70°C for 2 min, and resolved by sedimentation at 20°C in a linear 15–30% sucrose gradient in LSB/0.1% SDS. Pools of gradient fractions may be fingerprinted directly. However, resedimentation of each gradient fraction pool may be necessary to obtain greater size homogeneity within each sedimentation fraction (Nomoto et al., 1979b). In some procedures, oligo(dT) selection is the final step before fingerprinting, with each size class pool selected in small Pasteur pipet columns containing approximately 100 mg of oligo(dT).

Oligonucleotide maps may be constructed by visual comparison of the fingerprints of each size class and recording the order in which oligonucleotide spots become weak and then absent with decreasing fragment size. More quantitative comparisons are made by fingerprinting digests of each fragment size class mixed with differentially labeled full-length RNA. The RNA fragments are labeled with [³H]uridine and the full-length reference RNA with ^{32}P. The unique oligonucleotides are excised from the fingerprints, and their radioactivity is determined by scintillation counting. The order of the oligonucleotides is deduced by plotting the ^{3}H/^{32}P ratios for each oligonucleotide as a function of the size fractionation range of the gradients (Coffin et al., 1978).

Because of the limited resolving power of sedimentation gradients and the nonuniform distribution of unique oligonucleotides along the genome, the maps obtained by the above procedures are only approximate and are not a precise measure of the physical distances between oligonucleotides. The largest fragments are least efficiently resolved because they have the most similar sedimentation values; therefore the largest errors in assigning the oligonucleotide order are expected to occur for the 5'-terminal region. Oligonucleotide maps have been useful in identifying and characterizing recombinant genomes produced by intramolecular exchange (Beemon et al., 1974), in correlating biological properties with the presence of certain mapped oligonucleotides (Wang, 1978), in mapping deletion mutants of viral genomes (Kennedy, 1976), and in localizing some mutational differences between closely related strains (Nomoto et al., 1981).

2. Negative-Stranded Viruses

Oligonucleotides of the negative-stranded viruses vesicular stomatitis virus (Freeman *et al.*, 1979; Clewley and Bishop, 1979a) and Sendai virus (Amesse and Kingsbury, 1982) have been assigned to specific structural genes by identifying the set of oligonucleotides protected from nuclease digestion in mRNA–virion RNA duplexes. Freeman *et al.* (1979) separately hybridized ^{32}P-labeled virion RNA to saturating amounts of each viral mRNA. The partial duplexes were digested with RNase T1 and fingerprinted, and the oligonucleotides appearing in the fingerprints were assigned to the gene encoding the missing mRNA. Clewley and Bishop (1979a) applied the reciprocal approach and produced separate duplexes with each mRNA, then digested the single-stranded regions with a nuclease mixture containing S1 nuclease and RNases A, T1, and T2. The protected duplexes were purified on polyacrylamide gels and fingerprinted. The observed oligonucleotides were assigned to the gene encoding the protecting mRNA. Because the structural genes had been ordered by independent means, a partial oligonucleotide map could be constructed. Further ordering of oligonucleotides within each gene would require partial hydrolysis and fractionation of the poly(A)-containing mRNAs. This approach was useful in mapping deletions in defective-interfering mutants (Freeman *et al.*, 1979; Amesse *et al.*, 1982) and in comparing the genes of closely related rhabdoviruses (Clewley and Bishop, 1979a).

H. GENOME COMPLEXITY ANALYSIS

Fingerprinting has been used to provide a measure of the number of nucleotide residues per unique genome. The analysis is based on the principle that, for uniformly labeled RNA, the specific activity of any unique oligonucleotide is equivalent to that of the complete RNA molecule (Sinha *et al.*, 1965; Fiers *et al.*, 1965). Thus the chain length per unique genome can be calculated from the relationship

$$\text{chain length (RNA)} = \frac{\text{chain length (oligonucleotide)} \times \text{cpm (RNA)}}{\text{cpm (oligonucleotide)}}$$

In practice, uniformly labeled viral RNA is fingerprinted and several unique oligonucleotides are analyzed for radioactivity and chain length, determined from either composition data (Billeter *et al.*, 1974) or sequence data (Lee *et al.*, 1979). Several oligonucleotides are examined because of variable recoveries of oligonucleotides from the fingerprints. Some losses of the input RNA counts occur during fingerprinting, and correction for losses has been approached by the addition of oligonucleotides of known specific activities (internal standards), by parallel analyses of RNAs of known size and complexities (external standards) (Billeter *et al.*, 1974), or by counting the entire fingerprint gel (Lee *et al.*, 1979).

The complexity, or level of ploidy, of an RNA viral genome can be established by comparing the minimum genome size with other physical measures of the size of the viral RNA. Such analyses have been widely used to determine the complexities of the polyploid retroviral genomes (Billeter *et al.*, 1974; Beemon *et al.*, 1976; Friedrich *et al.*, 1976; Vigne *et al.*, 1977). Complexity analysis has also been used to determine the chain lengths of the genomic RNAs of Semliki Forest virus (Lomniczi and Kennedy, 1977), avian infectious bronchitis virus (Lomniczi and Kennedy, 1977), poliovirus types 1 and 2 (Lee *et al.*, 1979), and the genome segments of the bunyaviruses Uukuniemi virus (Petterson *et al.*, 1977) and snowshoe hare virus (Clewley *et al.*, 1977b).

I. Oligonucleotide Mapping in Genome Sequence Analysis

The ultimate identification of a virus strain is the determination of its complete genomic sequence. In studies to determine the total genome sequence of the Mahoney strain of type 1 poliovirus (Kitamura *et al.*, 1981), RNase T1- and RNase A-resistant oligonucleotides were used to prime the synthesis of DNA chains in the DNA polymerase I-mediated dideoxy sequencing system (Kitamura and Wimmer, 1980). Because the viral genome and the oligonucleotide primers are of the same polarity, the template was cDNA synthesized from virion RNA. The resultant sequence, unlike those determined from cloned DNA copies, represents the average genome sequence of the viral population (see the next section). Oligonucleotides have also been used in sequence studies with cloned DNA, serving as hybridization probes to characterize clones (Boothroyd *et al.*, 1981; van der Werf *et al.*, 1981).

IV. Applications

A. Fingerprinting as a Means of Studying Large Virus Populations

The fingerprint obtained experimentally represents the average genotype of a large virus population. Because of the extremely high inherent mutabilities of RNA genomes (Holland *et al.*, 1982), substantial genetic polymorphism may exist in the virus population. Model fingerprinting studies with phage Qβ revealed that the genome of most individual members of the population differed from the average genome at one or more sites (Domingo *et al.*, 1978). Further, more than one average genome may exist for the population, particularly if the virus population is in a state of disequilibrium (Holland *et al.*, 1982). If the alternative average genomes have different fingerprints, then the variant oligonucleotides will be detected as lighter, submolar spots. It is the experience of many laboratories that virus stocks maintained in the laboratory by low-multiplicity passage in cell culture appear to maintain a stable genetic equilibrium, such that their fingerprints remain unaltered with increased passage.

B. ANALYSIS OF NATURAL VIRUS ISOLATES

During natural transmission a virus population may encounter numerous complex conditions that favor disequilibrium and the selection of new variants. Given a succession of selective conditions, genome evolution is continual and is detected as cumulative changes in the fingerprints of successive isolates. The nature of the selective pressures is poorly defined, but appears to involve factors in addition to host antibodies, as natural genomic evolution can involve changes in sequences encoding nonstructural proteins (Young *et al.*, 1979; Domingo *et al.*, 1980; Nottay *et al.*, 1981) and can occur during persistent infection of cultured cells (Holland *et al.*, 1979; Meinkoth and Kennedy, 1980) and during infection of persons with severe immunodeficiencies (Yoneyama *et al.*, 1982).

In fact, the observed rates of genome evolution in nature vary widely among RNA viruses. Presumably, this is due to varying degrees of selective pressure encountered during the natural life cycles, as well as to possible differences in the basic mutation rate (Pringle *et al.*, 1981).

For some viruses, e.g., influenza A virus (Young *et al.*, 1979; Ortin *et al.*, 1980), foot-and-mouth disease virus (Domingo *et al.*, 1980), and poliovirus (Nottay *et al.*, 1981), the rate of natural evolution is so rapid that it is possible to reconstruct the general pathways of epidemiological transmission based upon the pattern of fixation of changes into the fingerprints. Such applications of fingerprint analysis have sometimes been described as "molecular epidemiology."

Many other viruses, including vesicular stomatitis virus (Clewley *et al.*, 1977a) and Western equine encephalitis virus (Trent and Grant, 1980), evolve so slowly that viruses with similar fingerprints can be isolated over a period of many years. In such cases, transmission pathways cannot be determined by fingerprint analysis, although the distribution of a genotype over space and time may be followed. Many virus strains appear to be restricted geographically, possibly because of the limited range of their reservoirs and vectors, and disappear from the environment as a genotype identifiable by fingerprinting, either by continued evolution or by displacement by strains of the same group but with distinct fingerprints. Virus genotypes, recognized by fingerprinting and geographic distribution, are called "topotypes."

Representatives of nearly every RNA virus group have been fingerprinted. The available data constitute a growing fingerprint catalog of RNA virus genotypes. For some viruses, the catalog may require frequent updating, because of the rapid evolution, whereas for others updating may not be necessary for decades.

1. Molecular Epidemiology

a. Influenza Viruses. Fingerprint analysis of influenza A (H1N1) isolates from the 1977–1979 pandemic revealed their close genetic relationship to H1N1 strains isolated in 1950 (Nakajima *et al.,* 1978). The 1977–1979 isolates showed a pattern of continual evolution by fixation of mutations, allowing construction of an evolutionary tree based upon fingerprint changes (Young *et al.,* 1979). In view of the rapid evolution of the pandemic virus, the great similarities in the fingerprints of isolates obtained 27 years apart suggested that the 1950 virus had reemerged after being maintained in a state of genetic dormancy, the exact nature of which is unknown. The evolution of H3N2 viruses has also been studied by fingerprinting (Ortin *et al.,* 1980; Nakajima *et al.,* 1982).

Influenza C viruses appear to be much more stable in nature, as isolates from four continents obtained over a 32-year period had generally similar fingerprints (Meier-Ewert *et al.,* 1981).

b. Enteroviruses. We use fingerprinting in our laboratory as a routine method for identifying the poliovirus genotype associated with an epidemic. The results of an epidemiological study are shown in Fig. 6 (Nottay *et al.,* 1981). It has been possible, on the basis of fingerprint data alone, to reveal unsuspected epidemiological links between cases or outbreaks that occur at different locations and at different times (Nottay *et al.,* 1981). An important fraction of isolates from the few remaining cases of paralytic poliomyelitis in developed countries has been shown, by fingerprinting, to be genetically related to the live poliovaccines (see, e.g., Fig. 2) (Minor, 1980; Nottay *et al.,* 1981; Kew *et al.,* 1981). Case isolates of enterovirus 70 from the 1980–1981 pandemic of acute hemorrhagic conjunctivitis were shown by fingerprinting to be closely related to each other and to some prepandemic isolates (Kew *et al.,* 1983).

c. Foot-and-Mouth Disease Viruses (FMDV). Isolates of type C FMDV obtained from diseased swine and cattle in Spain in 1979 were shown to be genetic variants of a 1970 strain from the same region (Domingo *et al.,* 1980). Similar findings were presented by this group for a type O strain. FMDV isolates from a 1981 outbreak in England and France were shown

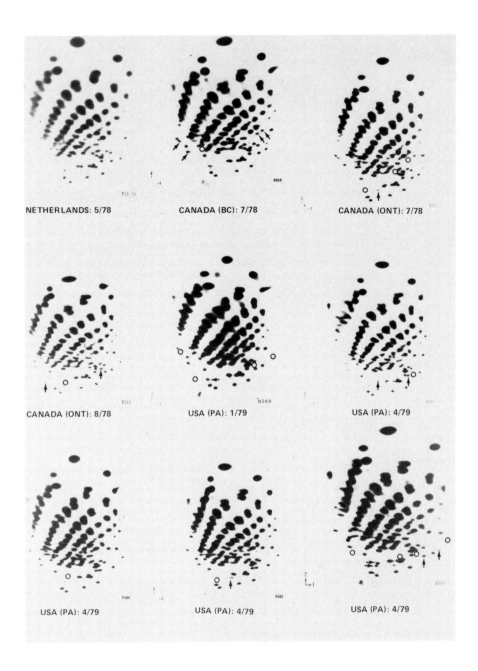

NETHERLANDS: 5/78

CANADA (BC): 7/78

CANADA (ONT): 7/78

CANADA (ONT): 8/78

USA (PA): 1/79

USA (PA): 4/79

USA (PA): 4/79

USA (PA): 4/79

USA (PA): 4/79

FIG. 6. Fingerprints of type 1 poliovirus isolates from the 1978–1979 Netherlands–Canada–United States epidemic. Fingerprints are arranged according to date of isolation. Note the general similarities among the fingerprints of isolates from this epidemic, and the obvious differences between this group and the fingerprint of a wild-type 1 poliovirus isolate from a different epidemic (Fig. 2). Note also evidence of evolution of the virus during epidemic transmission. Arrows (new spots) and open circles (missing spots) indicate differences in each pattern from the next earlier isolate's fingerprint. From Fig. 5 of Nottay et al. (1981).

to have fingerprints nearly identical to that of a vaccine strain isolated in 1965. Because of the high mutability of FMDV during epizootic transmission, unaltered survival of the 1965 strain in the field is unlikely, strongly suggesting that the outbreak was caused by accidental reintroduction of the virus into the field (King *et al.,* 1981).

2. Comparison of Natural Isolates Obtained at Different Times and Locations

a. Picornaviruses. Frisby *et al.* (1976) fingerprinted representatives of three of the four major picornavirus subgroups. Except for the cardioviruses encephalomyocarditis virus and Mengo virus, which had similar fingerprints, different viruses within and across subgroups had dissimilar fingerprints. The genetic relationships between isolates of swine vesicular disease virus and coxsackievirus B5 were studied by fingerprinting and competition hybridization (Harris *et al.,* 1977). Fingerprint analysis showed that the eight independent isolates of Theiler's murine encephalomyelitis virus belong to two distinct genetic subgroups (Lorch *et al.,* 1981). Attenuated strains of FMDV (Harris and Brown, 1977) and poliovirus (Nomoto *et al.,* 1981; Nottay *et al.,* 1981) have been compared by fingerprinting with their virulent parents.

b. Alphaviruses. The alphaviruses have undergone extensive evolutionary divergence, as little sequence homology is detected by hybridization across the three serocomplexes, Eastern equine encephalitis (EEE), Western equine encephalitis (WEE), and Venezuelan equine encephalitis (VEE) (Wengler *et al.,* 1977). For both the VEE (Wengler *et al.,* 1977; Trent *et al.,* 1979) and WEE (Trent and Grant, 1980) groups, fingerprint analysis generally corroborated serological classification. Isolates that were very close serologically had similar fingerprints, clearly distinguishable from those of close, but distinct, serological relatives. Individual alphavirus strains appear to evolve slowly, as very similar fingerprints were characteristic of WEE isolates widely separated temporally and geographically (Trent and Grant, 1980).

c. Flaviviruses. The flaviviruses are divided into seven subgroups, with very limited genetic homology across subgroups. Fingerprints of viruses of different subgroups (Wengler *et al.,* 1978) or different viruses within a subgroup (the prototypes of dengue virus types 1, 2, 3, and 4) (Vezza *et al.,* 1980a) showed no similarities. However, when comparisons were restricted to independent isolates of the same flavivirus, substantial genetic homology among representatives of a strain was evident. Trent *et al.* (1981) surveyed numerous North and Central American isolates of St. Louis encephalitis (SLE) virus from humans, birds, rodents, and mosquitoes obtained over a 47-year period. Fingerprint similarities were seen among all isolates, with differences among genotypic variants primarily correlating with differences in geographic region of isolation.

d. Rhabdoviruses. Fingerprints of the genetically divergent rhabdoviruses vesicular stomatitis virus (VSV), rabies, Cocal, Chandipura, and spring viremia of carp virus were, as expected, quite distinct (Clewley and Bishop, 1979b). However, independent isolates of the VSV Indiana group, obtained from diverse biological sources at different locations over a 50-year period, shared many similarities in their fingerprints (Clewley *et al.*, 1977a). The existence of two genetic subtypes of VSV New Jersey was demonstrated by fingerprint analysis. Three different laboratory strains (Clewley and Bishop, 1979b) of rabies have been fingerprinted.

e. Bunyaviruses. The bunyavirus group contains over 200 accepted or proposed members, divided among four genera. Members of different serogroups have clearly unique fingerprints for each L, M, and S genome segment (Pettersson *et al.*, 1977; Ushijima *et al.*, 1980; Clerx and Bishop, 1981). Within the California serogroup, individual strains, distinguishable on serological grounds, were also shown to have dissimilar fingerprints (El Said *et al.*, 1979). Fingerprint similarities were evident for independent isolates of La Crosse virus (California serogroup), although differences were observed among isolates. Variants existed as topotypes, as isolates obtained over 16 years from the same region were more closely related to each other than isolates obtained at the same times from different regions. The genome segments of La Crosse appeared to evolve at different rates, $M > L > S$ (El Said *et al.*, 1979).

f. Arenaviruses. Two variants of Pichinde virus (Tacaribe serocomplex) isolated 5 years apart in Colombia had clearly distinguishable fingerprints for their respective L and S segments. Vezza *et al.* (1980a) utilized these differences to provide physical evidence for independent assortment among arenaviruses.

g. Coronaviruses. The fingerprints of seven isolates of mouse hepatitis virus, obtained at different times and locations, indicated that all members of the group were closely related, having from 40 to 95% of their oligonucleotides in common (Lai and Stohlman, 1981). Independent isolates of infectious bronchitis viruses (IBV) had many distinct fingerprints, indicating a potential for substantial genetic variation even within a serotype. At the same time, some genotypes of IBV appeared to have wide geographic and temporal distributions (Clewley *et al.*, 1981).

h. Measles Virus. Fingerprints of measles virus isolated from cases of acute disease and from cases of subacute sclerosing panencephalitis revealed distinguishable patterns for each, with some large oligonucleotides shared (Stephenson and ter Meulen, 1982).

i. Ebola Virus. Coetaneous isolates of Ebola virus, from patients with hemorrhagic fevers, obtained in different locations (Zaire and Sudan) had very distinct fingerprints. Subsequent isolates had fingerprints similar to the earlier strains found in each location (Cox *et al.*, 1983), suggesting a slow rate of evolution for each Ebola topotype.

j. Orbiviruses. Fingerprint comparisons of isolates of bluetongue virus type 11 (BTV-11), obtained over a 12-year period in the United States, indicated that evolution in the natural environment occurred by both mutational drift and reassortment, and that several BTV-11 genotypes could be in circulation in the United States at the same time (Sugiyama *et al.*, 1982). Two Australian orbiviruses have also been compared by fingerprinting (Walker *et al.*, 1980).

k. Retroviruses. Fingerprinting has found wide application in the study of retroviruses. The large literature on this subject was separately and comprehensively reviewed by Beemon (1978) and Wang (1978). More recently, fingerprinting was used to reveal the genetic similarity of viruses from human leukemia patients to simian and gibbon ape retroviruses (Sahagan and Haseltine, 1980), to demonstrate that retroviral sequences from lymphomas of AKR mice are produced by recombination among at least three different endogenous retroviruses (Pedersen *et al.*, 1981; Thomas and Coffin, 1982), to document the participation of retroviral mRNA molecules in the recombination process (Wang and Stacey, 1982), and to show that retroviruses isolated from wild mice in different geographic locations had divergent sequences (Lai *et al.*, 1982). Using a combination of fingerprinting and oligonucleotide sequencing, Clements *et al.* (1982) demonstrated that common nucleotide sequence changes occurred in antigenic variants of visna virus isolated from different sheep.

C. OTHER APPLICATIONS RELATING TO VIRUS IDENTIFICATION

1. *Analysis of Recombinants and Reassortants*

Recombination by intramolecular exchange was demonstrated for retroviruses (Beemon *et al.*, 1974) and picornaviruses (King *et al.*, 1982) by fingerprint analysis of those progeny from mixed infections having recombinant genetic markers. In both systems, parental strains having easily distinguishable fingerprints and well-defined oligonucleotide maps were used. The approximate sites of crossover could be mapped because, for any region along the genome, only one parent contributed oligonucleotides to the recombinant RNA molecule. Natural intertypic recombinants of the live poliovaccines, isolated from humans exposed to trivalent oral vaccine, have been characterized by oligonucleotide mapping (Kew and Nottay, 1984).

Genetic exchange in viruses with segmented genomes occurs by independent assortment. Although reassortment can be confirmed by other physical means, such as one-dimensional electrophoresis (Young and Palese, 1979; Walker et al., 1980), fingerprinting of individual segments is much more sensitive and can clearly indicate whether each segment is unique. Fingerprinting has been used to characterize natural reassortants of influenza A viruses (Young and Palese, 1979; Nakajima et al., 1982), bluetongue virus (Sugiyama et al., 1982), and the bunyavirus Patois (Ushijima et al., 1980).

2. Analysis of Defective-Interfering Viral Genomes

Defective-interfering (DI) particles containing genome deletions are produced by nearly all RNA viruses. The presence of DI RNAs may complicate comparative studies among viruses because they can contribute substantial heterogeneity to the virus RNA population. Deletions may be located by comparison of the fingerprints of DI RNAs with the mapped oligonucleotides of full-length genome. Defective RNA genomes of avian retroviruses (Wang et al., 1975; Beemon, 1978), Semliki Forest virus (Kennedy, 1976; Pettersson, 1981; Soderlund et al., 1981), Sindbis virus (Dohner et al., 1979), poliovirus (Nomoto et al., 1979b), influenza virus (Davis and Nayak, 1979; Nakajima et al., 1979), vesicular stomatitis virus (Clerx-van Haaster et al., 1980; Hagen and Huang, 1981), and Sendai virus (Amesse et al., 1982) have been characterized by fingerprinting.

3. Mapping Subgenomic Intracellular RNAs

Subgenomic RNA molecules of genome polarity are produced intracellularly during the infectious cycles of alphaviruses, retroviruses, and coronaviruses. Fingerprinting has been used to map the subgenomic RNAs to the full-length genome for Semliki Forest virus (Kennedy, 1976; Wengler and Wengler, 1976), Rous sarcoma virus (Mellon and Duesberg, 1977), infectious bronchitis virus (Stern and Kennedy, 1980), and mouse hepatitis virus (Lai et al., 1981).

V. Summary and Conclusions

Fingerprinting is a definitive method of identifying RNA viruses according to their genotypes. It is not subject to the problems of antigenic drift or antigenic convergence that complicate serological identification. Furthermore, it provides a semiquantitative means of following the evolution of viral genomes in nature. Because all regions of the genome are represented by the large diagnostic oligonucleotides, a survey of the total genomic changes can be monitored.

Fingerprinting has two limitations as a diagnostic tool. First, although highly definitive, fingerprinting is not as rapid or inexpensive as serological techniques and cannot be as easily scaled up for routine identification of a large number of samples. Second, the evolutionary range of fingerprinting is short, and relationships may not be evident for isolates of rapidly evolving viruses obtained over long intervals. However, these limitations are not large, compared to the full benefits offered the virologist by the fingerprinting method.

ACKNOWLEDGMENTS

We are grateful to Drs. Fred Brown, Tim Harris, Yuan Fon Lee, and Eckard Wimmer for introducing our laboratory to the fingerprinting technique. We thank Marie Knox for careful preparation of the manuscript.

REFERENCES

Aaronson, R. P., Young, J. F., and Palese, P. (1982). *Nucleic Acids Res.* **10**, 237–246.
Adesnik, M. (1971). *Methods Virol.* **5**, 126–177.
Amesse, L. S., and Kingsbury, D. W. (1982). *Virology* **118**, 8–16.
Amesse, L. S., Pridgen, C. L., and Kingsbury, D. W. (1982). *Virology* 118, 17–27.
Aviv, H., and Leder, P. (1972). *Proc. Natl. Acad. Sci. U.S.A.* **69**, 1408–1412.
Banerjee, A. K. (1980). *Microbiol. Rev.* **44**, 175–205.
Barrell, B. G. (1971). *In* "Procedures in Nucleic Acids Research" (G. L. Cantoni and D. R. Davies, eds.), Vol. 2, pp. 751–779. Harper & Row, New York.
Beemon, K. L. (1978). *Curr. Top. Microbiol. Immunol.* **79**, 73–110.
Beemon, K. L., Duesberg, P. H., and Vogt, P. K. (1974). *Proc. Natl. Acad. Sci. U.S.A.* **71**, 4254–4258.
Beemon, K., Faras, A. J., Haase, A. T., Duesberg, P. H., and Maisel, J. E. (1976). *J. Virol.* **17**, 525–537.
Billeter, M. A., Parsons, J. T., and Coffin, J. M. (1974). *Proc. Natl. Acad. Sci. U.S.A.* **71**, 3560–3564.
Boothroyd, J. C., Highfield, P. E., Cross, G. A. M., Rowlands, D. J., Lowe, P. A., Brown, F., and Harris, T. J. R. (1981). *Nature (London)* **290**, 800–802.
Brown, F. (1982). *Ann. Neurol.* **9**, (Suppl.), 39–43.
Brown, W. M., George, M., and Wilson, A. C. (1979). *Proc. Natl. Acad. Sci. U.S.A.* **76**, 1967–1971.
Brownlee, G. G., and Sanger, F. (1967). *J. Mol. Biol.* **23**, 337–353.
Brownlee, G. G., and Sanger, F. (1969). *Eur. J. Biochem.* **11**, 395–399.
Chaconas, G., and van de Sande, J. H. (1980). *Methods Enzymol.* **65**, 75–85.
Clements, J. E., D'Antonio, N., and Narayan, O. (1982). *J. Mol. Biol.* **158**, 415–434.
Clerx, J. P. M., and Bishop, D. H. L. (1981). *Virology* **108**, 361–372.
Clerx-van Haaster, C. M., Clewley, J. P., and Bishop, D. H. L. (1980). *J. Virol.* **33**, 807–817.
Clewley, J. P., and Bishop, D. H. L. (1979a). *J. Virol.* **30**, 116–123.
Clewley, J. P., and Bishop, D. H. L. (1979b). *In* "Rhabdoviruses" (D. H. L. Bishop, ed.), Vol. 1, pp. 119–168. CRC Press, Boca Raton, Florida.
Clewley, J. P., and Bishop, D. H. L. (1982). *In* "New Developments in Practical Virology" (C. R. Howard, ed.), pp. 231–277. Liss, New York.
Clewley, J. P., Bishop, D. H., Kang, C.-Y., Coffin, J., Schnitzlein, W. M., Reichmann, M. E., and Shope, R. E. (1977a). *J. Virol.* **23**, 152–166.

Clewley, J., Gentsch, J., and Bishop, D. H. L. (1977b). *J. Virol.* **22**, 459-468.

Clewley, J. P., Morser, J., and Lomniczi, B. (1981). In "Biochemistry and Biology of Coronaviruses" (V. ter Meulen, S. Siddell, and H. Wege, eds.), pp. 143-153. Plenum, New York.

Coffin, J. M., Champion, M., and Chabot, F. (1978). *J. Virol.* **28**, 972-991.

Commerford, S. L. (1971). *Biochemistry* **10**, 1993-1999.

Commerford, S. L. (1980). *Methods Enzymol.* **70**, 247-252.

Cox, N. J., McCormick, J. B., Johnson, K. M., and Kiley, M. P. (1983). *J. Infect. Dis.* **147**, 272-275.

Darlix, J. L., Levray, M., Bromley, P. A., and Spahr, P. F. (1979) *Nucleic Acids Res.* **6**, 471-485.

Davis, A. R., and Nayak, D. P. (1979). *Proc. Natl. Acad. Sci. U.S.A.* **76**, 3092-3096.

De Wachter, R., and Fiers, W. (1972). *Anal. Biochem.* **49**, 184-197.

De Wachter, R., and Fiers, W. (1982). In "Gel Electrophoresis of Nucleic Acids: A Practical Approach" (D. Rickwood and B. D. Hames, eds.), pp. 77-116. IRL Press, Oxford.

Dohner, D., Monroe, S., Weiss, B., and Schlesinger, S. (1979). *J. Virol.* **29**, 794-798.

Domingo, E., Sabo, D., Taniguchi, T., and Weissman, C. (1978). *Cell* **13**, 735-744.

Domingo, E., Davila, M., and Ortin, J. (1980). *Gene* **11**, 333-346.

Donis-Keller, H. (1980). *Nucleic Acids Res.* **8**, 3133-3142.

Donis-Keller, H., Maxam, A., and Gilbert, W. (1977). *Nucleic Acids Res.* **4**, 2527-2538.

Duesberg, P. H., and Vogt, P. K. (1973). *J. Virol.* **12**, 594-599.

El Said, L. H., Vorndam, V., Gentsch, J. R., Clewley, J. P., Calisher, C. H., Klimas, R. D., Thompson, W. H., Grayson, M., Trent, D. W., and Bishop, D. H. L. (1979). *Am. J. Trop. Med. Hyg.* **28**, 364-386.

England, T. E., and Uhlenbeck, O. C. (1978). *Nature (London)* **275**, 560-561.

Faller, D. V., and Hopkins, N. (1978). *J. Virol.* **26**, 143-152.

Fiers, W., Lepoutre, L., and Vandendriessche, L. (1965). *J. Mol. Biol.* **13**, 432-450.

Freeman, G. J., Rao, D. D., and Huang, A. S. (1979). *Gene* **5**, 141-157.

Friedrich, R., Morris, V. L., Goodman, H. M., Bishop, J. M., and Varmus, H. E. (1976). *Virology* **72**, 330-340.

Frisby, D. (1977). *Nucleic Acids Res.* **4**, 2975-2996.

Frisby, D. P., Newton, C., Casey, N. H., Fellner, P., Newman, J. F. E., Harris, T. J. R., and Brown, F. (1976). *Virology* **71**, 379-388.

Hagen, F., and Huang, A. S. (1981). *J. Virol.* **37**, 363-371.

Harris, T. J. R., and Brown, F. (1977). *J. Gen. Virol.* **34**, 87-105.

Harris, T. J. R., Robson, K., and Brown, F. (1977). *J. Gen. Virol.* **35**, 299-315.

Harris, T. J. R., Robson, K., and Brown, F. (1980). *J. Gen. Virol.* **50**, 403-418.

Henry, C. (1967). *Methods Virol.* **2**, 427-462.

Holland, J. J., Grabau, E., Jones, C. L., and Semler, B. (1979). *Cell* **16**, 495-504.

Holland, J. J., Spindler, K., Horodyski, F., Grabau, E., Nichol, S., and Vande Pol, S. (1982). *Science* **215**, 1577-1585.

Horst, J., Kieth, J., and Fraenkel-Conrat, H. (1972). *Nature (London) New Biol.* **240**, 105-109.

Huang, E. S., and Pagano, J. S. (1977). *Methods Virol.* **6**, 457-497.

Ito, Y., and Joklik, W. K. (1972). *Virology* **50**, 189-201.

Jordan, E. M., and Raymond, S. (1969). *Anal. Biochem.* **27**, 205-211.

Kennedy, S. I. T. (1976). *J. Mol. Biol.* **108**, 491-511.

Kew, O. M., and Nottay, B. K. (1984). In "Modern Approaches to Vaccines" (R. M. Chanock and R. A. Lerner, eds.), pp. 357-362. Cold Spring Harbor Laboratory, New York.

Kew, O. M., Nottay, B. K., Hatch, M. H., Nakano, J. H., and Obijeski, J. F. (1981). *J. Gen. Virol.* **56**, 337-347.

Kew, O. M., Nottay, B. K., Hatch, M. H., Hierholzer, J. C., and Obijeski, J. F. (1983). *Infect. Immun.* **41**, 631–635.

King, A. M. Q., Underwood, B., McCahon, D., Newman, J. W. I., and Brown, F. (1981). *Nature (London)* **293**, 479–480.

King, A. M. Q., McCahon, D., Slade, W. R., and Newman, J. W. I. (1982). *Cell* **29**, 921–928.

Kitamura, N., and Wimmer, E. (1980). *Proc. Natl. Acad. Sci. U.S.A.* **77**, 3196–3200.

Kitamura, N., Semler, B. L., Rothberg, P. G., Larsen, G. L., Adler, C. J., Dorner, A. J., Emini, E. A., Hanecak, R., Lee, J. J., van der Werf, S., Anderson, C. W., and Wimmer, E. (1981). *Nature (London)* **291**, 547–553.

Laemmli, U. K. (1970). *Nature (London)* **227**, 680–685.

Lai, M. M. C., and Stohlman, S. A. (1981). *J. Virol.* **38**, 661–670.

Lai, M. M. C., Brayton, P. R., Armen, R. C., Patton, C. D., Pugh, C. H., and Stohlman, S. A. (1981). *J. Virol.* **39**, 823–834.

Lai, M. M. C., Shimuzu, C. S., Rasheed, S., Pal, B. K., and Gardner, M. B. (1982). *J. Virol.* **41**, 605–614.

Laskey, R. A. (1980). *Methods Enzymol.* **65**, 363–371.

La Torre, J. L., Underwood, B. O., Lebendiker, M., Gorman, B. M., and Brown, F. (1982). *Infect. Immun.* **36**, 142–147.

Lee, Y. F., and Fowlks, E. (1982). *Anal. Biochem.* **119**, 224–235.

Lee, Y. F., and Wimmer, E. (1976). *Nucleic Acids Res.* **3**, 1647–1658.

Lee, Y. F., Kitamura, N., Nomoto, A., and Wimmer, E. (1979). *J. Gen. Virol.* **44**, 311–322.

Lehrach, H., Diamond, D., Wozney, J. M., and Boedtker, H. (1977). *Biochemistry* **16**, 4743–4751.

Lindberg, V., and Persson, T. (1972). *Eur. J. Biochem.* **31**, 246–254.

Loening, U. E. (1969). *Biochem. J.* **113**, 131–138.

Lomniczi, B., and Kennedy, I. (1977). *J. Virol.* **24**, 99–107.

Lorch, Y., Friedmann, A., Lipton, H. L., and Kotler, M. (1981). *J. Virol.* **40**, 560–567.

McDonell, M. W., Simon, M. N., and Studier, F. W. (1977). *J. Mol. Biol.* **110**, 119–146.

Markham, R., and Smith, J. D. (1952). *Biochem. J.* **52**, 558–565.

Maxam, A., and Gilbert, W. (1980). *Methods Enzymol.* **65**, 499–560.

Meier-Ewert, H., Petri, T., and Bishop, D. H. L. (1981). *Arch. Virol.* **67**, 141–147.

Meinkoth, J., and Kennedy, S. I. T. (1980). *Virology* **100**, 141–155.

Mellon, P., and Duesberg, P. H. (1977). *Nature (London)* **270**, 631–634.

Merregaert, J., Barbacid, M., and Aaronson, S. A. (1981). *J. Virol.* **39**, 219–228.

Minor, P. D. (1980). *J. Virol.* **34**, 73–84.

Minson, A. C., and Darby, G. (1982). *In* "New Developments in Practical Virology" (C. R. Howard, ed.), pp. 185–229. Liss, New York.

Nakajima, K., Desselberger, U., and Palese, P. (1978). *Nature (London)* **274**, 334–339.

Nakajima, K., Ueda, M., and Suguira, A. (1979). *J. Virol.* **29**, 1142–1148.

Nakajima, K., Nakajima, S., and Suguira, A. (1982). *Virology* **120**, 504–509.

Nakajima, S., Cox, N. J., and Kendal, A. P. (1981). *Infect. Immun.* **32**, 287–294.

Nomoto, A., Kajigaya, S., Suzuki, K., and Imura, N. (1979a). *J. Gen. Virol.* **45**, 107–117.

Nomoto, A., Lee, Y. F., Babich, A., Jacobson, A., Dunn, J. J., and Wimmer, E. (1979b). *J. Mol. Biol.* **128**, 165–177.

Nomoto, A., Kitamura, N., Lee, J. J., Rothberg, P. G., Imura, N., and Wimmer, E. (1981). *Virology* **112**, 217–227.

Nottay, B. K., Kew, O. M., Hatch, M. H., Heyward, J. T., and Obijeski, J. F. (1981). *Virology* **108**, 405–423.

Ogra, P. L., Kargon, D. T., Righthand, F., and MacGillivray, M. (1968). *N. Engl. J. Med.* **279**, 893–900.

Ortin, H., Najera, R., Lopez, C., Davila, M., and Domingo, E. (1980). *Gene* 11, 319-332.

Peattie, D. A. (1979). *Proc. Natl. Acad. Sci. U.S.A.* 76, 1760-1764.

Pedersen, F. S., and Haseltine, W. A. (1980a). *J. Virol.* 33, 349-365.

Pedersen, F. S., and Haseltine, W. A. (1980b). *Methods Enzymol.* 65, 680-687.

Pedersen, F. S., Crowther, R. L., Tenney, D. Y., Reimold, A. M., and Haseltine, W. A. (1981). *Nature (London)* 292, 167-170.

Pettersson, R. F. (1981). *Proc. Natl. Acad. Sci. U.S.A.* 78, 115-119.

Pettersson, R. F., Hewlett, M. J., Baltimore, D., and Coffin, J. M. (1977). *Cell* 11, 51-63.

Pringle, C. R., Wilkie, D. M., Preston, C. M., Dolan, A., and McGeoch, D. (1981). *J. Virol.* 39, 377-389.

Ralph, R. K., and Bergquist, P. L. (1967). *Methods Virol.* 2, 463-545.

Rensing, V. F. E., and Schoenmakers, J. G. G. (1973). *Eur. J. Biochem.* 33, 8-18.

Robertson, H. D., Dickson, E., Plotch, S. J., and Krug, R. M. (1980). *Nucleic Acids Res.* 8, 925-942.

Robson, K. J. H., Crowther, J. R., King, A. M. Q., and Brown, F. (1979). *J. Gen. Virol.* 45, 579-590.

Rommelaere, J., Donis-Keller, H., and Hopkins, N. (1979). *Cell* 16, 43-50.

Sahagan, B. G., and Haseltine, W. A. (1980). *J. Virol.* 34, 390-401.

Sanger, F., and Coulson, A. R. (1978). *FEBS Lett.* 87, 107-110.

Sanger, F., Brownlee, G. G., and Barrell, B. G. (1965). *J. Mol. Biol.* 13, 373-398.

Schuerch, A. R., Mitchell, A. R., and Joklik, W. K. (1975). *Anal. Biochem.* 65, 331-345.

Simoncsits, A., Brownlee, G. G., Brown, R. S., Rubin, R. S., and Guilley, H. (1977). *Nature* 269, 833-836.

Sinha, N. K., Fujimura, R. K., and Kaesberg, P. (1965). *J. Mol. Biol.* 11, 84-89.

Soderlund, H., Keranen, S., Lehtovaara, P., Palva, I., Pettersson, R. F., and Kaarianen, L. (1981). *Nucleic Acids Res.* 9, 3403-3417.

Spirin, A. S. (1963). *Prog. Nucleic Acid Res. Mol. Biol.* 1, 301-345.

Stephenson, J. R., and ter Meulen, V. (1982). *Arch. Virol.* 71, 279-290.

Stern, D. F., and Kennedy, S. I. T. (1980). *J. Virol.* 36, 440-449.

Stewart, M. L., and Crouch, R. J. (1981). *Anal. Biochem.* 111, 203-211.

Stewart, M. L., Crouch, R. J., and Maizel, J. V. (1980). *Virology* 104, 375-397.

Sugiyama, K., Bishop, D. H. L., and Roy, P. (1982). *Am. J. Epidemiol.* 115, 332-347.

Szekely, N., and Sanger, F. (1969). *J. Mol. Biol.* 43, 607-617.

Takahashi, K., and Moore, S. (1982). *In* "The Enzymes" (P. D. Boyer, ed.), 3rd ed., Vol. 15, pp. 435-468. Academic Press, New York.

Thomas, C. Y., and Coffin, J. M. (1982). *J. Virol.* 43, 416-426.

Trent, D. W., and Grant, J. A. (1980). *J. Gen. Virol.* 47, 261-282.

Trent, D. W., Clewley, J. P., France, J. K., and Bishop, D. H. L. (1979). *J. Gen. Virol.* 43, 365-381.

Trent, D. W., Grant, J. A., Vorndam, A. V., and Monath, T. P. (1981). *Virology* 114, 319-332.

Ushijima, H., Klimas, R., Kim, S., Cash, P., and Bishop, D. H. L. (1980). *Am. J. Trop. Med. Hyg.* 29, 1441-1452.

van der Werf, S., Bregegere, F., Kopecka, H., Kitamura, N., Rothberg, P. G., Kourilsky, P., Wimmer, E., and Girard, M. (1981). *Proc. Natl. Acad. Sci. U.S.A.* 78, 5983-5987.

Vezza, A. C., Cash, P., Jahrling, P., Eddy, G., and Bishop, D. H. L. (1980a). *Virology* 106, 250-260.

Vezza, A. C., Rosen, L., Repik, P., Dalrymple, J., and Bishop, D. H. L. (1980b). *Am. J. Trop. Med. Hyg.* 29, 643-652.

Vigne, R., Brahic, M., Filippi, P., and Tamalet, J. (1977). *J. Virol.* 21, 386-395.

Volckaert, F., and Fiers, W. (1977). *Anal. Biochem.* 83, 222-227.

Volckaert, F., Min-Jou, W., and Fiers, W. (1976). *Anal. Biochem.* **72**, 433-446.

Walker, P. J., Mansbridge, J. N., and Gorman, B. M. (1980). *J. Virol.* **34**, 583-591.

Wang, L.-H. (1978). *Annu. Rev. Microbiol.* **32**, 561-592.

Wang, L.-H., and Stacey, D. W. (1982). *J. Virol.* **41**, 919-930.

Wang, L.-H., Duesberg, P., Beemon, K., and Vogt, P. K. (1975). *J. Virol.* **16**, 1051-1070.

Weislander, L. (1979). *Anal. Biochem.* **98**, 305-309.

Wengler, G., and Wengler, G. (1976). *Virology,* **73**, 190-199.

Wengler, G., Wengler, G., and Filipe, A. R. (1977). *Virology* **78**, 124-134.

Wengler, G., Wengler, G., and Gross, H. J. (1978). *Virology* **89**, 423-437.

Wimmer, E. (1972). *J. Mol. Biol.* **68**, 537-540.

Wimmer, E. (1982). *Cell* **28**, 199-201.

Yewdell, J. W., and Gerhard, W. (1981). *Annu. Rev. Microbiol.* **35**, 185-206.

Yoneyama, T., Hagiwara, H., Hara, M., and Shimojo, H. (1982). *Infect. Immun.* **37**, 46-53.

Young, J. F., and Palese, P. (1979). *Proc. Natl. Acad. Sci. U.S.A.* **76**, 6547-6551.

Young, J. F., Desselberger, U., and Palese, P. (1979). *Cell* **16**, 73-83.

3 Immunosorbent Electron Microscopy in Plant Virus Studies

R. G. Milne and D.-E. Lesemann

I. Introduction

A little over 10 years ago, Derrick (1972, 1973) introduced the method under discussion, and, after a slow start, its usefulness is now widely recognized (Kurstak, 1981; Torrance and Jones, 1981; van Regenmortel, 1982; Lesemann, 1983). Its essence is that an electron microscope (EM) support film, prepared on a specimen grid, is coated with antibody, and this coated film is used to trap the antigen (virus particles) from the suspension placed in contact with it. A 2- to 10,000-fold increase in particle numbers per grid area may be obtained, as compared with the results of methods not using antisera. After trapping, the preparation is rinsed and given contrast by shadowing, positive staining, or negative staining. The antigen is recognized in the EM by its morphology and staining properties and can be further labeled by a second incubation with a more concentrated antibody preparation, thus "decorating" each individual particle with an easily recognized halo of γ-globulin molecules (Milne and Luisoni, 1977).

Some advantages of the method are as follows. (1) The nature of the result is very direct; characteristic particles are seen in the microscope. (2) As a result of (1), false positives are minimal or absent. (3) Sensitivity is often high (about the same as with ELISA). (4) The "dirt" background is lowered as a result of coating the support film with antibody, and the virus particles are more easily seen. (5) Early bleedings or antisera of low titer may be used. (6) Antisera are used directly, without fractionation or conjugation; this permits greater reproducibility and longer shelf life of reagents. (7) Only very small volumes of dilute antisera are required (e.g., 5–10 μl). (8) Very small volumes of virus extract (e.g., 1 μl) can be used. (9) Results may be obtained rapidly (within 20 min to 2 hr). (10) A relatively wide spectrum of serological variants can be detected with one antiserum. (11) Contaminant antibodies, e.g., those directed against host material, can be tolerated since the virus particles can be morphologically distinguished from other adsorbed or trapped materials. (12) Using "decoration," contaminant or unexpected virus particles can be detected in preparations.

Some disadvantages of the method are that (1) it does not work, or works poorly, under some circumstances (which are not yet clear); (2) it does not detect "soluble antigens"; (3) it involves costly equipment (EM, vacuum coating unit) (4) it is labor intensive and requires EM skills; and (5) as a result of (4), it is not suited for handling large numbers of samples.

Terminology. Roberts *et al.* (1982) attempted the task of establishing uniform terminology in the area of interaction of electron microscopy and serology of particulate virus preparations. They suggested that two general terms were appropriate: immunoelectron microscopy (IEM) and electron microscope serology (EM serology).

For the particular technique introduced by K. S. Derrick, the term serologically specific electron microscopy (SSEM) has been used (e.g., by Derrick, 1973; Paliwal, 1977; Beier and Shepherd, 1978; Brlansky and Derrick, 1979). However, SSEM seems inappropriate, since all techniques in which serology and electron microscopy interact are "serologically specific electron microscopy"—for example, the use of ferritin-labeled antibodies in work with thin sections. Roberts *et al.* (1982) therefore argued for substitution of the term immunosorbent electron microscopy (ISEM) introduced by Roberts and Harrison (1979). Meanwhile, a rival term, solid-phase immune electron microscopy (SPIEM), has gained a certain currency among animal virologists (e.g., Almeida *et al.,* 1980; Kjeldsberg and Mortensen-Egnund, 1982), but to us, ISEM seems a much more apt term than SPIEM.

Also, in accord with Roberts *et al.* (1982), we would endorse the term coating for the process of attaching antibody to the grid, and the term trapping for what happens to the virus particles when they become specifically attached to the antibody on the grid.

II. Parameters of the Method

A. Support Films

Standard methods for making support films are widely available and will not be redescribed. The support films generally used for ISEM are made of Formvar or a similar plastic, or of Parlodion, strengthened with a layer of evaporated carbon. The carbon face of the film is used. Alternatively, support films made of carbon alone can be used, and may allow entrapment of more virus particles (Roberts and Harrison, 1979; Roberts, 1981b), but the films tend to be brittle for routine use. Unfortunately, support films, when made in a "standard" way, are not at all standard from one laboratory to another, owing to factors such as the characteristics of the carbon evaporating unit, the quality of the carbon, perhaps the backing plastic, and even the atmosphere of the laboratory. Changing the vacuum unit can, for example, change the suitability of the resulting films for ISEM.

Experience in our laboratories and elsewhere (see, e.g., Roberts and Harrison, 1979; Cohen *et al.* 1982) has suggested that the most consistent results are obtained if films are used for ISEM immediately after they are coated with carbon. An alternative is to "standardize" the surface properties of the films by subjecting them to glow-discharge (Lesemann *et al.,* 1980). The glow-discharge technique is generally carried out by exposing filmed grids for 1–30 sec to a high-voltage discharge in air at a pressure near 0.1 torr (Choppin and Stoeckenius, 1964; Reissig and Orrel, 1970).

van Balen (1982) investigated more systematically the effects on particle trapping of pretreating the support films. She used glow-discharge for 5 min in argon at 0.1–0.6 torr to induce a negative charge on the support, and floated grids on ethidium bromide (30 μg/ml in distilled water) (note that ethidium bromide is a carcinogen) for 15 min to induce a positive charge (Sogo *et al.,* 1979). In addition, either a thick or thin layer of carbon was used over a Formvar base. Clarified sap from potato leafroll virus-infected plants and the γ-globulin fraction of the homologous antiserum were used as test materials. Results indicated that the standard ISEM technique was more effective using a thick carbon film that was positively charged. On the other hand, in cases in which precoating with protein A was used (see Section II, J), a negatively charged film was more effective. When protein A was omitted, a negative charge on the film led to less efficient trapping than that achieved on untreated films.

B. Suitable Types of Antiserum or Antibody Preparation

So far, ISEM seems to have been used only with rabbit antisera, but there is no reason why antibodies from other sources should not function equally well. Early bleedings do not appear to be at a disadvantage, as they are with at least the simplest form of ELISA. Antisera of relatively low titer can also be used if necessary. These findings may well be related to the fact that no conjugation step is necessary, and thus the γ-globulin molecules remain undamaged and retain their full avidity (Koenig, 1978; Korpraditskul *et al.*, 1979; van Regenmortel, 1982). Fractionated γ-globulins have been used successfully for ISEM (e.g., Nicolaieff and van Regenmortel, 1980; Bovey *et al.*, 1982; Luisoni *et al.*, 1982), as well as F(ab')$_2$ fragments (Louro and Lesemann, 1984).

C. Antiserum Dilutions, Buffers, pH

As first shown by Derrick and Brlansky (1976) and Paliwal (1977), and later amply confirmed, the optimal dilution of a crude antiserum is around 1:1000 to 1:5000. This is a rule of thumb that requires confirmation or modification when optimizing any particular assay. At lower dilutions, especially at dilutions below about 1:100, marked inhibition of trapping occurs (Fig. 1), probably because of competition for attachment sites on the support film, between the IgG and other serum proteins (Paliwal, 1977; Milne and Lesemann, 1978; Lesemann *et al.*, 1980). Lesemann *et al.* (1980) suggested that the inhibitory effect was not related to antiserum titers, since it was observed equally with two sera having titers of 1:4096 and 1:256. Nicolaieff and van Regenmortel (1980) noted that the amount of inhibition at low dilution varies considerably with different antisera and viruses. Inhibition was not observed by Lesemann (1983) when γ-globulins were used at concentrations corresponding to strongly inhibitory concentrations of complete antisera.

Originally, 0.05 *M* Tris-HCl buffer, pH 7.2, was employed for diluting both antisera and virus preparations (Derrick, 1973), and this buffer has continued to be used satisfactorily (Derrick and Brlansky, 1976; Harville

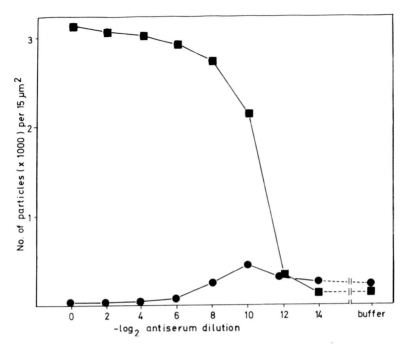

Fig. 1. Influence of the concentration of homologous antiserum and of pretreating the grids with 10 μg/ml protein A, on the trapping of virus particles from eggplant mottled crinkle virus-infected *Nicotiana clevelandii*. (■) Counts from protein A-pretreated grids; (●) counts from grids coated with antiserum only. Taken from Lesemann and Paul (1980).

and Derrick, 1978; Brlansky and Derrick, 1979). Beier and Shepherd (1978) used the same buffer, but at pH 7.5. Alternatively, 0.1 *M* phosphate buffer, pH 7.0 (Milne and Luisoni, 1977; Lesemann *et al.,* 1980; Luisoni *et al.,* 1982) or 0.06 *M* phosphate buffer, pH 6.5 (Roberts and Harrison, 1979; Roberts *et al.,* 1980) have been successful. Thomas (1980) and Korpradit-skul (1979) obtained excellent ISEM results, diluting the antisera in sodium carbonate buffer, pH 9.6 (the commonly used ELISA coating buffer). It is generally agreed that the incorporation of saline in the buffer is of no advantage. Cohen *et al.* (1982) examined the effect of pH conditions on ISEM, using a number of viruses and antisera. The pH of the antisera was varied between 5.0 and 8.0 using 0.1 *M* phosphate buffer. Although pH 5.0 generally gave the worst results, the greatest difference between the best and worst results in terms of trapping efficiency was only a factor of 2.6. Within the pH range 6.0–8.0, different viruses and antisera appeared to have different optima.

The conclusion may be drawn that, where differences in particle counts of the order of 3-fold are thought to be important, the best type of buffer, pH, and ionic strength should be sought empirically for each virus–antiserum combination.

D. MIXED ANTISERA

It has been shown that antisera can be mixed and used to trap simultaneously at least two different viruses on the same ISEM grid (Derrick and Brlansky, 1976; Harville and Derrick, 1978). Provided that serum dilutions of 1:1000 are used, there appears to be little or no inhibition of particle trapping by the respective heterologous sera.

E. SERUM COATING CONDITIONS

It is now reasonably well established that an antiserum coating time of 5 min at room temperature is sufficient to obtain essentially all the IgG adsorption that is going to occur (Milne and Luisoni, 1977; Roberts, 1981a). Bovey *et al.* (1982) found that there was little increase in the number of particles trapped when the coating period was increased from 1 min to 8 hr. Periods of less than 1 min gave uneven coating, but even with the shortest times (1 sec) an appreciable number of particles were trapped. Accotto (1982) confirmed that a coating time of 2 min was highly effective and that times up to 80 min gave no improvement over a 5-min incubation. Lesemann (1983) showed that trapping after coating times of 0.5 min was as effective as with 128 min, irrespective of low or high virus concentrations and of trapping times between 15 min and 4 hr. Longer coating times, if convenient for the work schedule, do not appear to offer any disadvantage. Incubation at 37°C rather than at room temperature is reported to yield more firmly attached virus particles and a better spread of negative stain (Roberts, 1981a).

F. RINSING THE GRIDS

After each phase of ISEM, except the final negative-staining step, the grids must be carefully rinsed with either buffer or water, as appropriate. For example, ISEM including protein A precoating (Section II,J) and decoration (Section II,K) would involve the following steps: protein A precoating, buffer rinse, antibody coating, buffer rinse, antigen incubation, buffer rinse, second antibody coating, water rinse, negative staining, drying, examination. Rinsing must be done accurately and consistently, for good results to be obtained; often rinsing is insufficiently thorough from the fear that the grid will be washed clean, but generally, more rinsing means better results.

Basically two rinsing systems are in current use. Individual grids may be retained in forceps throughout the process (Milne and Luisoni, 1975, 1977), and the vertically held face of the grid is rinsed with drops of buffer or water from a Pasteur pipet held almost in contact. Alternatively, as in most laboratories, the grids are floated on or immersed in a series of rinsing baths, which can simply be drops placed on a clean surface such as Parafilm.

G. Virus Preparations: Buffers, Protectants, Incubation Times

Virus preparations to be incubated on antibody-coated grids have been prepared in a variety of buffers at different ionic strengths and pH values. Often 0.05 M Tris-HCl (pH 7.2) or 0.1 M phosphate (pH 7.0) are used. Cohen et al. (1982) investigated the effect of preparing extracts of virus-infected plants in 0.1 M phosphate at pH values from 5.0 to 8.0. The preparations were of cucumber mosaic virus in tobacco, carnation mottle virus in carnation, lily symptomless virus in lily, and potato virus Y in tobacco, in tomato, or in Physalis floridana. Quite large effects of pH (up to 20-fold differences in particle counts) were reported, but the optimal pH was different for each virus–host combination and, with potato virus Y, varied according to the host plant. Thus no neat conclusion emerged, except that each system must be individually optimized; if it is not, substantially suboptimal conditions may inadvertently come to be employed.

Extracts of animal vectors such as aphids or nematodes (Roberts and Harrison, 1979; Roberts and Brown, 1980) can be studied as well as extracts from infected plants. Duncan and Roberts (1981) described a method of virus extraction from very small samples.

An example of specific requirements for virus extraction was observed during trapping tests with turnip mosaic virus isolates in Chinese cabbage (D.-E Lesemann and J. Vetten, unpublished observation). No trapping occurred after extraction with 0.1 M phosphate buffer (pH 7.0), but high trapping occurred when the same buffer was supplemented with 0.1 M EDTA. Similar effects of EDTA were experienced in ELISA tests with papaya ringspot virus (Gonsalves and Ishii, 1980) and turnip mosaic virus (J. Vetten, personal communication).

Nicolaieff and van Regenmortel (1980) tested whether trapping of purified tobacco mosaic and turnip yellow mosaic virus particles was influenced by the presence of sap from healthy tobacco and Chinese cabbage plants. It was found that crude sap constituents inhibited the adsorption of virus to uncoated grids but did not influence the trapping of virus particles on serum-coated grids. Luisoni et al. (1982) similarly reported, with rice ragged stunt virus, that the sensitivity of ISEM was not influenced by dilution of crude preparations either in healthy sap or in buffer.

On the other hand, it is also clear that the saps of certain plants contain materials that prevent the detection of virus particles by ISEM. Noel *et al.* (1978) noted that detection of plum pox virus from plum or apricot required the presence of 2.5% nicotine added to the sap. Incorporation of 2% polyvinylpyrrolidone (PVP) into saps of plants such as members of the Rosaceae, grapevine, or poplar was recommended by Milne (1981). Bovey *et al.* (1982) and Russo *et al.* (1982) found it necessary to grind grapevine tissues in the presence of 2.5% nicotine, before further dilution in phosphate buffer, when assaying for fanleaf virus by ISEM. Accotto (1982) tested the effects of several potential protectants on similar material; he confirmed that nicotine was an effective protectant but found it not as good as 2% PVP or 1% polyethylene glycol (MW 6000–7500); 1% sodium sulfite was ineffective, as was phosphate buffer alone. For detection of closteroviruses in grapevine, 2% PVP has been used successfully (Milne *et al.*, 1984).

The period for which the virus extract is incubated with the grid strongly influences the number of particles trapped. The relation between time and particle numbers may be essentially linear over a number of hours, and two incubation strategies therefore emerge. One is to use short incubation times (15 min) in order to trap sufficient virus for diagnosis, but still get a rapid result (Milne and Luisoni, 1977). The other is to use longer incubations (conveniently overnight, but possibly up to a week) in order to increase sensitivity (Derrick, 1973; Roberts and Harrison, 1979; Roberts, 1981a; Bovey *et al.*, 1982; Russo *et al.*, 1982). When long incubation times are used at room temperature or at 37°C, one part in 5000 of sodium azide may be incorporated in order to inhibit bacterial growth. Alternatively, incubation can be done in the cold, but this probably slows the kinetics of trapping.

It should be noted that, under some conditions, certain viruses are unstable or become detached from the grid, so that long incubations (for example, over 3 hr) may be counterproductive (Milne, 1981; Roberts, 1981a; Lesemann, 1983).

It has been observed that, under short incubation conditions, homologous serological reactions are favored, whereas with long incubations, more distantly related particles may also be trapped. Thus the specificity of the reaction can to some extent be manipulated by controlling incubation time (Lesemann and Paul, 1980; Lesemann, 1983). Protein A combined with little diluted antiserum can also be used to alter the degree of specificity (see Section II,J).

H. Applying Contrast to the Preparation

Derrick (1973) originally shadowed his preparations with platinum/palladium, and later Derrick and Brlansky (1976) used positive staining with ethanolic uranyl acetate. Although both these methods reliably provide good contrast, the great advantages of high resolution and good particle preservation obtainable with negative staining are sacrificed. Negative staining is strongly recommended over other methods of imparting contrast, because identification of the virus particles, especially when there are few, may hinge on seeing their morphology clearly.

Of various negative stains, both the authors routinely use 2% aqueous uranyl acetate, but 2% sodium phosphotungstate (PTA), either neutral or at pH 6.5, is equally good and perhaps easier to handle, so long as the viruses concerned are not labile in PTA. Some common labile viruses are alfalfa mosaic, ilarviruses, cucumoviruses, tomato spotted wilt, some rhabdoviruses, and some geminiviruses. For staining of labile viruses 2% ammonium molybdate at pH 6.0–7.0 may be very useful, as an alternative to uranyl acetate (Roberts, 1981a).

I. Assessment of Results: Particle Counts; Control Preparations

When ISEM is used simply as a detection method, all that is required is the presence of enough virus particles to make diagnosis certain. If the particles are sufficiently characteristic or have been labeled by "decoration," one particle per 400-mesh grid square may well suffice.

When quantitative comparisons are desired, particle counting becomes laborious, especially as the variance in particle numbers per unit area may be quite high on different areas of one grid square, on different squares, on duplicate grids, and with different batches of grids or different virus preparations (Nicolaieff and van Regenmortel, 1980). We would recommend counting at least one randomly chosen field in each of 10 grid squares on each of 2 grids (20 fields in all) and converting the results to a standard 1000 μm^2 area, as suggested by Roberts and Harrison (1979) and Roberts (1980). Counting directly from the EM screen is faster than counting from micrographs.

The degree of trapping can be related to one of two types of control preparation, each usually giving different results. One type of control is a grid coated with normal serum diluted to the same degree as the antiserum used on the test grid (Derrick, 1973; Lesemann et al., 1980; Nicolaieff and van Regenmortel, 1980). The other type of control is a grid "coated" with buffer only, or an untreated grid (Milne, 1980). The virus preparation is incubated with control grids under the same conditions as the test grids. Usually, particle counts are higher on untreated (or buffer-treated) grids than on normal serum-coated grids (Lesemann *et al.*, 1980; Cohen *et al.*, 1982). On untreated or buffer-treated grids counts greatly depend on the constituents of the virus samples. Thus, crude saps yield much lower counts than purified preparations of identical virus concentrations (Lesemann *et al.*, 1980). However, on normal serum-treated grids the counts are much more similar with both types of sample. Therefore, normal serum-coated grids should be preferred as controls to yield a more uniform basis for the comparison of trapping results. For critical interpretation of small quantitative differences in trapping, both kinds of controls should be used (Roberts, 1981b).

Some antisera may contain viruslike isometric particles as contaminants. Therefore, a grid incubated with the diluted antiserum alone should be prepared as a control if isometric virus particles are to be trapped at very low concentrations. A "decoration" (see Section II,K) following the incubation for trapping would avoid this problem.

J. USE OF PROTEIN A

Protein A, obtainable as a purified product, is a *Staphylococcus aureus* bacterial wall protein having the property of binding specifically to the Fc portion of IgG molecules. Two IgG molecules can be bound to one of protein A (Forsgren and Sjöquist, 1966; Langone, 1982). Shukla and Gough (1979) and Gough and Shukla (1980) were the first to use protein A for precoating EM grids, prior to using them as in the normal ISEM procedure.

Although Shukla and Gough (1979) and Gough and Shukla (1980) claimed rather dramatic increases in numbers of particles trapped using protein A, it is now realized that much more modest gains (2- to 6-fold increases) are obtained if ISEM results with and without protein A are compared rigorously (Fig. 1) (Lesemann and Paul, 1980; Milne, 1980; Nicolaieff and van Regenmortel, 1980; Roberts, 1981b). The reasons for the discrepancy between the results of Shukla and Gough (1979) and Gough and Shukla (1980), and those of later workers, are two, and they are worth discussing.

1. In the absence of protein A, the optimal antiserum dilution (Section II,C) may be in the region of 1:1000–1:5000. Higher concentrations of serum are inhibitory (Fig. 1). If the grids are precoated with protein A, the IgG is much less subject to competition for attachment sites from other serum proteins, and therefore lower dilutions of antiserum (even undiluted serum)

are effective (Fig. 1). Thus the optimal serum dilutions for the two methods may differ by more than five 2-fold steps (Lesemann and Paul, 1980; Milne, 1980; Nicolaieff *et al.*, 1982; van Regenmortel, 1982; Lesemann, 1983). This must be carefully borne in mind when comparisons are made.

2. When virus concentration is not limiting, the richer IgG layer obtainable with protein A allows trapping of more virus particles. However, as virus concentrations are decreased, the benefit of protein A appears to diminish: when ISEM is used precisely because virus concentrations are minimal, the amount of virus rather than the number of adsorbed IgG molecules may limit the number of particles trapped (Milne, 1980; Lesemann and Paul, 1980).

There are, however, circumstances in which the use of protein A may offer clear advantages. One is when the antiserum is of low titer (say, 1:32 in gel diffusion tests). In this case, precoating with protein A allows the effective use of undiluted or only moderately diluted sera that might not be effective when diluted to 1:500, for example, as required for "straight" ISEM (Lesemann and Paul, 1980; Lesemann, 1983).

Lesemann and Paul (1980), Makkouk *et al.* (1981), and Lesemann (1983) also noted that the use of protein A, combined with higher serum concentrations for coating, widens the range of serological variants that can be trapped. Thus, protein A–ISEM can be used under these conditions to detect distant serological relationships that would escape detection by normal ISEM.

In practice, protein A is usually diluted in buffer to a final concentration of 100 μg/ml (Shukla and Gough, 1979) or 10 μg/ml (Lesemann and Paul, 1980). It is applied to the grid for 5 min at room temperature, and the excess is then rinsed off before the coating antiserum is applied.

K. USE OF "DECORATION"

In the "decoration" step, virus particles already adsorbed or trapped on the grid are incubated with an antiserum (the same as used previously, or another) at a dilution that coats or decorates each individual particle in a halo of IgG molecules. The dilution used is generally between 1:10 and 1:100, and the incubation time is conveniently 15 min at room temperature (Milne and Luisoni, 1975, 1977; Milne, 1981).

Decoration identifies the particles very positively, and renders them highly visible, so that in practice the sensitivity of the ISEM technique is notably increased (Kerlan *et al.*, 1981).

III. *Varying Success with ISEM*

A. SENSITIVITY

The sensitivity of ISEM in detecting small amounts of virus has been discussed by van Regenmortel (1982) and Lesemann (1983). Although sensitivity varies widely, it can be as good as or better than that of ELISA (Beier and Shepherd, 1978; Hamilton and Nichols, 1978; Noel *et al.*, 1978; Thomas, 1980; Bovey *et al.*, 1982; Gillet *et al.*, 1982; Luisoni *et al.*, 1982; Sequeira and Harrison, 1982). In absolute terms, the limits of detection can be as low as 0.1–10 ng/ml, representing 10^7–10^9 particles per milliliter, for a small isometric virus (Derrick, 1973; Roberts *et al.*, 1978; Beier and Shepherd, 1978; Paliwal, 1979; Gillet *et al.*, 1982; van Regenmortel, 1982).

B. LOWERED SENSITIVITY OR FAILURE

As noted by Milne (1980), Cohen *et al.* (1982), and Lesemann (1983), ISEM does not always lead to trapping ratios of the order of 10–1000 times better than particle counts on various kinds of control grids. Almeida *et al.* (1980) also noted that ISEM did not increase numbers of particles either of bovine wart virus or of hepatitis B antigen.

As successes rather than failures are published, the failures or relative failures have not received their due, and the reasons for them remain unknown. Apart from trivial failures of technique or inadequacies of the antisera, two possible reasons come to mind: (1) Adsorption of the virus particles to the untreated grid may be so efficient that a layer of the specific antibody is of no further benefit. (2) The virus preparation may contain not only virus particles, but also free subunits or oligomers of the viral coat protein. If these are antigenically similar to the intact capsid, they are likely to bind to the antibody layer and inhibit trapping of the virus. This possibility has never been properly investigated, but it seems plausible because the authors have experienced instances in which fresh preparations of virus could be trapped effectively whereas frozen–thawed preparations or those stored for a long time could not.

IV. *Applications*

A. DIAGNOSTICS

As we have seen, ISEM can achieve a sensitivity of detection similar to that of ELISA and might in that case be considered a rival technique. In fact, it is complementary.

The merits and demerits of ISEM listed in Section II,I indicate that the method can often provide rapid and accurate answers using small amounts of materials, and using antisera that may be suboptimal in that they are from early bleedings, have low titers, or contain unwanted (e.g., host plant-directed) antibodies. Especially when combined with decoration, ISEM is useful for troubleshooting, e.g., detecting contaminant antigens in virus preparations, or contaminant antibodies in (your own and other people's) sera. ISEM is invaluable as an aid to developing a reliable ELISA, as, with a new virus, diagnosis by ISEM can usually be achieved 2 or 3 months earlier than by ELISA.

If large-scale routine diagnosis is needed, ELISA is the clear final goal, but even when running routinely, ELISA should always be subject to checks by ISEM, which is less prone to giving spurious positive results. Cases also often arise in which diagnosis may be required on a smaller scale or at unpredictable intervals. Here, it may not be worthwhile to prepare and maintain the ELISA reagents, and the labor-intensive nature of ISEM is less of a disadvantage.

B. Measuring Serological Relationships

As noted by Lesemann (1983), experience with the study of heterologous reactions using ISEM is still limited. In some cases, ISEM has given results qualitatively similar but quantitatively different from those of other methods (Paliwal, 1979; Lesemann et al., 1980; Waterhouse and Murant, 1981; Sequeira and Harrison, 1982). In other ISEM studies, a general agreement with relationships as revealed by other methods has been reported (Nicolaieff and van Regenmortel, 1980; Roberts et al., 1980; Rajeshwari et al., 1983).

As noted in Sections II,G and J, if it is desired to differentiate more sharply between serological variants, ISEM can be set up without protein A, using dilute antisera (e.g., 1:5000), and employing short incubation times (e.g., 15 min). Using no protein A, a serum diluted 1:3000, and a virus incubation time of 80 min, van Regenmortel et al. (1980) found that tobacco mosaic virus (TMV) antiserum did not significantly trap cucumber virus 4 (a tobamovirus differing from TMV by 4.2 serological differentiation index units) though TMV-U2 (difference from TMV, 2.7 units) was trapped.

In cases which require recognition of a wide spectrum of serotypes or the detection of very distant relationships, protein A can be used, with higher serum concentrations and long incubation times (Lesemann and Paul, 1980; Lesemann, 1983).

V. Detection of Double-Stranded RNA

Double-stranded (ds) RNA is immunogenic, although the antisera produced react with little specificity to any dsRNA, natural or synthetic, and also, to some extent, with DNA–RNA hybrids (Stollar, 1973).

Derrick (1978) showed that ISEM can be used to trap dsRNA molecules from crude virus-infected plant extracts. His technique was to coat filmed grids with a 1:1000 dilution of an antiserum produced against a complex of methylated bovine serum albumin and polyinosinic–polycytidylic acid [poly(I-C)] (Moffitt and Lister, 1975). The serum was diluted in 0.05 M Tris-HCl (pH 7.2). Coated grids were incubated with the plant extracts for 2 hr then rinsed with the buffer and floated on 0.1 mg/ml cytochrome c in the same buffer for 15 min (this step served presumably to add thickness and contrast to the RNA molecules). The grids were then washed with water, stained in ethanolic uranyl acetate (UA), dried, and rotary shadowed.

An anti-poly(I-C) serum has also been used with ISEM to trap dsRNA-containing viruslike particles from the mushroom *Agaricus bisporus* (del Vecchio *et al.,* 1978). In this case, presumably, at least part of the dsRNA had escaped from within the particles but had remained attached or adsorbed to the outside.

Luisoni *et al.* (1975) attempted to use the decoration technique to identify dsRNA from fijiviruses. However, at least with unfractionated serum, any specific attachment of the anti-dsRNA antibodies to the RNA molecules was obscured by heavy nonspecific adsorption of other serum proteins. It is possible that a purified IgG fraction or an IgG produced monoclonally would decorate dsRNA molecules more specifically, and thus form the basis for rapid identification.

VI. Preparation of Grids for Later Use

Any IEM process is best carried through as a whole in one laboratory, from preparation to observation. However, grids negatively stained with uranyl acetate are stable for long periods if kept desiccated, and it is therefore possible to send them through the mail, for example in small gelatin capsules surrounded by silica gel in a sealed polyethylene envelope. Grids stained with phosphotungstate or ammonium molybdate may, however, quickly deteriorate during storage.

Derrick and Brlansky (1976), Paliwal (1977), Milne and Luisoni (1977), and Milne (1981) have noted that, in the ISEM procedure, grids can also be dried and stored after the antiserum coating step and can be used up to 6 weeks later for trapping appropriate virus particles. The best method is probably to rinse the already coated grids with water just before drying them, store them desiccated, and rewet them with buffer before they are used for trapping. It is clear that, in this procedure, the grids might be prepared and coated at point A, mailed, incubated with virus at point B, negatively stained, and returned to A for examination.

Another possibility is opened up by the observation (Dr. G. I. Mink, personal communication) that grids bearing virus particles negatively stained in UA may be rinsed with water and then buffer and can then be incubated with antiserum to achieve decoration, which is followed by restaining in UA. Even grids that have been examined in the electron microscope can be decorated subsequently, but only on those grid squares that have not been directly in the electron beam.

Our experience shows that the above procedure can be successful but is not always so. In certain cases it, too, might form the basis of exchange between laboratories, if the facilities required are not all in one place.

VII. Problems and Prospects for the Future

The problems that beset ISEM are still formidable. First, there is as yet little understanding of the factors that determine attachment of protein A, γ-globulins, virus particles, and other substances to EM support films, and of the competition of these various elements for sites on the surface. We are still some way from the rational design of films with particular adhesive characteristics. A beginning has been made in understanding how to manipulate charges on the film surface, but there is still no good way to measure this charge, and its effect on "stickiness" is still in the realm of anecdote.

Second, it is clear that some virus preparations, in some laboratories, fail to be trapped efficiently by grids apparently coated appropriately. Even when trapping is considered satisfactory, its efficiency varies widely. It is not at present clear where the problem lies, but a careful analysis of the possible factors involved would be interesting and might lead to improvement.

Third, probably connected with the above points, is the problem of uniformity in particle counts per unit area. Variability is sometimes high despite efforts to produce uniform support films; if quantitative results are sought, this means counting many fields on several squares on different replicate grids.

One prospect for the future is the elimination or moderation of these difficulties through greater understanding and control of the different phases of ISEM. A second prospect is that experience with the study of heterologous reactions will increase and may enable distant serological relationships to be detected more easily than at present. This is important because virus taxonomy and epidemiology are always in need of different measures of relationship between viruses and virus strains.

Finally, it seems clear that ISEM is a technique that will be in use for some time, even though other methods of detection (of viruses, and of relationships), such as nucleic acid hybridization, continue to advance rapidly. The reason is that the high-technology moiety of ISEM lies with the electron microscope, whereas the rest can be done rapidly with very small amounts of unsophisticated materials. There are few links in the chain leading to the result, so that the kind of evidence obtained is very direct. These advantages mean that in novel or difficult situations, ISEM can often give results well before other methods have built up the necessary momentum.

REFERENCES

Accotto, G.-P. (1982). *Phytopathol. Mediterr.* **21**, 75.
Almeida, J. D., Stannard, L. M., and Shersby, A. S. M. (1980). *J. Virol. Methods* **1**, 325.
Beier, H., and Shepherd, R. J. (1978). *Phytopathology* **68**, 533.
Bovey, R., Brugger, J. J., and Gugerli, P. (1982). *Proc. 7th Meeting Int. Council for the Study of Viruses and Virus-like Diseases of the Grapevine, Niagara Falls, 1980* (A. T. McGinnis, ed.), p. 259.
Brlansky, R. H., and Derrick, K. S. (1979). *Phytopathology* **69**, 96.
Choppin, P. W., and Stoeckenius, W. (1964). *Virology* **22**, 482.
Cohen, T., Loebenstein, G., and Milne, R. G. (1982). *J. Virol. Methods* **4**, 323.
del Vecchio, V. G., Dixon, C., and Lemke, P. A. (1978). *Exp. Mycol.* **2**, 138.
Derrick, K. S. (1972). *Phytopathology* **62**, 753 (abstract).
Derrick, K. S. (1973). *Virology* **56**, 652.
Derrick, K. S. (1978). *Science* **199**, 538.
Derrick, K. S., and Brlansky, R. H. (1976). *Phytopathology* **66**, 815.
Duncan, G. H., and Roberts, I. M. (1981). *Micron* **12**, 171.
Forsgren, A., Sjöquist, J. (1966). *J. Immunol.* **97**, 822.
Gillet, J. M., Morimoto, K. M., Ramsdell, D. C., Chaney, W. G., Baker, K. K., and Esselmann, W. J. (1982). *Phytopathology* **72**, 937 (abstract).
Gonsalves, D., and Ishii, M. (1980). *Phytopathology* **70**, 1028.
Gough, K. H., and Shukla, D. D. (1980). *J. Gen. Virol.* **51**, 415.
Hamilton, R. I., and Nichols, C. (1978). *Phytopathology* **68**, 539.
Harville, B. G., and Derrick, K. S. (1978). *Plant Dis. Rep.* **62**, 290.
Kerlan, C., Mille, B., and Dunez, J. (1981). *Phytopathology* **71**, 400.
Kjeldsberg, E. and Mortensen-Egnund, K. (1982). *J. Virol. Methods* **4**, 45.
Koenig, R. (1978). *J. Gen. Virol.* **40**, 309.
Korpraditskul, P. (1979). Ph.D. Thesis, University of Göttingen, Federal Republic of Germany.
Korpraditskul, P., Casper, R., and Lesemann, D.-E. (1979). *Phytopathol. Z.* **96**, 281.
Kurstak, F., ed. (1981). "Handbook of Plant Virus Infections Comparative Diagnosis." Elsevier/North-Holland, Amsterdam.
Langone, J. J. (1982). *J. Immunol. Methods* **55**, 277.
Lesemann, D.-E. (1983). *Acta Hortic.* **127**, 159.

Lesemann, D.-E., and Paul, H. L. (1980). *Acta Hortic.* **110**, 119.

Lesemann, D.-E., Bozarth, R. F., and Koenig, R. (1980). *J. Gen. Virol.* **48**, 257.

Louro, D., and Lesemann, D.-E. (1984). *J. Virol. Methods* **9**, in press.

Luisoni, E., Milne, R. G., and Boccardo, G. (1975). *Virology* **68**, 86.

Luisoni, E., Milne, R. G., and Roggero, P. (1982). *Plant Dis.* **66**, 929.

Makkouk, K. M., Koenig, R., and Lesemann, D.-E. (1981). *Phytopathology* **71**, 572.

Milne, R. G. (1980). *Acta Hortic.* **110**, 129.

Milne, R. G. (1981). Notes for the Course on Immunoelectron Microscopy of Plant Viruses. Association of Applied Biologists Workshop in Electron Microscope Serology. John Innes Institute, England, 1981.

Milne, R. G., and Lesemann, D.-E. (1978). *Virology* **90**, 299.

Milne, R. G., and Luisoni, E. (1975). *Virology* **68**, 270.

Milne, R. G., and Luisoni, E. (1977) *Methods Virol.* **6**, 265.

Milne, R. G., Conti, M., Lesemann, D.-E., Stellmach, G., Tanne, E. and Cohen, J. (1984). *Phytopathol. Z., in press.*

Moffitt, E., and Lister, R. M. (1975). *Phytopathology* **65**, 851.

Nicolaieff, A., and van Regenmortel, M. H. V. (1980). *Ann. Virol. (Inst. Pasteur)* **131 E**, 95.

Nicolaieff, A., Katz, D., and van Regenmortel, M. H. V. (1982) *J. Virol. Methods* **4**, 155.

Noel, M.-C., Kerlan, C., Garnier, M., and Dunez, J. (1978). *Ann. Phytopathol.* **10**, 381.

Paliwal, Y. C. (1977). *Phytopathol. Z.* **89**, 25.

Paliwal, Y. C. (1979). *Phytopathol. Z.* **94**, 8.

Rajeshwari, R., Iizuka, N., Nolt, B. L., and Reddy, D. V. R. (1983). *Plant Pathology* **32**, 197.

Reissig, M., and Orrel, S. A. (1970). *J. Ultrastruct. Res.* **32**, 107.

Roberts, I. M. (1980). *J. Microsc.* **118**, 241.

Roberts, I. M. (1981a). Notes for the Course on Immunoelectron Microscopy of Plant Viruses. Association of Applied Biologists Workshop in Electron Microscope Serology. John Innes Institute, England, 1981.

Roberts, I. M. (1981b). Scottish Crop Research Inst. Annual Report for 1980, p. 106.

Roberts, I. M., and Brown, D. J. F. (1980). *Ann. Appl. Biol.* **96**, 187.

Roberts, I. M., and Harrison, B. D. (1979). *Ann. Appl. Biol.* **93**, 289.

Roberts, I. M., Torrance, L., Harrison, B. D., and Salazar, L. F. (1978). Scottish Crop Research Inst. Annual Report for 1977, p. 105.

Roberts, I. M., Tamada, T., and Harrison, B. D. (1980). *J. Gen. Virol.* **47**, 209.

Roberts, I. M., Milne, R. G., and van Regenmortel, M. H. V. (1982). *Intervirology* **18**, 147.

Russo, M., Martelli, G. P., and Savino, V. (1982). *Proc. 7th Meeting Int. Council for the Study of Viruses and Virus-like Diseases of the Grapevine, Niagara Falls, 1980* (A. J. McGinnis, ed.), p. 251.

Sequeira, J. C., and Harrison, B. D. (1982). *Ann. Appl. Biol.* **101**, 33.

Shukla, D. D., and Gough, K. H. (1979). *J. Gen. Virol.* **45**, 533.

Sogo, J. M., Rodeno, P., Köller, T., Venuela, E., and Salos, M. (1979). *Nucleic Acids Res.* **7**, 107.

Stollar, B. D. (1973). *In* "The Antigens" (M. Sela, ed.), Vol. 1, p. 1. Academic Press, New York.

Thomas, B. J. (1980). *Ann. Appl. Biol.* **94**, 91.

Torrance, L., and Jones, R. A. C. (1981). *Plant Pathol.* **30**, 1.

van Balen, E. (1982). *Neth. J. Plant Pathol.* **88**, 33.

van Regenmortel, M. H. V. (1982). "Serology and Immunochemistry of Plant Viruses." Academic Press, New York.

Waterhouse, P. M., and Murant, A. F. (1981). *Ann. Appl. Biol.* **97**, 191.

4 Quantitative Transmission Electron Microscopy for the Determination of Mass-Molecular Weight of Viruses

H. M. Mazzone, W. F. Engler, and
G. F. Bahr

I. Introduction

In the study of viruses, the determination of mass aids in the identification and classification of these microorganisms. Moreover, the mass value, or, as it is expressed in virology, molecular weight, is a prime prop-

103

erty of viruses. Because of its importance, the mass determination should be undertaken and evaluated with the utmost care and precision. In this regard, some of the methods used to provide data needed in the calculation of the molecular weight of viruses operate under the disadvantages that (1) some methods are indirect, requiring information from other parameters involving the use of extra equipment or data obtained from other laboratories; (2) some entail extensive time periods, thereby subjecting the virus particles to possible denaturation or degradation effects; and (3) some are constrained by the size of the virus, the relatively larger viruses being beyond the resolution of the methods.

The determination of the total dry mass of viruses by quantitative transmission electron microscopy (QTEM) has none of the above limitations. Indeed, the determination of the mass of a virus particle obtained after exposing it untreated in the electron microscope adds another important dimension to the usually observed features of size and shape. The mass equivalent of every virus particle whose image is recorded on an electron micrograph, plate, or film, may be analyzed directly, and the researcher is provided with mass values of individual virus particles and an average mass value of the virus population under study.

Viruses isolated from animals, bacteria, plants, and insects have been subjected to QTEM and their mass values determined (Bahr *et al.*, 1976). Moreover, insect viral inclusion bodies (Mazzone *et al.*, 1980a,b, 1981) and associative structures, such as the various maturation forms of herpes simplex virus (Lampert *et al.*, 1969), have been studied. The simplicity of the method, requiring relatively small quantities of material while yielding directly measurements of mass, aids in the characterization of the viruses. This is particularly important for laboratories studying viruses that infect man because, although these viruses have been serologically typed, appreciable titers of virus have not generally been available for further characterization, including the molecular weight determination. The quantitative electron microscopy procedure for mass determination should circumvent such restrictions for these and other viruses.

With the property of mass, in addition to size and shape, to consider for a given virus, the investigator may be better able to discern the degree of heterogeneity of a virus population. Is there now better evidence for the presence of a structural variant, or perhaps, a different type of virus particle? Can the extra or unusual symptoms observed at times in virus pathology be explained on this basis? These questions, we believe, are better answered, if at all, by the use of QTEM.

In this review we discuss, principally, studies in which mass determinations on viruses were derived by QTEM. Since its introduction as a research procedure, mass determination with the electron microscope has been extended through the use of quantitative scanning transmission electron microscopy (QSTEM). We discuss QSTEM, along with other methods, as a comparative procedure in the determination of the mass–molecular weight of viruses.

II. Theory

A detailed analysis of the theory of quantitative mass determination may be obtained from previous reports (Zeitler and Bahr, 1962, 1965; Bahr and Zeitler, 1965). The essential features are (1) the relationship between mass per area of a particle and the transmission recorded for its image on an electron micrograph negative, plate, or film; and (2) through integrating photometry, the conversion of photometric measurement to mass.

A. LINEAR RELATIONSHIP BETWEEN MASS PER AREA OF A PARTICLE AND TRANSMISSION

In the transmission electron microscope, a thin object, e.g., a virus particle, transradiated by electrons alters their initial energy and direction (Hall, 1951; Burge and Sylvester, 1960).

$$I = I_0 \, e^{-\alpha S w} \qquad (1)$$

where I_0 is the unscattered beam intensity; I is the intensity of the beam after it has passed through the object; α is an instrumental factor determined by the operating parameters and design of the microscope; S, the scattering cross section (cm^2) per gram of substance, is analogous to the extinction coefficient of light absorption measurements; and w, the mass per unit area—thickness of the object—is directly related to the net change in direction of electrons in the object.

In order to use Eq. (1) to derive the mass of an object transradiated in the electron microscope, it would be necessary to measure I, point by point, in the electron beam and then sum the logarithms of the individual measurements. It is simpler to convert I to the photographic density D, using the photographic emulsion as an analog converter of I. This conversion is linear for density values below 1.2 (Bahr, 1973) and is given by

$$D = Eb + \delta \qquad (2)$$

where E is exposure of the emulsion to the electron beam (I in amperes $\times$ the time); b is a constant governed by the response of the emulsion to electrons and by the conditions of development; and δ is the fog produced by the emulsion by diffusely scattered electrons and by the chemical development procedure. Linearity for electrons makes it possible to convert electron intensity [Eq. (1)] directly to photographic density D [Eq. (2)] and to use the relationship of photographic density to transmission:

$$D = - \log T \qquad (3)$$

In considering Eqs. (1) – (3), Bahr (1973) proposed the following relationship:

$$(T - T_0)/T_\infty \cong \alpha S w \qquad (4)$$

where T is the transmission of an image point in the electron micrograph of the object plus its underlying support film, T_0 is the corresponding transmission through the supporting film (background), and T_∞ is the transmission of the unexposed photographic material, which is chiefly determined by the fog level δ [Eq. (2)]. The transmission difference becomes proportional to the total mass W:

$$\overline{T} - \overline{T}_0/T_\infty \cong W \qquad (5)$$

where $\overline{T}$ is the mean transmission of the entire image area of an object and includes the mean background transmission, $\overline{T}_0$.

It was found experimentally by Hall (1966) and Reimer (1967) that Eq. (1) and, thus, Eq. (4) are largely independent of the elemental composition of the object. In Fig. 1, the increase of contrast in the negative was compared with increase of mass for such diverse objects as nylon crystals and carbon, silver, and beryllium layers. A linear relationship existed.

B. Integrating Photometry—Conversion of Photometric Measurement to Mass

1. The Integrating Photometer

After recording the image of the viral particle on the electron micrograph negative, the plate or film is analyzed by means of an integrating photometer. One integrating photometer, the IPM-2,* modeled after the prototype constructed by Bahr and co-workers (Bahr *et al.*, 1961; Zeitler and Bahr, 1962, 1965; Bahr and Zeitler 1965) is shown in Figs. 2 and 3. The optical path is partially analogous to that for Kohler illumination in a light microscope. For measuring transmission in the IPM-2 (Fig. 2), the measuring aperture (4) is illuminated by the light source (1) through a collecting lens (3), which forms a 1:1 image of the aperture through the condenser (6) in

---

*Mention of a commercial or proprietary product does not constitute an endorsement by the U.S. Department of Agriculture or by the Department of Defense.

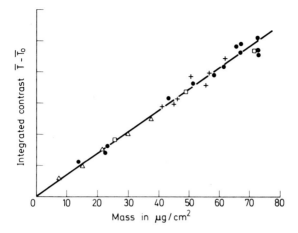

FIG 1. Linear relationship of integrated transmission of various substances vs their mass: nylon crystal (Δ); evaporate carbon (●); silver (□); and beryllium (+). The mass of steps was determined by independent methods such as multiple interference measurements, or by comparison with objects of known mass, and was plotted against integrated transmissions obtained with appropriate apertures in the Zeiss IPM-2 integrating photometer. From Bahr (1973) with permission.

the plane of the electron micrograph negative. The light passing through the aperture and not deflected by the beam splitter (12) produces a ×5 magnified image of the measuring area on a ground-glass screen (20). Simultaneously, the light source is imaged by the collecting lens (3) into the plane of the selector wheel (5), and from there via the beam splitter and lens (13) into the photomultiplier (16). The electrical signal generated is amplified, rectified (21), and fed to a suitable meter, in this case a galvanometer (22), and simultaneously to an analog output (25).

The light source is a 50-W photometer lamp, supplied by a separate, precisely stabilized dc power supply. Uniform illumination is achieved by having opal glass in front of the lamp (2) and in front of the photomultiplier (15).

Ten circular apertures (4) are arranged on a turret so that sizes of 1.5- to 18-mm diameter can be selected, and brought individually, precentered, into the beam. In order to compensate for the considerable difference in luminous flux between the smallest and largest apertures, a set of neutral density filters is changed automatically in front of the photomultiplier. Exact compensation is attained by automatic slight modification of the amplification factor. If 100% transmission has been set for unexposed emulsion with one aperture, changing to an aperture of a different size will have little or no effect on this setting.

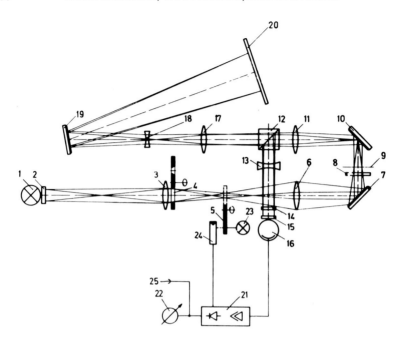

FIG 2. Optical path and components for the Zeiss IPM-2 integrating photometer. (1) Light source; (2) opal glass screen; (3) light collector; (4) measuring aperture; (5) sector wheel; (6) condenser; (7) deflecting mirror; (8) plane cover plate; (9) electron micrograph negative; (10) deflecting mirror; (11) objective lens; (12) beam splitter; (13) adjusting lens; (14) neutral filter; (15) opal glass screen; (16) photomultiplier; (17, 18) projecting lenses; (19) deflecting mirror; (20) projection screen; (21) amplifier and linear rectifier; (22) mirror galvanometer; (23) modulator lamp; (24) photo cell; (25) analog output. By courtesy of Carl Zeiss, Oberkochen, Federal Republic of Germany.

The integrating procedure may be used with noncircular apertures. For this purpose a second aperture turret is employed into which free-form apertures can be inserted. The IPM-2 is designed so that one can choose to read all values into a calculator–printer or into computer-compatible data storage, such as magnetic tape. Use of the integrating photometer provides the fastest, most accurate, and most convenient method of making mass determinations from electron micrographs. When measurements are recorded for computer processing, one can trigger them in rapid sequence with a foot switch. We have routinely made 150–250 measurements per hour, suggesting that the procedure is not limited by the time required for photometric evaluation, but that electron microscopy, and photoprocessing, set the pace.

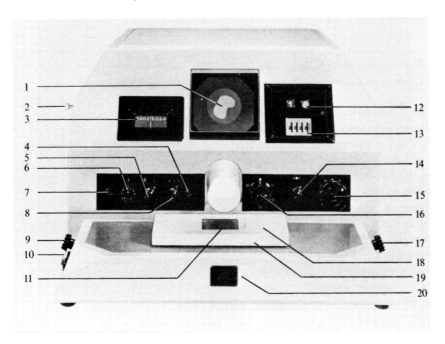

FIG 3. Front view of Zeiss IPM-2 integrating photometer. (1) Projection screen; (2) mechanical zero control; (3) galvanometer scale; (4) signal lamp indicating power on/off; (5) short-circuiting switch for galvanometer; (6) electric zero control; (7) 100-point control for analog output (not shown); (8) power switch; (9) background illumination control; (10) pushbuttons for printer control; (11) electron micrograph negative; (12) aperture number indication; (13) four-digit switch; (14) changeover between circular and noncircular apertures; (15) selector switch for circular apertures; (16) selector switch for noncircular apertures; (17) 100-point control for galvanometer readout; (18) holder for plates or cut film; (19) stage for negative; (20) desk. By courtesy of Carl Zeiss, Oberkochen, Federal Republic of Germany.

2. Measurement of Photometric Transmission—Conversion to Mass

The exponential Eq. (1) requires that every point of the object be measured, followed by transformation of the exponential values and addition to give the total mass, but no such complications are encountered when the measured value is related linearly to the mass. This is the case for the transmission over the background to at least a useful approximation. In practice, linearity holds for mass vs transmission readings for negatives of photographic density ranging from 0.5 to 0.75 (Fig. 4).

All values of contrast, $T - T_0$, for all image points can thus be determined by only two measurements, one over the image area, the other in its immediate vicinity (Fig. 5). The difference of these two measurements is proportional to the total mass. The transmission T of light through the plate

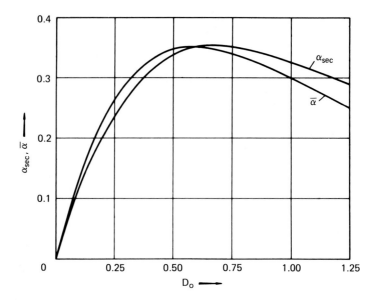

Fig 4. Dependence of the proportionality factor α as a function of background density, D_0. On the electron micrograph negative, the background D_0 can influence the mass determination unless it is at a constant value. The curves show a pronounced maximum in the range of $D_0 = 0.5$–0.75. This density range is recommended for practical work because the density variations of successive micrographs would affect the results very little. From Zeitler and Bahr (1962) with permission.

or film is directly related to the electron opacity and hence to the mass thickness w of the object. The more electron opaque the object, the higher the resulting light transmission. Thus, the transmission through the photographic plate, T_0, produced by the unaltered electron intensity I_0, which is the background, will always be at the lowest value.

A photometer whose homogeneous light beam of cross section A covers the entire image area of the object will measure a mean transmission $\overline{T}$. Provided the background transmission T_0 is homogeneous, the same photometer would measure a mean background transmission $\overline{T}_0$ equal to T_0. Because of the linear relationship between T and w, the difference ΔT of both measurements is a mean mass thickness $\overline{w}$ over the entire area A of the measuring light beam.

$$\frac{\Delta T}{T_\infty} = \frac{\overline{T} - T_0}{T_\infty} = \alpha S \overline{w} \qquad (6)$$

S is presumed not to vary for different parts of the object. This is especially unlikely for relatively small objects such as virus particles. To the inte-

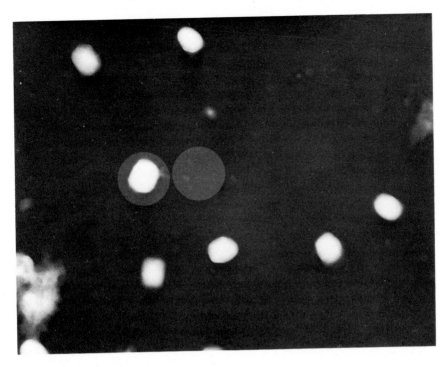

Fig 5. Vaccinia virions in the process of having their integrated transmission ($\overline{T} - T_0$) values measured. An aperture (left) was selected to cover as closely as possible the area of a vaccinia virion. In this manner the total transmission, $\overline{T}$, representing the transmission through the virion, T, plus that through the background, T_0, is measured. The background transmission T_0 alone is measured by the second aperture (right) which is identical in area to that of the first aperture. The difference of the two transmission readings is ΔT, which when multiplied by the area of the aperture gives the normalized transmission reading (R) for the virion. Original magnification ×60,000. Not shown is the shielded circular area for setting the transmission to 100%, thereby correcting for the effect of photographic fog, T_∞, on transmission (refer to Fig. 6.)

grating photometer, every object appears to have a mean mass thickness $\overline{w}$ uniformly spread over the area regardless of actual variations in mass thickness, w, or in the shape of the object. The treatment of the proportionality factors and S is discussed in Section II,C.

The apertures of the integrating photometer are chosen to fit the image as tightly as possible so that the contribution of the background T_0 to the total transmission measured $\overline{T}$ will be as small as possible, and the difference, $\overline{T} - T_0$, at a maximum (Fig. 5). The difference in transmission readings is normalized by multiplying it by the area A of the aperture used in the integrating photometer. This product, for calculation of mass, is referred to as R.

$$R = \frac{\overline{T} - T_0}{T_\infty} A \qquad (7)$$

The value of the photographic fog, T_∞, is present in Eqs. (4)–(7). It can be corrected by setting the transmission to 100% in an unexposed area of the electron micrograph, an area expressing only photographic fog. To produce such an area on the negative, a grounded metal disk about 15 mm in diameter is mounted to a suitable part of the microscope camera, as close as possible to the plane of the emulsion, in order to shield a circular area of the negative from exposure. In electron micrographs, the shielded circular area appears transparent on the negative and black on the positive.

An attractive alternative to integrating photometry with the IPM is scanning–digitizing of the negative. For this purpose a flat-bed scanner densitometer, a TV-based digitizer, or a digitizing drum scanner is a prerequisite. In most instruments of this type it is possible to define the area of scan by setting a scan window around the contours of the object image. The window is comparable to the object's feret diameters projected onto the x and y movements of the scanning instrument. Beyond these mechanical means there are software alternatives. For example, if the entire negative has been digitized, it can be displayed on a monitor and the object delineated with a light pen or cursor. Or, the object can be delineated by driving the scanning stage so that a fixed point, e.g., the cross hair in an ocular, appears to move around the image of the object, thus indicating to the computer the area to be scanned and reducing the digitizing to the object area only. An advanced step is the identification of the object area entirely by software through either thresholding or scene segmentation.

The first procedural step is to set the scanner on an unexposed portion of the photographic emulsion and adjust transmission to 100%, thus compensating subsequent measurements of photographic fog.

Assuming then that a digitized image is at hand, a histogram of transmission vs frequency will assist in separating background from object transmissions, whereupon the sum of the object transmission is calculated as well as the number of object pixels noted.

In order to correct for background the (average) transmission of one background pixel is calculated from 50 or 100 lowest transmission values. This is multiplied by the number of object pixels, and the sum is subtracted from the total object transmission.

If the object is the image of a polystyrene sphere (subjected to electron microscopy at the same magnification and processed under the same conditions as the object), the above procedure gives a transmission value that can be related to an absolute mass value calculated from the diameter of the sphere, corrected for magnification, i.e., the true volume of the sphere multiplied by the specific gravity of polystyrene (1.05 g/cm^3). Successive determinations of both total transmission and mass for spheres of different sizes produce a calibration curve of transmission vs mass, the slope of which may be used in calculations of the absolute object mass.

An outstanding advantage of the scanning–densitometric assessment of mass by the contrast method is that object images filling the area of the electron microscopic plate can be measured while the integrating photometer is limited to circular areas 2.0 cm in diameter or smaller. Measuring the mass of an entire cell or of cell nuclei can best be done with the scanning approach. A flat object 10 μm in diameter will become in an electron micrograph at $\times$ 6000 magnification a disk 6.0 cm in diameter fitting well on a 8.3 $\times$ 10.2-cm plate. A calibration sphere 0.756 μm in diameter, however, will be a disk 4.54 mm in diameter. Many nuclei have a size of 50 μm. At a magnification of $\times$ 1500 they will be 7.5 cm on the negative. A sphere safely within the range of the penetration of electrons is 0.756 μm in diameter. It will appear to be only 1.13 mm at the same magnification. This size is just at the limit of mass determination by contrast evaluation (Zeitler and Bahr, 1962).

C. STANDARDIZATION

The procedure for integrating photometric weight determination must be standardized:

$$W_{\text{unknown}} : W_{\text{standard}} :: R_{\text{unknown}} : R_{\text{standard}} \tag{8}$$

1. External Standards

When using standards that are not electrographed on the same film or plate as the unknown, polystyrene latex (psl) spheres (Fig. 6) are the choice in our laboratory for mass determinations. Because of their shape, the actual mass is determined from their volume and the knowledge of the density (g/cm^3) of polystyrene. In attempting to use psl spheres as internal standards, we have observed that they tend to form aggregates with virus particles (Bahr et al., 1980). Therefore, it is necessary to use the latex spheres as external standards. When exposed for a brief time to the electron beam, the psl spheres do not lose their shape, and they also have an additional advantage in that they can be purchased in an assortment of sizes representing a variety of mass values.

$$W_{\text{psl spheres}} = \text{volume} \times \text{density} \tag{9}$$

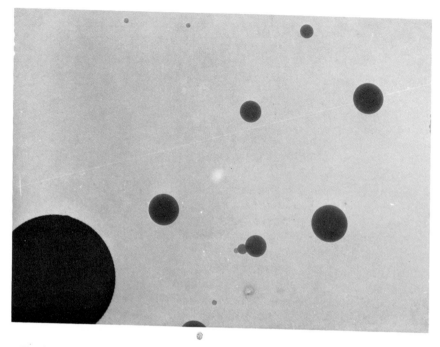

Fig 6. Electron micrograph of polystyrene latex (psl) spheres used as external standards in mass determinations of viruses. The black circular areas in the lower left corner represents the shielded areas of the electron micrograph (transparent on the negative). It is used to set the integrating photometer to 100% before measuring the two transmission values, $\bar{T}$ and T_0. Whatever the level of photographic fog, T_∞, it is now subtracted from every measurement of transmission in the emulsion. The psl spheres are used for calibration of the determination of dry mass in absolute units (gram).

The density of polystyrene is taken as 1.05 g/cm³. In practice, it is easier to measure the diameter of a sphere, d, than to attempt to pinpoint the center of the sphere for the radius measurement:

$$W_{\text{single psl sphere}} = \tfrac{4}{3} \pi \left(\tfrac{1}{2} d\right)^3 \times 1.05 \times 10^{-3}$$

and

$$\tfrac{4}{3} \pi \times \tfrac{1}{8} \times 1.05 = 0.5495 \qquad (10)$$

$$W_{\text{single psl sphere}} = 0.5495 \, d^3 \times 10^{-3}$$

(Since d was measured in millimeters, it must be multiplied by 10^{-3}.)

Experimentally, a mixture of latex spheres is electrographed separately under the same conditions as the unknown virus. Exposure must be controlled so that the density of the background on the negative is between 0.5 and 0.75, the ideal being 0.60. For experiments kept in this range, transmission of the image of objects, e.g., virus particles, are not affected by density differences (Fig. 4).

The negatives of the standard and those of the unknown *must* be developed together under standard photographic conditions. In our mass studies, development of electron micrograph negatives was done in Kodak D 19 solutions diluted 1:2 for 5 min, with intermittent agitation every 30 sec, at a temperature of 21°C (70°F).

Each psl sphere is measured for its $\overline{T} - T_0$, and multiplied by the area of the aperture used in the integrating photometer [Eq. (7)] to obtain $R_{psl\ sphere}$. All of the R vs d^3 values for the assorted sizes of psl spheres can be added and averaged or the R vs d^3 for each sphere size can be plotted. A straight line is obtained whose slope is proportional to αS, the terms defined in Eq. (1). We refer to the average value of R/d^3 or the slope value as β. The working equation for the determination of mass W, when using psl spheres, becomes

$$W_{unk} = \frac{0.5495 \times 10^{-3}}{\beta \times M^3} \times R_{unk} \tag{11}$$

In Eq. (11), the magnification M was raised to the third power because the diameter for each psl sphere size was raised to the third power to derive β, which is also present in Eq. (11). An example of the mass determination of SV40 is given in Section III,A; psl spheres were used as external standards.

2. Internal Standards

If the standard used in mass determination can be electrographed on the same negative, as would be the case in using a known virus as an internal standard, the proportionality factors of mass, W, and corrected transmission value, R, for the known virus and R for the unknown virus are all that are required. The magnification term does not enter into the calculation:

$$W_{unk} = W_{std} R_{unk}/R_{std} \tag{12}$$

After development of the negative, it is analyzed in the integrating photometer. A statistical number of standard virus particles and unknown virus particles are measured for their transmission values ($\Delta T = \overline{T} - T_0$), and ΔT is multiplied by the area of the aperture used in the integrating photometer to obtain the corrected reading, R. By proportionalization, the mass of the unknown virus is calculated from Eq. (12). In Section III,B, an example is given for the average mass determination of tobacco mosaic virus (TMV) using SV40 as an internal standard.

D. Logarithmic Normal Distribution of Viral Mass in a Population and the Standard Deviation of the Mean

1. Logarithmic Normal Distribution of Viral Mass

In constructing histograms reflecting the frequencies of corrected photometric readings R, values of the corresponding viral mass exhibit a distinct tendency to be skewed toward the higher side. However, the mass of viruses is distributed normally when considered logarithmically to the base 10. When R is converted to log R and plotted vs percentage cumulative frequency F in a probability graph, a straight line is obtained, indicating the lognormal distribution of viral mass, as shown in Fig. 7.

Lognormal distributions are common for most biological measurements, including weight, length, height, volume, and/or circumference (Gaddum,

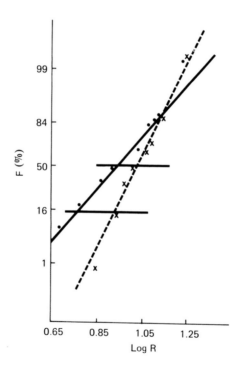

Fig 7. Lognormal distribution of virus particles in a plot of percentage cumulative frequency, F, vs corrected transmission, log R, in a probability chart. The solid line represents the plot for T2 virus, and the dashed line the plot for *Tipula* iridescent virus. At the 50% point, the value for the average log R can be obtained, and R is used in the calculation of the average mass of the virus population. In addition, log σ for distribution, the standard deviation, can be obtained as the difference of the readings at 50 and 16%.

1945a,b). In plotting lognormal distributions in the determination of mass by quantitative electron microscopy, one can observe mass distribution for the mode of the population. One can also observe the deviation in modality, indicating the heterogeneity in the virus population.

In their intensive analysis of the mass of vaccinia virus particles, Bahr *et al.* (1980) concluded that while genetic control in a presumably homogeneous virus population is strict with respect to quality, the quantity of viral components, presumably other than DNA, varies in a fashion that can be aptly described by a lognormal distribution.

2. Log σ for Distribution

As noted above the mass of viruses is given as the median of the distribution because the distribution is skewed toward higher mass values. In such cases, the more marked the skew, the more misleading is a mean, particularly the standard deviation of the mean, σ. If the distribution is normal, and log R is plotted against F, the percentage cumulative frequency, in a probability chart, the three distributional measures—median, mean, and mode—fall together. The standard deviation, log σ, is identified in each graph between the points of 16 and 50% cumulative frequency (Fig. 7).

III. Procedure and Treatment of Data to Obtain the Average Mass of Virus Particles

In this section the average mass value for two viruses is obtained using transmission data and employing standards whose mass is known or can easily be calculated. In the first example, an external standard is used, and in the second example, an internal standard.

A. AVERAGE MASS OF SV40 USING POLYSTYRENE LATEX SPHERES AS EXTERNAL STANDARDS

A drop containing virus particles was placed on Parlodion–carbon films (about 50 Å of carbon) and dehydrated through ascending concentrations of acetone to 100% acetone. The sample grids were critical-point dried from carbon dioxide according to Anderson (1951).

The grids were then placed in an oven at 180°C for 20 min to remove the Parlodion, leaving the virus particles supported only by the carbon film. The virus particles were electrographed in a Hitachi HU-12 electron microscope under the following conditions: magnification $\times 36,800$, a 100-μm condenser aperture, a 70-μm objective aperture, 75-kV accelerating voltage.

After being developed and dried, the plates were analyzed in the IPM-2 integrating photometer.

The plate was set emulsion-side up in the photometer. Electrical zero was obtained with the beam blocked by an opaque portion of the stage, and the image of a virus particle was centered on the ground-glass observation screen. An aperture was selected to cover conveniently the size of the image of the virus particle. The carriage was moved so that the bright disk of the aperture image fell clearly in the unexposed area of the plate and the transmission T_∞ was adjusted to 100% on the galvanometer. Next, the image of the virus particle was brought into the aperture, and the total transmission $\overline{T}$ over the image area was measured, i.e., T of virus image plus background T_0. The aperture image was then moved to a free background and T_0 was recorded. For 128 SV40 particles, readings of ΔT, representing $\overline{T} - T_0$, were obtained as demonstrated in Table I for 10 SV40 particles. As noted above ΔT multiplied by the area factor A for the aperture used in the integrating photometer yields R.

The virus particles were then grouped according to their R values. The range of R covered the values observed in measuring the particles. For the 10 SV40 particles noted in Table I, R had a range of 1.4–1.7. Assigning the number of particles to R values of 1.4, 1.5, 1.6, and 1.7, we would have 1 for R of 1.4, 4 for R of 1.5, 1 for R of 1.6, and 4 for R of 1.7.

After measurement of the virus particles, the analysis continued as shown in Table II, where n, $\Sigma\, n$, and $\%\, \Sigma\, n$ stand for number of particles, cumulative number, and percentage of cumulative number, respectively. In order to analyze a logarithmic distribution of the virus particles (Gaddum,

TABLE I

TRANSMISSION VALUES FOR 10 IMAGES OF SV40 PARTICLES,
OBTAINED WITH THE INTEGRATING PHOTOMETER[a]

Virus particle	$\overline{T}$	T_0	ΔT	R^b
1	345	302	43	1.68 ≅ 1.7
2	340	296	44	1.72 ≅ 1.7
3	335	296	39	1.52 ≅ 1.5
4	330	288	42	1.64 ≅ 1.6
5	335	296	39	1.52 ≅ 1.5
6	334	295	39	1.52 ≅ 1.5
7	330	292	38	1.48 ≅ 1.5
8	330	286	44	1.72 ≅ 1.7
9	320	276	44	1.72 ≅ 1.7
10	312	375	37	1.44 ≅ 1.4

[a] Taken at aperture number 2. Plate number 4910.

[b] $R = \Delta T\, A$ [see Eq. (7)]. A, the area factor for aperture number 2, is 3.9×10^{-2}.

TABLE II

PERCENTAGE CUMULATIVE FREQUENCY, F,
OF TRANSMISSION VALUES, R, FOR SV40 VIRUS PARTICLES

R	n	$\Sigma\,n$	F^a	Next higher $\log R^b$
1.0	0	0	0	0.04
1.1	3	3	2.3	0.08
1.2	11	14	10.9	0.11
1.3	20	34	26.6	0.15
1.4	38	72	56.3	0.18
1.5	22	94	73.4	0.20
1.6	20	114	89.1	0.23
1.7	9	123	96.1	0.26
1.8	2	125	97.7	0.29
1.9	3	128	100	0.30
2.0	—	—	—	—

^{a} % $\Sigma\,n = F$.
^{b} Log values are rounded.

1945a), log R is assigned to each R value, as shown in Table II. The % $\Sigma\,n$ is compared to the next higher log R. Thus % $\Sigma\,n$ of 2.3 is compared to the log of 1.2, or 0.08. This procedure is followed for the next % $\Sigma\,n$ in the series, and so on. In order to assign a log R to the last % $\Sigma\,n$, i.e., a % $\Sigma\,n$ of 100 in the series, the range of R was extended by 1 to give 2.0. The log of 2.0 is 0.301, rounded off to 0.3, which was used as the log R corresponding to a % $\Sigma\,n$ of 100.

1. Lognormal Distribution of SV40 Particles

A plot of % $\Sigma\,n$ (or F) vs its corresponding log R is made to obtain the log R at the 50% point, or the average log R. A lognormal distribution was obtained for SV40. From the graph, log R at a frequency F of 50% was 0.170. The value of R, was 1.479. The next step was to obtain R and w for the polystyrene latex spheres, the standards.

2. Standards

Various-sized psl spheres were electrographed separately but under the same conditions as the SV40 particles. The proportionalization is as shown in Eq. (8).

$$W_{\text{SV40}} = W_{\text{psl spheres}} \times (R_{\text{SV40}}/R_{\text{psl spheres}})$$

As noted in Section II, when using psl spheres as standards, the working equation for the average mass determination is Eq. (11).

$$W_{SV40} = \frac{0.5495 \times R_{SV40} \times 10^{-3}}{M^3 \times \beta}$$

where M, the magnification, is raised to the third power. The magnification used in the experiment was $\times 36,800$. The term β was obtained from an average of the ratio of R to d^3; d is the diameter raised to the third power of the psl spheres and R is the corresponding normalized transmission value. As noted above, the term β may also be obtained from the slope of the line representing a plot of R vs d^3 for each psl sphere size. Inserting the corresponding values in Eq. (11),

$$W_{SV40} = \frac{\overset{(R)}{0.5495 \times 1.479 \times 10^{-3}}}{\underset{(M^3)\qquad(\beta)}{49.8 \times 10^{12} \times 0.3}}$$

$$W_{SV40} = \quad 5.44 \times 10^{-17}\,g$$

The molecular weight (grams/mole) was obtained by multiplying the mass value by Avogadro's number, $5.44 \times 10^{-17} \times 6.02 \times 10^{23}$, or a molecular weight for SV40 of 32.7×10^6.

B. Average Mass of TMV Using SV40 Particles as Internal Standards

Electrographing a mixture of SV40 particles and TMV particles and analyzing their images on the negative (Fig. 8) allows a direct comparison by which the average mass value of TMV is obtained, according to Eq. (12).

The average mass of SV40 obtained in the preceding experiment was given as 5.44×10^{-17} g. What is needed is to obtain $\overline{T} - T_0$ transmission values for each of the two viruses, and normalize ΔT with respect to the area of the aperture used in the integrating photometer to obtain R values for SV40 and TMV, respectively, as demonstrated in Table III.

For SV40, a plot of % $\Sigma\,n$ or F vs log R gave the 50% value or average log R as 0.85; R is 7.08. In the same manner, a plot of % $\Sigma\,n$ or F vs log R gave the 50% value of log R for TMV particles as 1.01; R was 10.23.

The average mass of TMV was calculated from Eq. (12).

$$\begin{aligned} W_{TMV} &= W_{SV40} \times (R_{TMV}/R_{SV40}) \\ &= 5.44 \times 10^{-17} \times 10.23/7.08) \\ &= 7.61 \times 10^{-17}\,g \end{aligned}$$

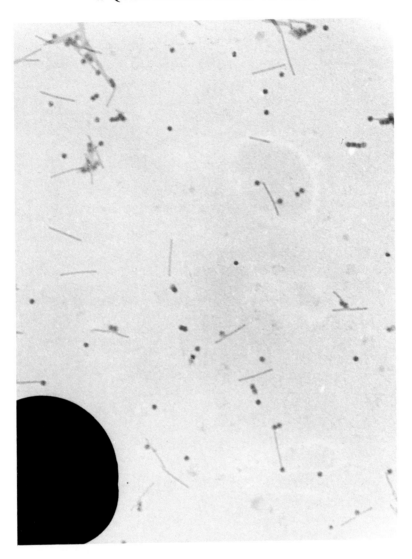

FIG 8. Electron micrograph of tobacco mosaic virus (TMV) particles and SV40 particles. The SV40 particles serve as internal standards to obtain the average mass of TMV. Note black circular area in lower left corner, which is transparent on the negative and is used to correct for photographic fog. No contrasting materials such as heavy metal stains were added to the virus particles, since, obviously, these materials would alter the true weight of the viruses. Original magnification on the negative ×30,000.

TABLE III
TREATMENT OF TRANSMISSION DATA FOR SV40 PARTICLES USED
AS AN INTERNAL STANDARD FOR TMV PARTICLES

Virus	R	n	$\Sigma\, n$	F	Next higher log R^a
SV40	5	7	7	5.7	0.78
	6	23	30	24.6	0.85
	7	54	84	68.9	0.90
	8	33	117	95.9	0.95
	9	4	121	99.2	1.00
	10	1	122	100	1.04
	11	—	—	—	—
TMV	5	1	1	1.5	0.78
	6	—	—	1.5	0.85
	7	6	7	10.9	0.90
	8	4	11	17.2	0.95
	9	12	23	35.9	1.00
	10	25	48	75.0	1.04
	11	—	—	—	1.08
	12	14	62	96.9	1.11
	13	—	—	—	1.15
	14	2	64	100	1.18
	15	—	—	—	—

[a] Log values are rounded.

or, on a molecular weight basis,

$$MW_{TMW} = (MW_{SV40} \times R_{TMV}/R_{SV40}$$
$$= (32.7 \times 10^6 \times 10.23)/7.08$$
$$= 45.8 \times 10^6 \text{ g/mol}$$

The reader should note that, when internal standards are used in the determination of mass, the magnification and β term are not considered.

Of the two procedures for determining mass, the use of external standards or of the use of internal standards, the virologist would be tempted to choose the latter. Such a choice is permitted provided that, in the mixing of the two viruses, there is not extensive aggregation to make transmission measurements difficult, nor are any physical or chemical alterations noted for the virus particles. If any of these factors is observed, the researcher can use the method of external standards to obtain the average mass value of the virus particles. Consider, also, that psl spheres store better than most viruses. Therefore, if there is any doubt concerning the structural integrity of the virus particles to be used as the internal standard, it would be prudent to use psl spheres, and the external standard procedure for mass determination of the unknown virus particles.

IV. Applications in Virology

The virus studies in which mass values were determined by quantitative transmission electron microscopy are presented in this section.

For the different laboratory procedures cited below, it should be clear to the reader that there is no strict methodology that must be used for mass determinations. Thus, for example, all types of transmission electron microscopes, conventional or high voltage, may be employed at various accelerating voltages. What must be recognized is the necessity for constancy in electron microscopy, photography, and integrating photometry when comparing an unknown virus to an external standard. Using an internal standard, in mass determinations, where unknown and known virus particles are analyzed together, such factors, obviously, are equalized.

It should be emphasized that the virus particles and standards were electrographed in the absence of heavy-metal stains or other substances that would alter the true mass of the particle. The electron microscope may be adjusted instrumentally to produce the desired contrast of the image by using a suitable objective aperture or a lower accelerating voltage.

Except for SV40 particles used as internal standards for the average mass determination of TMV, all the mass values for the viruses presented here were obtained using psl spheres as external standards. We have found psl spheres to be a reliable standard in our laboratory, using minimum exposure times in the electron microscope, as noted in Section II.

In converting mass values of viruses to molecular weights, we have expressed the latter in grams per mole. Some virologists choose to express molecular mass in daltons. One dalton is equal to 1.65×10^{-24} g. The expression of molecular weight in either one or the other units gives practically the same numerical value. Thus, for SV40 virus particles having an average mass of 5.44×10^{-17} g, the molecular weight (g/mol) would be the mass value multiplied by Avogadro's number: 5.44×10^{-17} g $\times 6.02 \times 10^{23}$ molecules per mole, is equal to 32.74×10^6 g/mol. In dalton units, we would obtain 5.44×10^{-17} g divided by 1.650×10^{-24} g, or 32.96×10^6 daltons. In each case, rounding off gives the same numerical value, 33 million. In the comparison of mass–molecular weight values that follows, the values in grams per mole and in daltons are compared directly.

We have attempted to describe the essential features of mass determinations of the viruses studied by quantitative transmission electron microscopy (QTEM). Such features include the source of the virus, some of the main points noted in the electron microscopy, and comparisons with other literature reports. It seemed unnecessary to include, for every virus presented, discussion of features such as the size of the objective aperture, condenser aperture(s), or the accelerating voltage level. The interested reader is urged to consult the original papers. On the other hand, where we have felt that an important concept should be made known to the reader, we have stated it in some detail.

A. Animal Viruses

Average mass values are listed in Table IV.

1. Herpes Simplex (Lampert and Bahr, 1969)

The study of herpes simplex virus was our first attempt to measure directly the mass of individual virus particles by quantitative transmission electron microscopy. The preparative method employed allows one to isolate viruses from small amounts of cellular material without further purification and to study the morphology of the different maturation stages without recourse to staining or other contrasting procedures.

TABLE IV
Average Mass and Molecular Weight Values of Animal Viruses

Virus	n	Dry mass (g)	Molecular weight (g/mol)
Herpes simplex maturation stage			
Core	13	2.07×10^{-16}	125×10^6
Empty capsid	17	5.22×10^{-16}	315×10^6
Full capsid	31	7.55×10^{-16}	455×10^6
Enveloped nucleocapsid	19	13.33×10^{-16}	804×10^6
Vaccinia strains			
NIH	176	6.54×10^{-15}	3.94×10^9
Hamburg	7300	5.26×10^{-15}	3.17×10^9
Lea (7N)	158	6.17×10^{-15}	3.72×10^9
W.R.	109	5.98×10^{-15}	3.61×10^9
W.R.	150	6.04×10^{-15}	3.64×10^9
Cowpox	156	6.50×10^{-15}	3.92×10^9
Rabbitpox	79	5.31×10^{-15}	3.20×10^9
Fowlpox—Australia	176	6.02×10^{-15}	3.63×10^9
Fowlpox—Mississippi	164	7.22×10^{-15}	4.35×10^9
Myxoma	135	6.22×10^{-15}	3.75×10^9
SV40	128	5.44×10^{-17}	33×10^6

Source of virus. Herpes simplex virus stain 11140 (Ashe and Scherp, 1963).

Preparation and electron microscopy. Infected cells were spread on a trough filled with distilled water at pH 6.4. Virus particles released from the cells were placed on Formvar grids coated with carbon, by touching the grids to the water surface. After drying the sample grids by the critical point method (Anderson, 1951), the virus particles were electrographed at 100 kV in a Siemens Elmiskop I electron microscope at a magnification of $\times 34,000$ to $\times 38,000$, depending upon the maturation form being analyzed.

Other literature reports. Some indication of the mass of herpes simplex can be gleaned from studies of its DNA. It is reported to have a molecular weight approximating 100 million (Becker *et al.,* 1968; Lee *et al.,* 1971) and is believed to be located exclusively within the core (Kaplan, 1969).

2. *Poxviruses: Vaccinia Strains, Cowpox, Rabbitpox, Fowlpox, Myxoma*

Bahr and co-workers have made intensive studies on the mass of poxviruses (Bahr *et al.,* 1976; 1980). The magnification used in analyzing the vaccinia (NIH) strain was $\times 22,896$. For other strains of vaccinia and other poxviruses, magnifications close to $\times 15,000$ were used. A comparison with other literature reports is presented at the end of the discussion on poxviruses.

a. Vaccinia (NIH)

Source of virus. Dr. B. Moss, National Institute of Allergy and Infectious Disease, National Institutes of Health, Bethesda, Maryland.

Preparation and electron microscopy. As noted for SV40, in Section III,A.

b. Vaccinia (Hamburg)

Sources of virus. Vaccinia virus was cultivated according to Hoagland *et al.* (1940) in rabbit skin as described by Peters (1960). Virus particles were prepared from homogenates of infected tissue in McIlvane buffer solution, diluted 1:40, pH 7.3, to which 50% (v/v) trifluorotrichloroethylene (Freon 113) had been added. Differential centrifugation in buffer to which increasing proportions of Freon were added produced viral preparations of high purity (Marquardt *et al.,* 1963).

Preparation and electron microscopy. A drop of virus suspension was placed on Formvar–carbon-coated grids, and the buffer salts were dialyzed out by floating the grids on distilled water (Bahr and Zeitler, 1965). Alternatively, grids were prepared from extensively dialyzed suspensions (Marquardt *et al.,* 1963; Peters, 1960). In either case, the sample grids were air dried from distilled water and examined in an RCA-4B or a Zeiss EM-10 electron microscope.

c. Vaccinia, Lea (7N), and Vaccinia, W.R.

Source of virus. The Tropeninstitut, Hamburg, Federal Republic of Germany.

Preparation and electron microscopy. These strains were studied in a fashion similar to that described for the vaccinia (Hamburg) strain.

d. Cowpox, Rabbitpox, Fowlpox, Myxoma

Source of viruses. The Tropeninstitut, Hamburg; Dr. W. R. Sobey, Division of Animal Genetics, C.S.I.R.O., Sidney, Australia; Dr. C. C. Randall, University of Mississippi Medical Center, Jackson, MS.

Preparation and electron microscopy. As described under Vaccinia (Hamburg).

For the poxviruses, Bahr *et al.* (1980) observed a spread of mass values greater than 2-fold. Vaccinia virions, for example, ranged in mass from 3.5 to 8×10^{-15} g. The variation was believed to result from mass changes in viral components other than DNA.

Other literature reports on the mass of poxviruses. Smadel *et al.* (1939) assumed a spherical shape for vaccinia and measured the diameter as 236×10^{-7} cm. A mass value of 5.34×10^{-15} g was obtained from a volume of 4.24×10^{15} cm^3 and a density of 1.26–1.27 g/cm^3. Joklik (1962) obtained essentially the same value from electron microscopic counting of particles and the amount of protein per particle. Marquardt *et al.* (1963) reported a mass value for vaccinia of 5.69×10^{-15} g from measurements of the particle, assuming a spherical shape.

3. Simian Vacuolating Virus (SV40) (Bahr et al., 1976)

Source of virus. Dr. M. Martin, National Institute of Allergy and Infectious Disease, National Institutes of Health, Bethesda, MD.

Preparation and electron microscopy. As described in Section III,A.

Other literature reports. Mayor *et al.* (1963) reported a molecular weight of 44 million based on a DNA content of 9% and a molecular weight of 4 million. The diameter of the particle was calculated as 45 nm. A molecular weight of 28.8 million for the virus can be calculated from a DNA content of 12.5% (Anderer *et al.,* 1967) and a molecular weight of 3.6 million (Tai *et al.* 1972). For virus particles averaging 41.1 nm, Koch *et al.* (1967) calculated a molecular weight of 17.3 million from sedimentation and diffusion coefficients and a partial specific volume determination.

B. Bacterial Viruses

Mass values for the viruses studied are listed in Table V.

1. T2 (Mazzone et al., 1980c)

Source of virus. Miles Laboratory, Inc., Elkhart, IN.

Preparation and electron microscopy. The virus (ATCC 11303 B2)

TABLE V

AVERAGE MASS AND MOLECULAR WEIGHT VALUES OF BACTERIAL VIRUSES

Virus	n	Dry mass (g)	Molecular Weight (g/mol)
T2	101	34.9×10^{-17}	211×10^6
T5	156	19.4×10^{-17}	114×10^6
T7	219	7.4×10^{-17}	45×10^6

was applied to Formvar–carbon-coated grids and critical-point dried. A Zeiss EM-10 electron microscope was used at a magnification of $\times 35,280$.

Other literature reports. Cummings and Kozloff (1960) calculated a molecular weight value of 215 million from data on sedimentation and diffusion coefficients and the partial specific volume of the virus particle. Giddings *et al.* (1975), employing the technique of sedimentation field flow fractionation, obtained an average molecular weight of 225 million from three experiments.

2. T5 (Mazzone et al., 1980c)

Source of virus. Dr. R. Fujimura, Oak Ridge National Laboratory, Oak Ridge, TN.

Preparation and electron microscopy. As described above for T2, using a magnification of $\times 35,856$.

Other literature reports. Dubin *et al.* (1970) calculated a molecular weight value of 109 million from sedimentation and diffusion coefficients, the latter obtained by the time-saving technique of laser light scattering (Dubin *et al.*, 1967). The value for the partial specific volume and the value for the sedimentation coefficient were obtained from studies in other laboratories.

3. T7 (Adolph and Haselkorn, 1972)

Source of virus. Dr. F. W. Studier, Brookhaven National Laboratory, Upton, NY.

Preparation and electron microscopy. Drops of virus suspension were placed on Formvar–carbon-coated grids and dehydrated through a graded series of aqueous ethanol solutions, then through absolute ethanol and Freon 113. The sample grids were dried by the critical-point method with Freon 13 as the transitional fluid (Cohen *et al.*, 1968). Electron micrographs were taken with an RCA-4B electron microscope operated at 100 kV. This study was combined with those for the plant viruses LPP-1 and LPP-2 (refer to Section IV,C). Depending upon the virus analyzed, magnifications from $\times 36,000$ to $\times 39,000$ were used.

Other literature reports. Bancroft and Freifelder (1970) reported a molecular weight value of 40.4 million from a sedimentation equilibrium experiment. Using laser light scattering to obtain the diffusion coefficient, Dubin *et al.* (1970) and Camerini-Otero *et al.* (1974) calculated molecular weights of 50.4 million and 50.9 million, respectively. In each case the value of the sedimentation coefficient or that of the partial specific volume was obtained from the literature.

C. Plant Viruses

Mass values for the viruses studied are listed in Table VI.

1. *Blue-Green Algal Virus, LPP-1M (Adolph and Haselkorn, 1972)*

Source. LPP-1 belongs to a class of viruses that infect the blue-green alga *Plectonema boyanum* (Sherman and Haselkorn, 1970a). The virus is similar to bacteriophage T7 in a number of characteristics: morphological features (Luftig and Haselkorn, 1967), DNA metabolism (Sherman and Haselkorn, 1970b), and protein synthesis (Sherman and Haselkorn, 1970c).
Preparation and electron microscopy. As described above for T7.
Other literature reports. There were none.

2. *Blue-Green Algal Virus, LPP-2, Isolate WA (Adolph and Haselkorn, 1972)*

Source. The LPP-2 virus, as noted above for LPP-1, resembles T7 bacteriophage. The host for LPP-2 WA, like that for LPP-1, is *P. boyanum;* however, the two algal viruses appear to be unrelated serologically (Safferman *et al.,* 1969)
Preparation and electron microscopy. As described above for T7.
Other literature reports. There were none.

3. *TMV (Bahr et al., 1976)*

Source. Dr. R. L. Steere, Plant Virology Laboratory, Agricultural Research Service, U.S. Department of Agriculture, Beltsville, MD.

TABLE VI
Average Mass and Molecular Weight Values for Plant Viruses

Virus	n	Dry mass (g)	Molecular weight (g/mol)
Blue-green algal virus LPP–1M	186	8.88×10^{-17}	53×10^6
Blue-green algal virus LPP–2 isolate WA	134	8.46×10^{-17}	51×10^6
Tobacco mosaic	64	7.61×10^{-17}	46×10^6

Preparation and electron microscopy. As noted for SV40 in Section III,A. The mass of TMV was determined using SV40 particles as an internal standard. Although the magnification is not a requisite in determining mass using an internal standard, the two viruses were analyzed at ×30,000. The length of TMV averaged 2900 Å.

Other reported values. Numerous molecular weight values have been reported for TMV, and some are noted in Table VII. The length of the TMV particle is believed to approximate 3000 Å. (Williams and Steere, 1951; Hall, 1958; Fraenkel-Conrat and Ramachandran, 1959; Haltner and Zimm, 1959; Steere, 1963).

Most of the values presented in the table were obtained by procedures, e.g., light scattering, with which only average values can be obtained. Thus, for such methods, all the particles contribute to the molecular weight value, even if their lengths are significantly longer or shorter than 3000 Å. Lauffer and Stevens (1968), in their critique on molecular weight determinations of TMV, noted that "many physical methods are not useful for particles this

TABLE VII

VALUES FOR MOLECULAR WEIGHT REPORTED FOR TOBACCO
MOSAIC VIRUS VS AVERAGE LENGTH OF VIRION

Method[a]	MW[b] (× 10⁶)	Length (Å)	Reference
EM counting	50	2980	Williams *et al.* (1951)
	42	3300	Polson *et al.* (1970)
EM, X ray, and chemical analysis	50	2890	Markham *et al.* (1964)
EM and X ray	40	2700	Lauffer (1944)
	45	3000	Watanabe and Kawade (1953)
QTEM	46	2900	Bahr *et al.* (1976)
s and *D*	32	2560	Lauffer (1944)
	42	3600	Watanabe and Kawade (1953)
s and *η*	33	2760	Lauffer (1944)
	37	3100	Watanabe and Kawade (1953)
D and *η*	36	2830	Lauffer (1944)
	53	3600	Watanabe and Kawade (1953)
Light scattering	40	3200	Boedtker and Simmons (1958)
	40	2700	Oster *et al.* (1947)
	50	3260	Jennings and Jerrard (1966)
	51	2650	Doty and Steiner (1950)

[a] EM, electron microscopy; QTEM, quantitative transmission electron microscopy; *s*, sedimentation coefficient; *D*, diffusion coefficient; *η*, viscosity.

[b] Molecular weight values were rounded off to the nearest whole number, or averaged for multiple experiments.

large. On the other hand, the particles are too small for many direct particle counting techniques. In addition the determination is complicated by the tendency of the virus particles to associate end to end. Thus, independent evidence of the homogeneity of the preparation is required.''

In the quantitative transmission electron microscopy procedure for determining mass, the molecular weight of a virus may be derived for every particle of a measured length. A range of molecular weight values, as well as accurate measurements of correlating length, diameter, and structural integrity, may be obtained.

D. Insect Viruses

Mass values for the viruses studied are listed in Table VIII.

1. *Tipula Iridescent Virus* (*TIV*) (Bahr *et al.*, 1976)

Source. Dr. R. L. Lowe, Entomology Research Division, Agricultural Research Service, U.S. Department of Agriculture, Gainesville, FL.

Preparation and electron microscopy. A drop of the sample was placed on Formvar grids and air dried. A Siemens Elmiskop IA was used at a magnification of $\times 2100$.

Other reported values. Thomas (1961) reported a molecular weight of 1.22×10^9 (diameter 1300 Å) by electron microscopy counting. Weber

TABLE VIII
AVERAGE MASS AND MOLECULAR WEIGHT VALUES OF INSECT VIRUSES

Virus	n	Dry mass (g)	Molecular weight (g/mol)
Tipula iridescent	137	1.37×10^{-15}	820×10^6
Single virus (rod form) of the gypsy moth NPV	22	1.55×10^{-15}	935×10^6
Single virus (rod form) of the European pine sawfly NPV	100	4.66×10^{-16}	281×10^6
Single virus (round form) of the European pine sawfly NPV	30	4.62×10^{-16}	279×10^6
Inclusion body of the gypsy moth NPV	241	3.44×10^{-12}	207×10^{10}
Inclusion body of the European pine sawfly NPV	157	0.04×10^{-12}	2.4×10^{10}
Inclusion body of the spruce budworm NPV	213	2.02×10^{-12}	122×10^{10}
Inclusion body of the cabbage moth NPV	133	2.49×10^{-12}	150×10^{10}

et al. (1963) calculated the molecular weight for TIV as 1.05×10^9 from sedimentation equilibrium by gravity, in an experiment lasting 18 days. Glitz *et al.* (1968) calculated the following molecular weights: 1.28×10^9 from sedimentation and diffusion coefficients, 1.19×10^9 from sedimentation and viscosity, and 1.30 from partial specific volume and particle diameters measured in electron micrographs. Kalmakoff and Tremaine (1968) calculated a molecular weight value of 0.55×10^9 from sedimentation and diffusion.

2. Baculoviruses: Nucleopolyhedrosis Viruses

The baculoviruses are rod-shaped viruses, which may be occluded within a structure, the inclusion body. The latter is composed of a carbohydrate envelope and matrix protein. The occluded baculoviruses are of two types: nucleopolyhedral viruses (NPVs), which contain a number of virus rod structures within the inclusion body; and the granulosis (capsular) viruses (GVs), which contain, generally, only one virus rod structure per granule or capsule inclusion body (Wildy, 1971).

The baculoviruses have great potential as biological insecticides (Podgwaite and Mazzone, 1981) and are being evaluated for this role in the environment.

a. Single Virus Rod Structures

i. Single virus rod structure of the gypsy moth NPV (Bahr et al., 1976)

Source. Inclusion bodies purified from infected gypsy moth larvae (Breillatt *et al.*, 1972) were lysed at pH 10.5 (Bergold, 1953). A virus rod fraction was obtained by differential centrifugation.

Preparation and electron microscopy. A drop of the sample was placed on Formvar grids and air dried. The virus rod structures were electrographed with an RCA EMU-4B at a magnification of ×22,570. The single virus rods measured 350–375 nm long × 75 nm wide.

Other reported values. Caldwell *et al.* (1980), using the procedure of sedimentation field-flow fractionation, calculated a molecular weight of 968×10^6 g/mol for the single virus rod of the gypsy moth NPV.

ii. Single virus rod structure of the European pine sawfly NPV (Bahr et al., 1976)

Source. Inclusion bodies were purified from infected larvae of the European pine sawfly (Mazzone, *et al.*, 1970). The virus rods were liberated from lysed (pH 10.5) inclusion bodies (Bergold, 1953), and a virus rod fraction was obtained by differential centrifugation.

Preparation and electron microscopy. As reported above under the single rod virus of the gypsy moth NPV, using a magnification of ×31,000. We observed that in the lysing procedure some of the rod forms tended to round up. These forms were also measured, and their mass was found to approximate very closely the normal rod-shaped virus forms (refer to Table VIII).

Other reported values. None were reported in the literature.

   b. Studies on the Inclusion Bodies of Baculoviruses (NPVs) Using a High-Voltage Electron Microscope (HVEM). The inclusion bodies of four NPVs were also measured for their mass values in order to satisfy registration requirements of the U.S. Environmental Protection Agency. The range of size of the four types of NPVs studied is from 0.1 to 10 μm in the widest dimension. In order for the electron beam to penetrate such structures, high-voltage electron microscopes had to be employed. It is a pleasure to acknowledge the support of the following facilities, which made this study possible: the Biotechnology Resources Program, Division of Research Resources, National Institutes of Health, Bethesda, MD; the High Voltage Electron Microscopy Laboratory of the U.S. Steel Research Center, Monroeville, PA; the High Voltage Electron Microscope Laboratory of the University of Colorado, Boulder, CO.

   i. Inclusion bodies of the gypsy moth NPV (Mazzone et al., 1980a)

Source. As noted above under single virus rod structure of the gypsy moth NPV.

Preparation and electron miscroscopy. A drop of the inclusion body suspension was allowed to air dry or Formvar grids. The U.S. Steel RCA HVEM was used at a magnification of ×1646 and an accelerating voltage of 800 kV.

   ii. Inclusion bodies of the NPV of the European pine sawfly (Mazzone et al., 1980b)

Source. As noted above under single virus rod of the NPV of the European pine sawfly.

Preparation and electron microscopy. As noted above under inclusion body of the NPV of the gypsy moth. A magnification of ×4725 was used, and an accelerating voltage of 800 kV.

   iii. Inclusion body of the NPV of the spruce budworm (Mazzone et al., 1980b)

Source. Dr. J. Cunningham, Canadian Forestry Service, Great Lakes Forest Research Centre, Box 490, Sault Ste. Marie, Ontario, Canada.

Preparation and electron microscopy. The University of Colorado HVEM was used at a magnification of ×4900 and an accelerating voltage of 1000 kV. Inclusion bodies were air dried on Formvar-coated grids.

   iv. Inclusion body of the NPV of the cabbage moth (Mazzone et al., 1981)

Source. Dr. A. Gröner, Institut für Biologische Schädlingsbekämpfung, Darmstadt, Federal Republic of Germany.

Preparation and electron microscopy. As noted above under inclusion bodies of the NPV of the spruce budworm. The HVEM was used at a magnification of ×1922 and an accelerating voltage of 1000 kV.

Other reported values for the mass of inclusion bodies of the insect nuclear polyhedrosis viruses studied. None were reported.

V. Discussion

A. SOURCES OF ERROR

Bahr and Zeitler (1965) analyzed the errors existing in the determination of mass by quantitative transmission electron microscopy. From their extensive assessment, these investigators concluded that when an external standard such as polystyrene latex spheres was used, the error in mass determination was ±7–9%. If only two objects, such as SV40 and TMV, described above, were compared with each other on the same electron micrograph, an error of +1–2% resulted.

In this section we cite the major factors contributing to sources of error in the method. For a more detailed description of errors, the reader is asked to consult the original study.

1. Focus, Astigmatism, and Magnification

Hall (1955), in analyzing point measurements on through-focus series of stained virus particles, reported that determinations of mass are little influenced by normal off-focus conditions. Bahr and Zeitler (1965) observed that the influence of focus on mass determinations is larger, the smaller the mass; a small degree of underfocusing has little influence on the mass (Klug and de Rosier, 1966). Errors are small, if not negligible, with minor degrees of astigmatism and are negligible with properly compensated electron microscopes. The magnification can be determined to ± 1.5% using grating replicas.

2. Contamination

In quantitative electron microscopy, contamination increases the mass per area linearly with the electron dose. The beam intensity, time of exposure to the beam, and accelerating voltage all contribute to the contamination rate at the object surface. In practice, one reduces the intensity of the illuminating beam to a minimum, compatible with visibility of the image at the fluorescent screen. The conditions of regular electron microscopy may not always permit reduction of the beam intensity to desirable levels. Under ordinary conditions, it is then necessary to use smaller condenser apertures and/or increase the distance between filament and grid cap in the electron gun assembly.

In working with objects that have a small mass per area, it is often necessary to increase contrast in the electron microscope. This can be achieved either by using a smaller objective aperture or by operating at lower voltages. At voltages from 40 to 100 kV, by substituting a 20-μm aperture for a 100-μm aperture, contrast is increased by a factor of 2. Lowering the accelerating voltage of the electron microscope from 100 to 40 kV increases contrast by a factor of 2.5. Thus, combining changes of aperture and voltage will increase contrast about five times. Although it appears from such consideration that lowering the voltage is slightly more effective, it is preferable to resort to smaller apertures first, since lower voltages are associated with higher contamination rates.

3. Radiation Changes

In transradiating an object, some of the electrons of the beam interact with it. The majority of reactions are elastic scatterings, essential to image formation. A small fraction suffers inelastic collisions, producing diverse effects that include ionization and generation of X rays and of visible radiation from ultraviolet to near infrared, as well as that of heat and far infrared (Bahr et al., 1965; Stenn and Bahr, 1970a,b; Glaeser, 1971; Beer et al., 1975). As a result of inelastic scattering events, the organic object changes the moment the electron beam hits it. At a stable current density of the beam, the object attains a steady state and changes little thereafter (Bahr et al., 1965; Stenn and Bahr 1970a,b; Glaeser, 1971). The physical consequences of inelastic scattering produce a loss of mass of the irradiated object. The larger and more cross-linked the molecules, the less is the loss of substance.

In spite of irradiation effects, mass determination by electron microscopy can be relied upon. Most of the molecules in biological objects are juxtaposed in a three-dimensional network. These molecules are quickly and randomly cross-linked as radicals are formed and new bonds are established. The entire object becomes one molecule having considerable radiation resistance. Moreover, with respect to electronic radiation the polystyrene standard behaves comparably to a biological object. In our experience, as noted in Section II, polystyrene molecules maintain their perfect shape, as do objects of biological origin. Polystyrene molecules are cross-linked to a stable meshwork and, upon irradiation, lose about the same amount of mass as albumin. If a known virus is used as an internal standard, one can safely assume that its mass loss would be comparable to that of the unknown virus. Last, mass determinations by electron microscopy compare favorably to results obtained by methods such as interference microscopy, microradiography, and actual weighing (Bahr and Zeitler, 1962; Glas and Bahr, 1966; Ruch and Bahr, 1970; Reedy et al., 1972).

The procedure of quantitative transmission electron microscopy (Zeitler and Bahr, 1962; Bahr and Zeitler, 1965) has been employed to obtain the mass of particles in the range 10^{-10} to 10^{-18} g. In addition to viruses it has been successfully applied in the study of mitochondria (Bahr et al., 1970; Glas and Bahr, 1966), chromosomes (Golomb and Bahr, 1971), erythrocytes (Bahr and Zeitler, 1962), thrombocytes (Ambs et al., 1971), muscle fibers (Reedy et al., 1972), spermatozoa (Bahr and Zeitler, 1964: Bahr and Wied, 1971), and malaria parasites (Bahr and Mikel, 1972).

4. Photometry

Errors in photometry were determined by measuring T 100 times on the same micrograph. For a sphere with an average diameter of 5.54 mm, the standard deviation was ± 1.5%, and for a sphere of 4.28 mm, ±2%. The inaccuracy of the average weight of a population is merely a statistical error. The larger the number of particles measured, the smaller the error. For 500 measured particles, the error would be $(500)^{-1/2}$ ±4.5%, and for 1000 particles, ±3.2%.

5. Errors of Standardization

In the use of external standards such as polystyrene latex spheres, the measurement of the diameters of the spheres is a source of error, since, as noted in Eq. (11), the diameter term is raised to the third power. The error of magnification enters adding ±1.5%. A further error of ±1% stems from the determination of the specific gravity of polystyrene latex (Burge and Sylvester, 1960), 1.05 g/cm^3.

The size of psl spheres is of crucial significance, as is the determination of the electron microscopically magnified size. The latter is subject to errors of determining magnification and of the measurement itself. We have redetermined the size of psl spheres for this review in view of the fact that others reported deviations of size from the manufacturer's nominal values (Daniels et al., 1978; Porstendoerfer and Heyder, 1972).

We used samples of Dow latex, both fresh ones and ones with a shelf life of 10 years. An optical comparator was used to determine the size of more than 100 particles in three size populations; 0.220 μm measured 0.228 μm, 0.364 μm measured 0.377 μm, and 0.620 μm measured 0.591 μm. The sizes were about 3.6% larger to 5% smaller than indicated by the manufacturer. The standard error of the mean of measurement, determined from repeat measurements, had a standard deviation of the mean of 3 Å in terms of the unmagnified sphere; this was 20 times smaller than the standard deviation.

Two of the size classes showed a very slight skew to the left; the third was normally Gaussian.

It is known that radiation damage decreases with increasing acceleration voltage of the electron beam. We used throughout the determination described in this study 100-kV acceleration tension and also inserted a relatively small condenser aperture. Through measurement it was determined that the current at the fluorescent screen could be kept at 3×10^{-11} A.

In summary, our mass standard used for 20 years displays an excellent resistance to radiation damage and does not deviate significantly in size from those given then and now by the manufacturer. This agrees well with data of Rowell *et al.* (1979).

B. Comparison with Other Methods for Determining Mass–Molecular Weight of Viruses

The mass or molecular weight of a virus is generally determined by procedures involving sedimentation, light scattering, or electron microscopy. Let us discuss these procedures and compare them to the QTEM procedure.

1. Sedimentation

Earlier sedimentation procedures, such as sedimentation velocity and diffusion and sedimentation equilibrium, were limited by the size of the virus. Determining molecular weight in excess of 40 million g/mol was so time consuming that doubt was cast on the results obtained (Schmid and Mazzone, 1963). For sedimentation equilibrium in particular, the time required to achieve equilibrium was excessive. The greater the virus mass, the more slowly the rotor must revolve so as not to spin down the virus particles. Moreover, the rotor must spin at constant speed and maintain a constant temperature throughout the experiment, which could last for days (Mazzone, 1963). For sedimentation equilibrium in particular, the time required columns of sedimenting solutions (Schmid and Mazzone, 1963) or the meniscus depletion method (Yphantis, 1964), the larger viruses could not be analyzed in a convenient time frame. For example, the determination of the molecular weight of bacteriophage T7 by the meniscus depletion method required 5 days (Bancroft and Freifelder, 1970).

The sedimentation velocity procedure was similarly inhibited by virus size when the diffusion coefficient was sought. The time required to obtain the diffusion coefficient was reduced significantly with the introduction of self-beat laser spectroscopy (Dubin *et al.*, 1967). Still, the sedimentation and diffusion coefficients and the partial specific volume are seldom obtained in the same laboratory from the same virus preparation. Usually, one of the parameters is obtained, the other two coming from independent laboratories or from other virus preparations.

Of recent importance in sedimentation procedures for obtaining the mass–molecular weight of objects such as viruses was the introduction of sedimentation field-flow fractionation (sedimentation FFF) (Giddings *et al.*, 1975, Caldwell *et al.*, 1980, 1981; Kirkland and Yau, 1982). For some viruses, experimental time did not exceed 30 min.

The sedimentation FFF is restricted to giving average molecular weight-mass values. The QTEM procedure, as demonstrated by Bahr and co-workers (1980) in the case of vaccinia, analyzed 7300 virus particles individually.

2. Light Scattering

In conventional light scattering, the size of the virus limits the procedure (Lauffer and Stevens, 1968). Small viruses are amenable for analysis, but the virus preparation must be extremely pure, as the presence of other material adversely affects the scattering, giving false values for the average molecular weight calculated. As demonstrated by Lampert and Bahr (1969) in determining the mass of the maturation forms of herpes simplex virus by QTEM, one need only recognize the particle to be studied in order to set it apart from other material that may be present.

3. Electron Microscopy Procedures

The electron microscopy counting procedure mixes unknown virus particles with particles of a standard, generally polystyrene latex spheres, and the two types of particles are counted (Williams *et al.*, 1951). The counting procedure is subject to errors in viewing whole drops sprayed onto grids. The observer should be aware of the limits of the drop, for only then can all the virus and standard particles be counted. Particles obscured by the grid wire would give a false count, but this feature may possibly be circumvented by counting with the scanning electron microscope. Pure suspensions are needed in the electron microscopy counting procedure, since another parameter such as DNA content, protein content, or actual weight of the sample must be accurately known in order to correlate molecular weight measurements. Another drawback is the aggregation of particles, making it difficult to count all the particles in the aggregate. If the standard used is polystyrene latex spheres, our experience shows that aggregation is induced between the spheres and the unknown virus particles (Bahr *et al.*, 1980).

The scanning transmission electron microscope (STEM) has been used to determine the mass of objects, including viruses (Lamvik and Langmore, 1977; Wall, 1979; Wall and Hainfeld, 1980). There is a direct relationship between specimen mass thickness and annular detector signal (or energy loss signal) in the STEM. The number of electrons striking the annular detector, n_s, is given by Beer's law:

$$n_s = DA\,(1 - e^{-N\sigma/A}) \cong DN\sigma$$

where D is the dose (el/Å^2) incident on a picture element of area A containing N atoms of a certain type, and σ (Å^2) is the cross section for an atom in A to scatter an electron from the conical illuminating beam onto the annular detector. As with quantitative TEM, particles are measured in the STEM one at a time, so that the spread in size as well as the average can be determined accurately.

In determining the mass of viruses by QSTEM, either an external or internal standard may be used; similarly, as for QTEM, the external standard is that of polystyrene latex spheres and the internal standard, a virus of known mass–molecular weight. Wall (1979) obtained molecular weights in dalton units of 39 million and 16 million for tobacco mosaic virus and the filamentaous bacterial virus fd, respectively.

Freeman and Leonard (1981) reported the molecular weights of three additional viruses: pf1 phage, 166 million daltons; tomato bushy stunt virus, 7.8 million daltons; and the arbovirus Semliki Forest virus, 35 million daltons. Freeman and Leonard noted that when their molecular weight values were somewhat lower than other literature values, implying dissociation of the virus particles, this problem could be overcome by fixation of the virus particles, influencing the mass very little.

Wall and Hainfeld (1980) stated that the accuracy of mass measurements by QSTEM was determined by errors of two types. Random errors arise from statistical fluctuations in the thickness of the supporting substrate and the number of scattered electrons detected. Random errors could be minimized by using a high dose (at the expense of mass loss) or by averaging over many particles. Systematic errors arise from mass loss, sampling errors, faulty specimen preparation, and variation in electron collection. Systematic errors could be minimized by extrapolation to zero dose, freeze drying specimens frozen in deionized water, inclusion of internal controls, use of constant angular magnification optics between specimen and detector, and geometrical correction for sampling error. If particles being averaged are identical and no systematic errors are present, the mass error will decrease in proportion to the square root of the number of particles.

VI. Summary

The determination of the mass of microscopic objects by quantitative transmission electron microscopy is a simple but powerful procedure. In addition to allowing the observation of the general features of size and shape, the important property of mass is quickly and directly ascertained.

In assigning an average mass value to a virus, quantitative electron microscopy attains the same experimental goals as other procedures for determining molecular weight. However, QTEM, unlike other methods, is able to go further in analyzing virus populations. Each individual object, virus particle, can be analyzed for its mass and compared to the average mass value of the population. In this regard, the degree of heterogeneity, which can also be observed by other procedures, e.g., sedimentation velocity experiments, can for QTEM be assigned to definite particles. The virologist may, in the case of QTEM, have more certitude in answering questions on anomalies observed in viral infections—questions concerning the presence of structural variants as being responsible for pathological consequences in the infected host.

In deriving the mass–molecular weight of viruses, the electron microscope is used as an analytical balance covering the entire weight range of all the viruses thus far reported in the literature. Its ease of operation, coupled with the critical features of time and the requirement of only small quantities of virus, ranks the electron microscope as a premier research instrument for this important aspect of virology.

ACKNOWLEDGMENTS

For their discussion and helpful suggestions, we are grateful to Dr. R. K. Fujimura, Oak Ridge National Laboratory, Oak Ridge, Tennessee, and to Drs. G. S. Walton and J. D. Podgwaite, United States Department of Agriculture, Hamden, Connecticut.

REFERENCES

Adolph, K. W., and Haselkorn, R. (1972). *Virology* **47**, 701–710.
Ambs, E., Bahr, G. F., and Zeitler, E. (1971). *Z. Gesemte Exp. Med. Einschl. Exp. Chir.* **154**, 187–197.
Anderer, F. A., Schlumberger, H. D., Koch, M. A., Frank, H., and Eggers, H. J. (1967). *Virology* **32**, 511–523.
Anderson, T. F. (1951). *Trans. N. Y. Acad. Sci. (s2)* **13**, 130–134.
Ashe, W. K., and Scherp, H. W. (1963). *J. Immunol.* **91**, 658–665.
Bahr, G. F. (1973). *In* "Molecular Biology, Biochemistry and Biophysics" (V. Neuhoff, ed.), Vol. 14, pp. 257–284. Springer-Verlag, Berlin and New York.
Bahr, G. F., and Mikel, U. (1972). *Proc. Helminthol. Soc. Wash.* **39**, 361–372.
Bahr, G. F., and Wied, G. L. (1971). *Acta Cytol.* **15**, 499–505.
Bahr, G. F., and Zeitler, E. (1962). *Lab. Invest.* **11**, 912–917.
Bahr, G. F., and Zeitler, E. (1964). *J. Cell Biol.* **21**, 175–189.
Bahr, G. F., and Zeitler, E. (1965). *Lab Invest.* **14**, 955–977.
Bahr, G. F., Carlson, L., and Zeitler, E. (1961). *Int. Biophys. Congr., Stockholm*, p. 327.
Bahr, G. F., Johnson, F. B., and Zeitler, E. (1965). *Lab. Invest.* **14**, 1115–1133.

Bahr, G. F., Pihl, E., Engler, W., and Mikel, U. (1970). *Cytobiologie* 2, 163-187.
Bahr, G. F., Engler, W. F., and Mazzone, H. M. (1976). *Q. Rev. Biophys.* 9, 459-489.
Bahr, G. F., Foster, W. D., Peters, D., and Zeitler, E. (1980). *Biophys. J.* 29, 305-314.
Bancroft, F. C., and Freifelder, D. (1970). *J. Mol. Biol.* 54, 537-546.
Becker, Y., Dym, H., and Sarov, I. (1968). *Virology* 36, 184-192.
Beer, M., Frank, J., Hanszen, K.-J, Kellenberger, E., and Williams, R. C. (1975). *Q. Rev. Biophys.* 7, 221-238.
Bergold, G. E. (1953). *Adv. Virus Res.* 1, 91-139.
Boedtker, H., and Simmons, N. S. (1958). *J. Am. Chem. Soc.* 80, 2550-2556.
Breillatt, J. P., Brantley, J. N., Mazzone, H. M., Martignoni, M. E., Franklin, J. E., and Anderson, N. C. (1972). *Appl. Microbiol.* 23, 923-930.
Burge, R. E., and Sylvester, N. R. (1960). *J. Biophys. Biochem. Cytol.* 8, 1-11.
Caldwell, K. D., Nguyen, T. T., Giddings, J. C., and Mazzone, H. M. (1980). *J. Virol. Methods* 1, 241-256.
Caldwell, K. D., Karaiskakis, G., and Giddings, J. C. (1981). *J. Chromatogr.* 215, 323-332.
Camerini-Otero, R. D., Pusey, P.N., Koppel, D. E., Schaefer, D. W., and Franklin, R. M. (1974). *Biochemistry* 13, 960-970.
Cohen, A. L., Marlow, D. P., and Garner, G. E. (1968). *J. Microsc. (Paris)* 7, 331-342.
Cummings, D. J., and Kozloff, L. M. (1960). *Biochim. Biophys. Acta* 44, 445-458.
Daniels, C. A., McDonald, S. A., and Davidson, T. A. (1978). *In* "Emulsions, Lattices, Dispersions" (P. Becker and M. N. Yudenfreund, eds.). Dekker, New York.
Doty, P., and Steiner, R. F. (1950). *J. Chem. Phys.* 18, 1211-1220.
Dubin, S. B., Lunacek, J. H., and Benedek, G. B. (1967). *Proc. Natl. Acad. Sci. U.S.A.* 57, 1164-1171.
Dubin, S. B., Benedek, G. B., Bancroft, F. C., and Freifelder, D. J. (1970). *J. Mol. Biol.* 54, 547-556.
Fraenkel-Conrat, H., and Ramachandran, L. K. (1959). *Adv. Protein Chem.* 14, 175-229.
Freeman, R., and Leonard, K. R. (1981). *J. Microsc. (Paris)* 122, 275-286.
Gaddum, J.H. (1945a). *Nature (London)* 156, 463.
Gaddum, J. H. (1945b). *Nature (London)* 156, 747.
Giddings, J. C., Yang, F. J. F., and Meyers, M. N. (1975). *Sep. Sci.* 10, 133-149.
Glaeser, R. M. (1971). *J. Ultrastruct. Res.* 36, 466-482.
Glas, U., and Bahr, G. F. (1966). *J. Cell Biol.* 29, 507-523.
Glitz, D. G., Hills, G. J., and Rivers, C. F. (1968). *J. Gen. Virol.* 3, 209-220.
Golomb, H. M., and Bahr, G. F. (1971). *Science* 171, 1024-1026.
Hall, C. E. (1951). *J. Appl. Phys.* 22, 655-662.
Hall, C. E. (1955). *J. Biophys. Biochem. Cytol.* 1, 1-12.
Hall, C. E. (1958). *J. Am. Chem. Soc.* 80, 2556-2557.
Hall, C. E. (1966). "Introduction to Electron Microscopy," 2nd ed. McGraw-Hill, New York.
Haltner, A. J., and Zimm, B. H. (1959). *Nature (London)* 184, 265-266.
Hoagland, C. L., Smadel, J. E., and Rivers, T. M. (1940). *J. Exp. Med.* 71, 737-750.
Jennings, B. R., and Jerrard, H. G. (1966). *J. Chem. Phys.* 44, 1291-1296.
Joklik, W. K. (1962). *Virology* 18, 9-18.
Kalmakoff, J., and Tremaine, J. H. (1968). *J. Virol.* 2, 738-744.
Kaplan, A. S. (1969). *Virol. Monog.* 5, 10.
Kirkland, J. J., and Yau, W. W. (1982). *Science* 218, 121-127.
Klug, A., and de Rosier, D. J. (1966). *Nature (London)* 238, 435-440.
Koch, M. A., Eggers, H. J., Anderer, F. A., Schlumberger, H. D., and Frank, H. (1967). *Virology* 32, 503-510.
Lampert, F., Bahr, G. F., and Rabson, A. S. (1969). *Science* 166, 1163-1165.

Lamvik, M. K., and Langmore, J. P. (1977). *In* "Scanning Electron Microscopy" (O. Johari, ed.), pp. 401–408. Chicago Press, Chicago.

Lauffer, M. A. (1944). *J. Am. Chem. Soc.* **66**, 1188–1194.

Lauffer, M. A., and Stevens, C. L. (1968). *Adv. Virus Res.* **13**, 1–63.

Lee, L. F., Kieff, E. D., Bachenheimer, S. L., Roizman, B., Spear, P. G., Burmester, B. R., and Nazerian, K. (1971). *J. Virol.* **7**, 289–294.

Luftig, R., and Haselkorn, R. (1967). *J. Virol.* **1**, 344–361.

Markham, R., Hitchborn, J. H., Hills, G. J., and Frey, S. (1964). *Virology* **22**, 342–359.

Marquardt, J., Geister, R., and Peters, D. (1963). *Arch. Gesamte Virusforsch.* **12**, 561–578.

Mayor, H. D., Jamison, R. M., and Jordan, L. E. (1963). *Virology* **19**, 359–366.

Mazzone, H. M. (1967). *Methods Virol.* **2**, 41–91.

Mazzone, H. M., Breillatt, J. P., and Anderson, N. G. (1970). *Proc. Int. Colloq. Insect Pathol. 4th* College Park, Maryland, pp. 371–379.

Mazzone, H. M., Wray, G., Engler, W. F., and Bahr, G. F. (1980a). *In* "Invertebrate Systems in Vitro" (E. Kurstak, K. Maramorosch, and A. Dübendorfer, eds.), pp. 511–515. Elsevier/North-Holland, Amsterdam.

Mazzone, H. M., Engler, W. F., Wray, G., Szirmae, A., Conroy, J., Zerillo, R., and Bahr, G. F. (1980b). *Proc. Annu. Meet. Electron Microsc.* Soc. Am. **38**, 486–487.

Mazzone, H. M., Engler, W. F., and Bahr, G. F. (1980c). *Curr. Microbiol.* **4**, 147–149.

Mazzone, H. M., Engler, W. F., Wray, G., Gröner, A., Zerillo, R., and Bahr, G. F. (1981). *Proc. Annu. Meet. Electron Microsc. Soc. Am.* **39**, 398–399.

Oster, G., Doty, P. M., and Zimm, B. H. (1947). *J. Am. Chem. Soc.* **69**, 1193–1197.

Peters, D. (1960). *Proc. Int. Congr. Electron Microsc. 4th,* Vol. 2, pp. 552–573.

Podgwaite, J. D., and Mazzone, H. M. (1981). *Protection Ecology* **3**, 219–227.

Polson, A., Stannard, L., and Tripconey, D. (1970). *Virology* **41**, 680–687.

Porstendoerfer, J., and Heyder, J. (1972). *Aerosol Sci.* **3**, 141–184.

Reedy, M., Bahr, G. F., and Fischman, D. A. (1972). *Cold Spring Harbor Symp. Quant. Biol.* **37**, 397–421.

Reimer, L. (1967). "Elektronenmikroscopische Unterschungs- und Präparationsmethoden," 2nd. Springer-Verlag, Berlin.

Rowell, L., Farinato, R. S., Parsons, J. W., Ford, J. R., Langley, K. H., Stone, J. R., Marshall, T. R., Parmenter, C. S., Seever, M., and Bradford, E. B. (1979). *J. Colloid Interface Sci.* **69**, 590–595.

Ruch, F., and Bahr, G. F. (1970). *Exp. Cell Res.* **60**, 470.

Safferman, R. S., Morris, M. E., Sherman, L. A., and Haselkorn, R. (1969). *Virology,* **39**, 775–780.

Schmid, K., and Mazzone, H. M. (1963). *Nature (London)* **197**, 671–673.

Sherman, L. A., and Haselkorn, R. (1970a). *J. Virol.* **6**, 820–833.

Sherman, L. A., and Haselkorn, R. (1970b). *J. Virol.* **6**, 834–840.

Sherman, L. A., and Haselkorn, R. (1970c). *J. Virol.* **6**, 841–845.

Smadel, J. E., Rivers, T. M., and Pickels, E. G. (1939). *J. Exp. Med.* **70**, 379–385.

Steere, R. L. (1963). *Science* **140**, 1089–1090.

Stenn, K. S., and Bahr, G. F. (1970a). *J. Histochem. Cytochem.* **18**, 574–580.

Steen, K. S., and Bahr, G. F. (1970b). *J. Ultrastruct. Res.* **31**, 526–550.

Tai, H. T., Smith, C. A., Sharp, P. A., and Vinograd, J. (1972). *J. Virol.* **9**, 317–325.

Thomas, R. S. (1961). *Virology,* **14**, 240–252.

Wall, J. (1979). *In* "Introduction to Analytical Electron Microscopy" (J. J. Hren, J. I. Goldstein, and D. C. Joy, eds.), pp. 333–342. Plenum, New York.

Wall, J., and Hainfeld, J. (1980). *Annu. Meet. Proc. Electron Microsc. Soc. Am.* **38**, 674–675.

Watanabe, I., and Kawade, Y. (1953). *Bull. Chem. Soc. Jpn.* **26**, 294–298.
Weber, F. N., Jr., Kupke, D. W., and Beams, J. W. (1963). *Science* **139**, 837–838.
Wildy, P. (1971). *Monogr. in Virol.* **5**, 17–32.
Williams, R. C., and Steere, R. L. (1951). *J. Am. Chem. Soc.* **73**, 2057–2061.
Williams, R. C., Backus, R. C., and Steere, R. L. (1951). *J. Am. Chem. Soc.* **73**, 2062–2066.
Yphantis, D. A. (1964). *Biochemistry* **3**, 297–317.
Zeitler, E., and Bahr, G. F. (1962). *J. Appl. Phys.* **33**, 847–853.
Zeitler, E., and Bahr, G. F. (1965). *Lab. Invest.* **14**, 946–954.

5 Use of Thin Sectioning for Visualization and Identification of Plant Viruses

Giovanni P. Martelli and Marcello Russo

I. Introduction

A thorough review of the technical aspects of electron microscopy methods used for studying biological materials would indeed constitute a task far beyond the scope of this presentation. This matter has been extensively treated in a number of books and review articles (Hayat, 1970; Milne, 1972; Glauert, 1974; Hall, 1978) to which the reader is referred for comprehensive information.

This chapter aims at illustrating (1) some of the current techniques used for investigating virus-infected plant cells at the ultrastructural level, and (2) the possible use of information gathered from these studies for virus identification and taxonomy. Since the early days of plant virology (Iwanowski, 1903), research workers have been intrigued by the cytological modifications induced in host plants by viral infections. These studies, notwithstanding the low resolution of the light microscope, have yielded a wealth of information (reviewed by McWhorter, 1965; Bald, 1966) that has provided a solid background for the cytopathological reappraisal of virus infections made possible by the development of the electron microscope.

Developments that have made an array of diversified cytochemical tests available have helped to overcome one of the major drawbacks of electron microscopy: the difficulty of correlating structure with function. Cytochemical electron microscopy is a field in which great advances can be expected in the years to come. However, the results obtained so far are of the utmost importance, as they have conferred physiological significance upon some of the cytopathic structures associated with infections by members of several taxonomic virus groups. For this reason a section of the chapter is devoted to a description of these structures, which, in addition to having specific physiological functions, may serve as diagnostic markers at the virus group level.

II. Technical Procedures for Processing Plant Tissues

A. Choice of Plant Material

When thin sectioning is to be used for the *in situ* study of a particular plant virus, the first problem is tissue sampling.

Only a small portion of tissue can be examined, and features important for diagnosis may easily escape attention if the sampling is not representative. Generally speaking, two situations may be encountered. In the first, only plants naturally infected with a virus (or reputed to be so) can be examined, because the disease agent is not transmissible to test plants; in the second, the virus is transferable to experimental hosts, and some of its properties may be known. Clearly, the first situation is the more difficult because it is not known how the virus is distributed in the plant and its concentration may be very low, as is often the case with natural infections. But this is also a situation in which observation of thin sections may be particularly useful for the identification of the causal agent of the disease. In these cases, the researcher can only proceed blindly, sampling tissues at random from different plant organs, and being guided by the nature of specific symptoms, if present.

The second situation, when one or more artificially inoculated test plants are available, is in a sense simpler. Sampling can be done from plants of different species at different times after inoculation, thus allowing for the study of the sequential events leading to the development of cytopathological structures.

In systemic infections, samples can be taken from the same plant using leaves of different ages including both the newly produced ones (even though they may not show symptoms) and the oldest, i.e., those that had been inoculated. Absence of visible symptoms does not imply that no cytological alterations are present. In fact, in many instances, the very early cytopathic events can be seen only in asymptomatic young leaves. Alternatively, a different batch of plants can be inoculated simultaneously and samples taken from leaves of the same age at various times after inoculation. To quantify leaf age, it may be useful to take into account the plastrochron index (Pennazio et al., 1978).

When infection is suspected to be restricted to the vascular bundles, as in the case of luteoviruses and closteroviruses, sampling may be limited to veins and/or petioles.

With localized infections it is advisable to examine lesions of different ages. When lesions do not exceed 1–2 mm in diameter, they can be processed whole, including part of the apparently unaffected surrounding tissue. Blocks up to 3–4 mm^2 can be satisfactorily fixed and embedded as they are. Thin sections can then be cut from the same block at set intervals from the periphery to the center of the lesion. In this way, the sequence of cytopathological events taking place along the infection gradient in the lesion can be established (Milne, 1966a,b, 1970; Russo and Martelli, 1981; Russo et al., 1981). Larger lesions can be cut into two or four pieces at right angles and processed and sectioned as above.

As a rule glasshouse-grown infected plants are used for ultrastructural observations. In most cases such plants are actively growing, and the tissues are soft and easy to handle. It is often convenient to keep the plants in the dark before sampling in order to reduce the starch content in chloroplasts, which may cause difficulties during microtoming. After removal, tissue pieces are fixed immediately to minimize unwanted modifications of the internal structure that may give rise to artifacts. Sometimes, however, it may be necessary to utilize material from field-grown plants. Here again, collection of prefixation of samples are best done on the spot, although, if large samples (whole shoots for instance) are taken and kept in a moist and cool environment, processing of leaf tissues can be initiated after many hours without appreciable alteration of the cellular structure.

B. FIXATION AND FIXATIVES

The purpose of fixation is to stabilize the cell structure so that it does not change with respect to the *in vivo* condition despite exposure to drastic treatments such as dehydration, embedding in resins, sectioning, staining, and electron irradiation.

As a general rule, it is best to use well-established procedures, known to yield good generalized fixation of the cells. However, conditions suitable for adequate cytological stabilization of tissues are not always suitable for preserving crystalline aggregates of virus particles or the structural organization of certain inclusion bodies, such as the laminated plates of some potyviruses. In these cases, specific fixing procedures can be adopted, as will be specified below. An outline of the commonest chemicals used for fixation of plant material follows.

1. *Aldehydes*

Until 1963, osmium tetroxide and potassium permanganate were the only fixatives used in electron microscopy. Sabatini *et al.* (1963) introduced aldehydes for fixing animal tissues. Shortly afterward aldehydes became widely used also in the fixation of plant tissues. However, aldehydes do not stabilize lipids, and therefore tissues must be postfixed with osmium tetroxide. Prefixation with aldehydes has several advantages: large fragments of tissue can be fixed, to be reduced afterward to smaller pieces before postfixation with osmium tetroxide; and inter- and intramolecular cross-links with protein molecules are more stable and cytochemical tests can be carried out prior to fixation with osmium. This is possible because aldehydes permit the identification of enzymatic and immunological characteristics that are destroyed by osmium fixation. Among the aldehydes tested by Sabatini *et al.* (1963) three have gained and still enjoy considerable favor: glutaraldehyde, acrolein, and formaldehyde. The first two can be used alone, whereas formaldehyde is more commonly used in mixtures, especially with glutaraldehyde (Karnovsky, 1965).

a. Glutaraldehyde. Glutaraldehyde, a five-carbon dialdehyde, is the chemical most commonly used for prefixation of plant material. Tissues can be left in the fixative for many hours (even days when shipped from country to country) with negligible or no apparent harm because completion of fixation is fast and the cross-links formed are remarkably stable. Glutaraldehyde reacts with proteins that are stabilized and made insoluble by inter- and intramolecular cross-links. Its affinity for lipids, however, is almost nil, and this is a major reason for postfixation in osmium tetroxide.

Glutaraldehyde is also believed to be effective in preserving delicate structures such as microtubules, which are usually not seen after fixation with osmium tetroxide alone. However, the reason for failure of osmium tetroxide to retain microtubules may simply be that, at the concentrations normally used (1-2%), it acts too slowly to be effective (Hayat, 1970).

Glutaraldehyde penetrates tissues more slowly than formaldehyde and acrolein, but operates faster, especially when compared with formaldehyde.

Fixation with glutaraldehyde is generally carried out at room temperature (18-22°C), as the penetration rate increases with temperature. Moreover, low temperatures (e.g., 4°C) are not suited for preserving labile structures such as microtubules.

Cacodylate and phosphate buffers are routinely used to prepare glutaraldehyde fixatives. Both are used at 0.05-0.1 M and pH around neutrality. It would appear that pH and osmolality of a fixative should be chosen according to their relative values in plant cells. However, Milne (1972) pointed out that not much is known about the effect of osmotic pressure on plant cells, as pressure of the different cellular compartments is difficult to measure. He also reported that pH could be varied from 5 to 7.4 with no detriment to the quality of fixation. In any case, the pH and osmolality used are traditionally those developed for animal tissues, and are not necessarily the best for plants.

A detailed account of the influence of pH and tonicity of fixatives on the fine structure and stability of cellular components was also given by Salema and Brandaõ (1973). Preservation of cellular components (i.e., chlorophyll, phospholipids, and proteins) was most satisfactory when the osmolality of the glutaraldehyde solution used to fix mature, vacuolated cells was 800 mosM. The rinsing solutions and osmium fixative had the best tonicity between 650 and 700 mosM. The best pH was found to be 8.0 for glutaraldehyde and 6.8 for osmium, the difference being due to the different modes of action of these fixatives. Glutaraldehyde establishes mainly methylene bridges, and the production of polymers is favored by alkaline pH. Osmium, instead, acts primarily through formation of H bonds. Low pH values increase the number of active hydrogen atoms on protein chains, thus allowing for a more extensive bonding with the oxygen of osmium tetroxide.

In the same paper, Salema and Brandaõ (1973) showed that PIPES-NaOH buffer is better than phosphate and carbonate buffers in preserving and stabilizing the fine structure of cell components. This buffer, however, to our knowledge has not been used in ultrastructural studies of virus-infected plant cells.

The concentration of glutaraldehyde in the fixative can be varied safely over a fairly wide range, concentrations between 3 and 6% being the most often used.

A detailed study of the influence of aldehyde fixatives on the amenability of plant tissue to enzymatic digestion was made by Weintraub *et al.* (1969). Glutaraldehyde was found to be superior to other aldehydes tested in preserving the fine structure of cytoplasm and organelles, but it greatly reduced susceptibility to proteolytic enzymes and no digestion at all was possible with nucleases. Formalin or acrolein–formalin mixtures were the best fixing agents for enzymatic digestion, but proved to be not as good as glutaraldehyde for preserving cellular details.

Glutaraldehyde is commercially available as a 25 or 50% solution in water. It is advisable to purchase products that are especially made for electron microscopy. These do not require purification, have a pH between 5 and 6, and are rather stable if kept cold and well stoppered in the dark. Some commercial batches may contain impurities. Their use is not advisable without previous "cleaning," and they should be discarded if the pH falls below 3.5 and the color becomes dark yellow. A simple method for purifying glutaraldehyde consists in treatment with activated charcoal followed by vacuum distillation and dilution of the distillate to the desired concentration with hot distilled water (Fahimi and Drochmans, 1965).

b. Acrolein (Acrolaldehyde). Acrolein, an unsaturated monoaldehyde, penetrates tissues and fixes cellular structures quickly. It has a high specificity for proteins, but little is known of its ability to stabilize lipids. Weintraub *et al.* (1969) reported that plant cells exposed to acrolein were unsatisfactorily fixed and unsuitable for enzymatic digestion. Some workers, however, include acrolein in fixing fluids, usually mixed with glutaraldehyde (Lawson and Hearon, 1973; Appiano *et al.,* 1977) to take advantage of its high speed of penetration.

Acrolein has an unpleasant odor, is highly irritating, and must be handled under a fume hood. It deteriorates upon storage unless it contains an oxidation inhibitor (usually 0.1% hydroquinone). Fresh preparations (5–10% in buffer) should be used for fixation.

c. Formaldehyde. Among the three aldehydes mentioned, formaldehyde is perhaps the least satisfactory as a fixing agent for electron microscopy. It does not, for instance, stabilize certain virus structures sufficiently to withstand the electrom beam (Morgan and Rose, 1967). Furthermore, most of the reactions it undergoes, leading to formation of methylene bridges whereby proteins are fixed, are reversible and the chemical itself is readily washed with water. Therefore, formaldehyde is seldom used alone for plant material unless digestion tests with DNase are to be performed (Weintraub *et al.,* 1969). However, mixed with glutaraldehyde, it makes an excellent fixative for animal and plant tissues (Karnovsky, 1965).

It is advisable to prepare formaldehyde solution from solid paraformaldehyde rather than from formalin, which if used, should be methanol free.

2. *Osmium Tetroxide (OsO₄, Osmic Acid)*

Osmium tetroxide was used alone as a fixing agent for electron microscopy before the introduction of aldehydes. It is current practice, adopted because osmic acid penetrates tissues slowly, to postfix tissues with osmium after prefixation with aldehydes (usually glutaraldehyde or Karnovsky's mixture). When cell structures are already largely stabilized by prefixation, the slow penetrating action is no longer a drawback and all advantages of using osmium are maintained. These advantages reside primarily in the reaction of osmium tetroxide with unsaturated lipid molecules, which are cross-linked and insolubilized. The ability of OsO_4 to cross-link proteins is much less than that of aldehydes, and the amount of cross-link introduced may be further reduced upon prolonged exposure to the fixative. Factors involved in the preservation of actin filaments *in vitro* for electron microscopy have been examined by Maupin-Szamier and Pollard (1978). These authors found the conditions that seemed the best for successful fixation of actin filaments *in vitro,* but left unsolved the problem of fixing actin filaments inside cells.

No reaction of osmium tetroxide is reported with native DNA and RNA.

When postfixation with oxmium is omitted, the lipid moiety of cell membranes is depleted, and the membranes become almost invisible. Organelles with extensive membrane systems, such as chloroplasts and mitochondria, appear as a negative image, with electron-lucent lamellae surrounded by a dense stroma. Hills and Plaskitt (1968) described a technique ("uranyl soak method") in which no postfixation with osmium tetroxide is used after glutaraldehyde fixation and the tissue is dehydrated in an ethanol series saturated with uranyl acetate. The method allows better visualization of isometric viruses because ribosomes are only lightly stained compared with virus particles. Withholding osmic acid after fixation has proved to be useful also for localizing lipid components in thin-sectioned particles of eggplant mottled dwarf virus (EMDV), a plant rhabdovirus (Martelli and Castellano, 1971).

Osmium tetroxide solutions may be prepared in cacodylate, phosphate, Veronal, or PIPES–NaOH buffer. The pH is usually maintained around 7 although slightly lower values may be useful (see the suggestions of Salema and Brandaõ, 1973), and the buffer molarity is chosen between 0.02 and 0.1 M. Osmium tetroxide is currently employed at concentrations of 1–2%, but concentrations as low as 0.2% (Harrison *et al.,* 1970) and 0.1% (Milne, 1970) have been used. Very dilute solutions must be used to preserve the structure of some virus crystals, such as tobacco mosaic virus crystals (Warmke and Christie, 1967).

Osmium tetroxide vapors are very dangerous, and solutions must be handled under a fume hood. Breathing of vapors or spilling of solution on the skin should be carefully avoided, and the eyes should be protected.

3. Potassium Permanganate

Potassium permanganate is a strong oxidant. It is reduced to manganese dioxide, a brown precipitate that confers a marked contrast upon structures that react with it. Permanganate fixes cell membranes extremely well, and they acquire high electron density. However, most cytoplasmic constituents including ribosomes and virus particles are lost. The fixative is no longer used except for a few special purposes. For instance, it has been useful for visualizing the connection of the outer envelope of EMDV particles with the nuclear membrane (Russo and Martelli, 1973) and for preserving the arrangement in a crystalline lattice of tobacco mosaic virus (TMV) particles (Warmke and Edwardson, 1966a,b).

Potassium permanganate can be used alone in 2–5% unbuffered aqueous solutions (Mollenhauer, 1959) or after prefixation with aldehydes (Hayat, 1968; Russo and Martelli, 1973). Fixation is done at room temperature for a few minutes to about 1 hr. Tissues fixed with permanganate do not require staining before observation.

C. Buffers and Buffered Fixing Solutions

A number of buffers for fixatives are available; the most popular ones are described below.

1. Buffers

a. Cacodylate-HCl (0.1 M)

Solution A: 0.2 M sodium cacodylate [4.28 g of $Na(CH_3)_2AsO_2 \cdot 3H_2O$ in 100 ml of distilled water]
Solution B: 0.2 N HCl
To 50 ml of solution A add solution B in the quantities specified in the tabulation below to obtain the desired pH and bring to a total volume of 100 ml with distilled water.

pH	Soln B (ml)
6.4	18.75
6.6	13.3
6.8	9.3
7.0	6.3
7.2	4.15
7.4	2.7

Cacodylate buffers of any molarity can also be prepared by adding drops of concentrated HCl to a solution of cacodylate of the desired molarity until the desired pH is reached and then bringing to volume.

b. *Sodium Phosphate (0.2 M)*

Solution A (monobasic sodium phosphate): $NaH_2PO_4 \cdot H_2O$, 27.6 g; $NaH_2PO_4 \cdot 2H_2O$, 31.2 g; distilled water to 1000 ml

Solution B: $Na_2HPO_4 \cdot 2H_2O$, 35.61 g; $Na_2HPO_4 \cdot 7H_2O$, 53.7 g; or $Na_2HPO_4 \cdot 12H_2O$, 71.64 g; distilled water to 1000 ml

Mix solutions A and B in the ratios as tabulated below to have the desired pH.

pH	Soln A (ml)	Soln B (ml)
6.4	73.5	26.5
6.6	62.5	37.5
6.8	51.0	49.0
7.0	39.0	61.0
7.2	28.0	72.0
7.4	19.0	81.0
7.6	13.0	87.0

c. *Millonig (1961a) Monosodium Phosphate.* The pH of this buffer can be adjusted by varying the quantity of alkali without changing tonicity. It was designed for use with mammalian tissues, as it is isotonic with blood plasma, but it may be used also for plant tissues.

Solution A (monosodium phosphate): $NaH_2PO_4 \cdot H_2O$, 22.6 g; or $NaH_2PO_4 \cdot 2H_2O$, 25.6 g; distilled water to 1000 ml

Solution B: NaOH, 25.2 g; distilled water to 1000 ml

Mix solutions A and B (see tabulation below) to the desired pH. Glucose may be added (5 ml of a 5.4 % solution to 45 ml of final buffer solution) to increase tonicity.

pH	Soln A (ml)	Soln B (ml)
6.4	92.5	7.5
6.8	87.9	12.1
7.0	85.8	14.2
7.2	83.9	16.1
7.4	82.5	17.5
7.6	81.6	18.4
7.8	80.8	19.2

d. Veronal–Acetate. This buffer is used only with osmium tetroxide, as it reacts with aldehydes, losing buffering capacity. Veronal–acetate enters the composition of OsO_4 fixatives such as Palade's which was formulated for animal tissues, and Caulfield's, which is an adaptation of the former to which sucrose is added to increase tonicity. Formulations for both these fixatives follow. Stock Veronal–acetate buffer solutions are prepared as follows: sodium Veronal, 29.5 g; sodium acetate anhydrous, 11.7 g; or sodium acetate · $3H_2O$, 19.4 g; distilled water to 1000 ml.

2. Fixatives

Fixatives are the fluids obtained by mixing a fixing agent with a buffer.

Some very efficient fixatives are readily made. For instance, glutaraldehyde is simply prepared by diluting a stock solution with appropriate buffer; e.g., to prepare 5% glutaraldehyde, 5 ml of a 25% stock solution is mixed with 20 ml of buffer.

Similarly, acrolein, which is commercially available as a pure chemical, is dissolved in buffer to 5–10%. Use of freshly made solutions is recommended. Waste acrolein is disposed of in 10% sodium bisulfite solutions. Acrolein–glutaraldehyde fixatives usually contain a 1:1 mixture, the combined concentration being in the range 1.25–6%.

Formaldehyde solutions are best prepared from solid paraformaldehyde by the addition of the appropriate amount of distilled water, heating at 60–70°C, and adding a few drops of 1 N NaOH until the solution clears. This solution is diluted with buffer before use. A widely used glutaraldehyde–formaldehyde fixative, designed by Karnovsky (1965), is prepared as follows:

1. Dissolve 2 g of paraformaldehyde in 25 ml of distilled water.
2. Heat to 60–70°C with stirring.
3. Add 1–3 drops of NaOH until the solution clears (persisting turbidity, if mild, is not harmful), and cool to room temperature.
4. Add 10 ml of 25% glutaraldehyde (or 5 ml of a 50% solution) and bring the volume to 50 ml with 0.2 M cacodylate or phosphate buffer, pH 7.4–7.6. Final pH is 7.2. Add 25 mg of anhydrous $CaCl_2$ if cacodylate is used.

Osmium tetroxide is sold in sealed glass ampules containing 100 mg to 1 g. As osmium tetroxide dissolves slowly, the solution is prepared several hours in advance of use. For protection, all operations must be carried out under a fume hood with the glass window lowered. All glassware must be cleaned thoroughly, for any trace of organic matter will induce a quick darkening of the solution.

Osmium tetroxide solutions are made usually in 0.05 or 0.1 M cacodylate or phosphate buffer, pH 6.8–7.2, or Veronal–acetate. Two formulations with the latter buffer follow.

Palade's (1952) Fixative. This is composed of stock Veronal–acetate buffer solution, 5 ml; 0.1 N HCl, 5 ml; 2% OsO$_4$, 12.5 ml; distilled water.

Caulfield's (1957) Fixative. The formulation is stock buffer solution, 1 part; 0.1 N HCl, 1 part; 2% OsO$_4$, 4 parts; distilled water, 2 parts; sucrose, 15 mg/ml.

The pH of both fixatives should be around 7.4. Small adjustments can be made by the addition of more stock buffer or drops of 0.1 N HCl before bringing to volume with distilled water.

D. Fixation Techniques

For thorough fixation, dehydration, and embedding, the tissue pieces should not exceed 1 × 1 × 1.5 mm. However, larger specimens (up to 1 cm^2 if they are thin) can be prefixed with aldehydes, to be subdivided or trimmed later in a drop of fixative or of buffer, during washing. Infiltration of large tissue samples with aldehyde fixatives under vacuum is recommended, for intercellular spaces of leaf mesophyll are filled with air that impairs a quick and uniform penetration of the fixing fluid inside the cells.

In our laboratory we carry out the following steps.

1. Large (0.5–1 cm^2) pieces of leaf lamina are cut with a corkborer or a razor blade.
2. Specimens are placed in glass tubes with a small amount of the appropriate fixative and are sunk by the aid of small pieces of glass tubing or small wire baskets.
3. Containers are placed in a desiccator, and vacuum is applied with a rotary or a water pump.
4. About 1 min after maximum vacuum is reached, the pump is switched off and normal pressure is slowly restored.
5. Tissue pieces appear water-soaked and remain sunken if well infiltrated. If not, steps 3 and 4 are repeated.

Fixation with aldehydes is reputed to be complete after 1–3 hr at room temperature. However, samples may be left longer in the fixative, especially glutaraldehyde, if they are collected in the field or must be shipped from one laboratory to another.

After infiltration with fixative, specimens are cut into smaller pieces and transferred to smaller containers, such as scintillation vials or screw-cap glass vials. This operation can also be done after the chosen fixation time, when the fixing fluid has been exchanged with the washing buffer.

It is usually recommended that excess aldehydes be washed from tissues prior to postfixation with osmium tetroxide. Traces of aldehyde are said to react with OsO_4, which would lead to the formation of precipitates in the tissue. Although washing between the two fixatives was not found necessary by several authors (Milne and De Zoeten, 1967; Milne, 1970; Langenberg and Schroeder, 1974, 1975; McMullen and Gardner, 1980), who did not detect precipitates in the materials examined, elimination of excess aldehyde is customary. Rinsing consists in a few changes, over a few minutes, or up to 2 hr, of the same buffer as that of the fixing fluid. This and all subsequent operations in which one liquid is exchanged with another can be conveniently done with a Pasteur pipet fitted with a rubber bulb.

After washing, tissue specimens are ready for fixation with osmium tetroxide. This is done at room temperature for 1–3 hr or overnight at 4°C. Overfixation should be avoided, as there may be some leaching of cell constituents. Finally, tissues are washed with distilled water (two or three changes in about 1 hr) and dehydrated or are bulk stained with uranyl acetate prior to dehydration. Samples fixed with permanganate are also washed with distilled water before dehydration.

Several fixing schedules are available in the literature. Among these, a few are reported that, in our experience, are fully exchangeable. References attached to the different methods are for the reader's convenience, so that the results can be evaluated with the original papers.

Method I (Esau and Hoefert, 1972a; Russo *et al.,* 1978)

1. Cut tissue pieces in a drop of Karnovsky's (1965) aldehyde fixative.
2. Infiltrate under vacuum.
3. Keep pieces in the fixing solution for 2–3 hr at room temperature.
4. Rinse samples with several changes of phosphate buffer, 0.1 M, pH 7.2 for 3–4 hr.
5. Postfix overnight at 4°C in 2% osmium tetroxide in the same buffer.
6. Rinse with three or four changes of distilled water.

Method II (Kim *et al.,* 1978; Martelli and Russo, 1981)

1. Fix tissues in 4% glutaraldehyde in 0.05 M cacodylate buffer, pH 7.0, for 2 hr at room temperature.
2. Rinse for 2 hr with 0.05 M cacodylate buffer, pH 7.0.
3. Fix for 2 hr with 1% OsO_4 in the same buffer at room temperature.
4. Rinse with distilled water.

Method III (Esau and Magyarosy, 1979)

1. Fix tissue for 2 hr in 4% glutaraldehyde in 0.1 M cacodylate buffer, pH 6.8.
2. Rinse briefly in buffer.

3. Fix for 1 hr in cold buffered 2% OsO_4.
4. Rinse with distilled water.

Method IV (Hatta and Matthews, 1974)

1. Fix tissues in 3-5% glutaraldehyde in Millonig's (1961a) phosphate buffer for 2 hr at room temperature.
2. Rinse in Millonig's buffer.
3. Fix with 1% OsO_4 in Millonig's buffer for 3 hr at room temperature.
4. Rinse with distilled water.

Method V (Milne, 1970)

1. Fix tissue in 2.5% glutaraldehyde in 0.1 M phosphate buffer, pH 6.8, for 1 hr at room temperature.
2. Blot tissue pieces and transfer them directly to 1 or 0.1% OsO_4 in the same buffer at room temperature for 3 hr.
3. Rinse with 50% aqueous acetone.

Method VI (Appiano et al., 1977)

1. Fix tissues in a 1:1 mixture of 3% glutaraldehyde and 3% acrolein in 0.1 M phosphate buffer, pH 7.2, for 1 hr at room temperature.
2. Rinse with buffer.
3. Fix with 1% OsO_4 in buffer for 1 hr.
4. Rinse with buffer or distilled water.

The sequential use of the two basic fixatives (aldehydes and osmium tetroxide) is recommended in all the above schedules, although these can also be applied in mixture. Contemporary glutaraldehyde–osmium fixation followed by further osmication was found to be satisfactory for both healthy (Franke et al., 1969) and virus-infected (Plumb and James, 1973) plant tissues. However, whether this procedure has any distinct advantage over standard fixing schedules is unclear. In fact, as shown by Langenberg (1978), contemporary glutaraldehyde–osmium fixation may be unsuitable, for it induces recognizable artifacts in the lamellar system of chloroplasts due to the faster rate of penetration of osmic acid.

Dalton's (1955) fixative, another mixture containing both chromic and osmic acid, proved to be equally detrimental to the integrity of intracellular accumulations of TMV and wheat streak mosaic virus (WSMV) particles (Langenberg and Schroeder, 1972). Instead, when the same two chemicals were used in succession (i.e., prefixation with buffered chromate followed by osmication), a much better preservation of cellular structures and viral inclusions was obtained. This led the authors (Langenberg and Schroeder, 1972) to concluded that osmic acid was responsible for the disrupting action. Individual TMV particles, however, are well preserved by the simultaneous treatment with chromate and osmium, as Milne and De Zoeten (1967) found no difference in their appearance after chrome–osmium fixation and glutaraldehyde plus osmium fixation. Also the length of fixation may, in certain cases, be critical in the maintenance of unaltered ground cytoplasm and virus aggregates. McMullen and Gardner (1980) found, for instance, that WSMV-infected cells that had been "short-fixed" according to Langenberg and Schroeder (1974) (i.e., fixation in cold glutaraldehyde for 5–8 min followed by postfixation in cold osmic acid for 15–16 min) looked better than comparable samples processed through longer procedures (i.e., 6 hr in glutaraldehyde and 16 hr in osmium tetroxide). It should be pointed out that in the latter schedule fixation times much longer than normal are used and comparison should perhaps have been made with more conventional procedures (about a 2-hr fixation period). However, WSMV-induced cylindrical inclusions, which are a diagnostic feature for this and other potyviruses (Edwardson, 1974), were not affected by the different fixing schedules (McMullen and Gardner, 1980).

Langenberg (1979) showed that prolonged (up to 5 hr) chilling at 5°C of tissue samples vacuum infiltrated with buffer before exposure to cold 2% glutaraldehyde greatly improved conservation of the crystalline structure of particle aggregates of five different plant viruses. Fixation at 25°C resulted in an almost invariable disruption of the crystalline lattice. In this case, however, it cannot be excluded that chilling itself induced the formation of crystals, which are then not present *in vivo*.

It is, then, evident that any of the fixing procedures outlined above should give satisfactory general preservation of the cell's fine structure, but not necessarily of certain constituents of infected tissues, which may be significant in a cytopathological or diagnostic context. In these instances, the basic techniques may be adapted to suit individual cases.

E. DEHYDRATION

The majority of resins used for embedding biological materials are water insoluble. Water in the tissue must therefore be exchanged for an organic solvent in which the resins are soluble. A graded series of acetone or ethanol dilutions serves this purpose. There are no definite arguments in favor of the use of one or the other solvent. Acetone seems to extract less material (in particular phospholipids) from tissues than ethanol, and, with it, a transitional solvent may not be necessary. For instance, epoxy resins are fully compatible with acetone, but not with ethanol, which requires substitution with a compatible intermediate solvent such as propylene oxide before embedding. On the other hand, Spurr's resin, a widely used epoxy embedding medium, is compatible with ethanol. We prefer ethanol as dehydrating agent.

Tissues are soaked in ethanol dilutions for a time varying between 10 and 30 min. It is not advisable to use shorter or longer periods because incomplete dehydration may occur or lipids may leach out. Also, no step should be omitted, as large, vacuolated plant cells are rather sensitive to damage during dehydration. If specimens were in the cold (for example, if they had been stained with 0.5% uranyl acetate at 4°C), they should be kept in the cold until 90–95% ethanol, then brought to room temperature.

We use the following dehydrating schedule: ethanol 20%, 30 min; ethanol 40%, 30 min; ethanol 60%, 30 min; ethanol 95%, 30 min; ethanol 100%, 2 × 30 min.

Water contamination (even traces) of pure solvents (100% ethanol) should be carefully avoided. This can be achieved by keeping absolute ethanol over anhydrous copper sulfate in the bottle and using oven-dry glass vials at the absolute ethanol stages that complete dehydration.

A new dehydrating agent (2,2-dimethoxypropane; DMP) has been suggested (Muller and Jacks, 1975), which allows a very rapid dehydration of various tissues for both light and electron microscopy. Thorpe and Harvey (1979) have made a search of the best conditions for the use of DMP in plant cytology. We have used it occasionally, and so have others working with virus-infected plants (see, for example, Esau and Magyarosy, 1979).

F. EMBEDDING

Embedding media may be divided in three groups: epoxy, polyester, and acrylic resins. Epoxy resins are widely used, and they will be described in some detail. Polyester resins, e.g., Vestopal W, are used less frequently and description of their use will not be given. Acrylic resins (methacrylates) in the original formulations (i.e., mixtures of butyl and methyl methacrylates) have been totally abandoned. However, a formulation consisting of butyl or methyl methacrylate and styrene (vinylbenzene) is often used in plant cytology. Water-soluble methacrylates are of widespread use in cytochemistry, and their use will be also be described.

1. *Epoxy Resins*

Epoxy resins are characterized by the presence of an epoxide group: C—O—C. They are polymerized by the addition of a cross-linking agent (accelerator). Final hardness is controlled by mixing the resin with a hardener and a plasticizer before adding the accelerator.

When liquid, epoxy resins are harmful. Disposable plastic gloves should be worn to prevent contact with the skin. In case of accidental contact, the parts must be promptly washed with soap and water, not with organic solvents. Inhaling of vapors is dangerous. When cured, epoxy resins are inert, but it is strongly advised to avoid creating or inhaling epoxy resin dust. A thorough account of dangers involved in the use of epoxy resins and all measures needed for their safe handling are described by Ringo *et al.* (1982). Epon, Araldite, and ERL 4206 are commonly used epoxy resins.

a. Epon. Epon 812 (Epikote) used to be a most widely employed resin. Some companies have replaced it with other formulations, such as Agar 100 or TAAB 812, which according to the manufacturers are less viscous and are produced to higher specification.

Luft's (1961) is still the preferred formulation.

Mixture A: Epon 812, 62 ml; dodecenyl succinic anhydride (DDSA), 100 ml

Mixture B: Epon 812, 100 ml; methyl nadic anhydride (MNA), 89 ml

Components are measured volumetrically, but care should be taken to let the viscous resin slide down completely from the walls of graduated cylinders. Handling is made easier if the chemicals are heated at about 45°C to reduce viscosity. After mixing they should be thoroughly stirred with a glass rod.

The two mixtures are stable for many months in a refrigerator. Just before use, they are blended and the accelerator is added. Hardness is controlled by varying the proportions of mixtures A and B: the final product will be harder if the proportion of mixture B is increased. A 1:1 mixture is suitable for many purposes, but the proportions may be changed. The accelerator 2,4,6-tri(dimethylaminomethyl)phenol (DMP-30) is used at a concentration of 1.5–2%. The DMP is added with a thoroughly dry syringe. The final mixture is stirred carefully to secure uniform distribution of the components. It is convenient to use plastic disposable beakers and cylinders for handling the resins and to dispose of them after use. The final mixture (with accelerator) can be kept frozen at −80° for several months in small disposable vials so that only the amount needed can be thawed. Epon can be cured overnight at 70°C or, apparently without detriment, at 90°C for 60 min.

b. Araldite. There are a few slightly different formulations of Araldite. Araldite M (Araldite CY 212) is produced in Europe. Araldite 502 seems to be very similar to Araldite M and is produced in the United States. Araldite 6005, which was used at the one time in the United States, has been discontinued. Araldite is mixed with the hardener DDSA and the accelerator DMP-30. Some recipes include also dibutyl phthalate as a plasticizer.

The following recipes are suggested.

Glauert and Glauert (1958) formulation: Araldite CY212, 10 ml; DDSA, 10 ml; dibutyl phthalate, 1 ml; DMP-30, 0.5 ml

Luft (1961) formulation: Araldite 502, 27 ml; DDSA, 23 ml; DMP-30, 0.75–1 ml. Addition of dibutyl phthalate is not necessary, as it is contained in Araldite 502.

Formulation suggested by Fluka for Durcupan ACM Kit: Durcupan ACM, 10 ml; DDSA, 10 ml; dibutyl phthalate, 0.1–0.2 ml; accelerator, 0.3 ml. Araldite is cured at 60°C for 2–3 days.

c. *Epon–Araldite Mixtures.* For plant tissues Mollenhauer (1964) proposed the following three formulations (A–C) containing Epon, Araldite, and a plasticizer (dibutyl phthalate) or a flexibilizer (Cardolite NC 513).

	Mixture (ml)		
Constituents	A	B	C
Epon	25	62	—
DDSA	55	100	—
Araldite M (or CY212)	15	—	—
Araldite 506	—	81	50
Cardolite NC 513	—	—	25
Dibutyl phthalate	2–4	4–7	1–2
Benzyldimethylamine or	3%	3%	3%
DMP-30	1.5%	1.5%	—

Blocks prepared with the above mixtures vary in hardness, image contrast, and tissue preservation as indicated in the following tabulation.

	Mixture		
Characteristic	A	B	C
Hardness	Medium	Soft	Medium
Image contrast	High	Medium	Low
Tissue preservation	Good	Excellent	Excellent

Blocks made with mixture B are more readily sectioned than those made with mixture A or mixture C. Mixture C is useful for tissues that do not bond well to the resin, such as leaves with a thick cuticle. Curing is done at 40–80°C overnight.

With Epon and/or Araldite, after dehydration with ethanol, tissue samples are given two changes of 30 min each of propylene oxide in the same glass vial. After the last change, a 3:1 mixture of propylene oxide and resin is poured on the specimens and the vial (unstoppered) is placed in an oven at about 35°C until the solvent is completely evaporated. The resin is poured out and a fresh batch without propylene oxide is added. The vial is placed back in the warm oven for about 4 hr. As this point embedding molds are filled with resin, and tissue pieces are poured with little resin from the vial on filter paper (several layers). The samples are picked up individually with a wooden stick, placed in the resin-containing molds, and allowed to sink through the resin. Moderate heat (35–40°C) may be applied to speed up sinking. Then the molds are placed in the oven for curing.

When orientation of samples is not required, gelatin capsules of different sizes can be used as molds. However, for botanical and virological work, in which orientation is usually necessary, flat embedding molds are preferred. Flat blocks require a special chuck for microtoming, but no further operations are needed after embedding.

When embedding is in gelatin capsules or any other mold requiring reorientation of the specimens, these must be cut away and glued back in the proper orientation on the same block from which they originated or on a wooden block. A drop of epoxy resin or of sealing wax melted with a hot scalpel is used for gluing.

Before use it is advisable to dry molds thoroughly in the oven overnight.

d. Spurr's Resin. Spurr (1969) introduced a new epoxy resin, vinylcyclohexene dioxide (VCD or ERL 4206), to improve infiltration of plant tissues. When cured in the presence of only a hardener, blocks are too hard; therefore a plasticizer (flexibilizer) is required. An anhydride hardener [(nonenyl) succinic anhydride; NSA] and a plasticizer (diglycidyl ether of polypropylene glycol; DER 736) were chosen because of their low viscosity. Hardness is controlled by varying slightly the amount of the flexibilizer (DER 736). Setting is done in the presence of the accelerator dimethylaminoethanol (DMAE, or S-1).

For general purposes, the following formulation yields firm blocks: resin (VCD), 10 g; flexibilizer (DER 736), 6 g; hardener (NSA), 26 g; accelerator (S-1), 0.4 g.

The components are measured gravimetrically. A plastic disposable bottle with wide neck is placed on a scale, and the scale is balanced. The first component is poured in the bottle, care being taken to avoid overweighing. The last drops are added with a Pasteur pipet. Flexbilizer and hardener are added in succession, pipet being changed each time. The mixture in shaken gently for thorough blending before the accelerator is added, then mixed again.

Dehydration for infiltrating with Spurr's resin can be done with several solvents, including acetone and ethanol. After the last change of the dehydrating agent, a small quantity of agent is left in the vial and an equal quantity of resin is added. The mixture is swirled and allowed to stand for 30 min, or the vial may be rotated on a motor-driven wheel inclined at an angle of 45° (1–5 rpm). An additional equal quantity of resin is then added to the mixture, which is swirled or rotated again for another 30 min. The mixture is then poured off and drained from the vial, and pure resin is substituted. After 2–3 hr, the resin is changed and tissue pieces are left in it overnight.

Embedding is carried out the next morning using the procedures described above, and curing is done at 70°C for 8 hr or longer.

2. Methacrylate–Styrene Mixtures

Mohr and Cocking (1968) suggested the use of methacrylate–styrene mixtures for embedding vacuolated, senescent plant tissue. They proposed the following recipe: n-butyl methacrylate (stabilizer removed), 70 ml; styrene (stabilizer not removed), 30 ml; benzoyl peroxide (anhydrous), 1 g.

The stabilizer contained in the n-butyl methacrylate is hydroquinone, which inhibits polymerization. It is removed with repeated washing in a separating funnel with a 30–50% sodium hydroxide solution. The alkali is in turn removed by washing with distilled water. The methacrylate is then filtered through filter paper to eliminate most of the water and is dried with anhydrous calcium chloride, which is finally eliminated by filtration (Pease, 1964).

Harrison and Roberts (1968) used a similar mixture with methyl methacrylate and styrene (6:1) and 1% benzoyl peroxide.

Dehydration is done in ethanol, and the tissue is infiltrated with increasing concentrations of the resin mixture and embedded in predried closed gelatin capsules. Butyl methacrylate–styrene blocks are polymerized for 44 hr at 55°C. For the methyl methacrylate mixture, Harrison et al. (1970) suggested incubation for 48 hr at 40°C, then for 24 hr at 60°C.

Methacrylate–styrene mixtures have been used in our laboratory for some time with satisfactory results (Martelli and Russo, 1973; Russo and Martelli, 1972, 1973). However, their utilization has been discontinued in favor of epoxy resins (mainly Spurr's medium), which are much less troublesome to prepare.

3. Glycol Methacrylate

Glycol methacrylate (GMA) is a water-soluble resin introduced by Leduc and Bernhard (1967). Specimens are not dehydrated with organic solvents but by passage through a graded series of resin and water. Polymerized GMA is still hydrophilic, so that sections in GMA can be treated with aqueous enzyme solutions. The technique was first used in plant virology by Weintraub and Ragetli (1968), who also established conditions best suited for preserving the amenability of cellular substances to enzymes.

Two stock solutions are prepared and used fresh:

Solution A: GMA monomer, 80 ml; water, 80 ml

Solution B: GMA monomer, 97 ml; water, 3 ml

Solution C: Solution B, 7 vol; butyl methacrylate (stabilizer not removed) with 2% Luperco,* 3 vol

Solution C is the embedding mixture and should be prepolymerized before infiltration of specimens. The use of prepolymerized methacrylate serves to prevent the so-called "polymerization damage" occurring when simple methacrylate monomer is used. Its causes and effects are not clear, but the end results are artifacts due to swelling of cells and organelles. For a thorough discussion of this point, see Pease (1964).

To prepare prepolymerized GMA, the following procedure was given by Leduc and Bernhard (1967).

1. Pour into an Erlenmeyer flask a small amount of solution C, so as to form a layer no thicker than 1 cm.
2. Cap it with aluminum foil and heat over a Bunsen flame with very rapid swirling until boiling just begins, which takes about 1 min.
3. Plunge the flask in ice water. Continue swirling until the mixture cools down to the temperature of the bath. The prepolymer should have the viscosity of a thick syrup. If not, the procedure should be repeated.
4. Store prepolymerized GMA in a freezer and warm it to 0–4°C when needed.

Specimens are infiltrated by passage through solution A (20 min), solution B (20 min), and prepolymer (overnight). Embedding is done in gelatin capsules. These are left without cap for about 30 min to eliminate air bubbles, then capped excluding as much air as possible (air inhibits polymerization). The capsules are held with a wire support and placed beneath a long-wavelength UV light source (> 315 min). The top of the capsules should not be farther than 5 cm from the light source. Curing is completed in 24–48 hr, depending on the viscosity of the prepolymer and the source of UV light.

Dehydration, embedding, and polymerization are carried out in the cold (0–4°C).

G. MICROTOMY

Several ultramicrotomes are commercially available having comparable technological levels. They are supplied with comprehensive instruction manuals containing both specifications for a proper use of the instrument and suggestions on how to operate to avoid problems and defects in sectioning. A few practical suggestions, however, will be given.

---

*Luperco is the trade name of an accelerator consisting of a mixture of dibutyl phthalate and the catalyst 2,4-dichlorobenzoyl peroxide.

1. *Knives*

Glass or diamond knives can be used for sectioning. Glass knives are made when needed from glass strips of standard size and discarded after use. Diamond knives are commercially available that have different sizes and angles. It is common practice today to use diamond knives. Although expensive, they are reliable and the user may ask the manufacturer to supply a knife preliminarily tested with the proposed material.

Glass knives can be made with pliers especially adapted to this purpose or with special machines, such as the LKB Knifemaker. The cutting quality of glass knives can be improved by coating the cutting edge with a thin film of evaporated tungsten metal. Knives treated in this way give up to 10 times as many good sections as uncoated knives (Roberts, 1975).

In our laboratory diamond knives are regularly used. However, glass knives are still used to cut semithin sections for light microscopy and also for trimming blocks into the wanted shape and size for ultrathin sectioning.

2. *Water Bath*

Sections are made to float on a water surface extending right from the cutting edge. A "boat" is then fixed to a knife to provide a water bath.

Two types of boats may be applied to glass knives. One, made of metal, is recovered after the used knife is discarded, and it is cleaned and used again. The second type is made with a strip of a water-insoluble tape wrapped around the back of the knife edge. Both types are sealed to the glass with wax melted with a warm glass spatula.

The boat is filled with either pure water or 10% ethanol or acetone. Acetone solutions have a lower surface tension, so that they may wet the knife edge more easily.

A third approach, for diamond and glass knives, is not to use a fixed water bath, but to hang a water drop on the sloping face of the knife. The drop is prevented from sliding down by using a small hydrophobic barrier, such as a strip of epoxy resin. The advantage of this system is that the water drop can be rapidly removed to clean the knife edge, or replaced if it becomes contaminated. Generally the cutting quality of the knife is superior and cleaner sections can be obtained (R. G. Milne, unpublished observation).

3. *Trimming of Blocks and Sectioning*

In our laboratory the block is first trimmed either with a one-edge razor blade or with half of a double-edge razor blade fitted in a holder supplied with the spares of the microtome, or a file. Trimming is then continued with the ultramicrotome until the final shape and size are reached. Usually, the block ready for sectioning has a pyramidal shape with the upper face rectangular or trapezoidal not exceeding 0.5 mm on each side. It should be kept in mind that, in general, the smaller the block face, the better the quality of the sections.

Once prepared, the block is mounted in the microtome as well as the knife (for a diamond knife, a clearance angle not higher than 1° and a knife angle of about 45° are used). The knife is then advanced manually until the block face almost touches the knife edge, the automatic advance is inserted, and the manual advance is continued (0.5 μm each time) until the first section is cut. Automatic advance (at cutting speed of 1–2 mm/sec) is adjusted at each stroke until a section is cut and the sections are of uniform thickness. Thickness is evaluated from interference colors shown by the sections floating on the water meniscus. Gray to silver sections are suitable. A ribbon of sections should form. If not, the cause of the trouble (most probably a misshapen pyramid) must be found, for sections floating freely in the water bath are very difficult to collect on the grids and many will be lost. Other troubles such as compression wrinkling, knife marks, and "chatter" (variation in thickness on the same section, showing as alternate bands parallel to the knife edge) will generally not appear if embedding was properly done, the knife edge is sharp, and the microtome works properly.

4. Collecting Sections on the Grids

In our laboratory, bare grids (400 mesh or 300 × 75 mesh) are routinely used for sections of plant material embedded in epoxy resins or cross-linked methacrylates that are stable under the electron beam. On the other hand, since part of the section can be obscured by the bars of grids of fine mesh, coarse-mesh grids (200 mesh) with film can be used (for film preparation, see Milne, 1972). For viewing serial sections, when a three-dimensional reconstruction of cellular structures is needed and each of the sections of a ribbon must be observed, still coarser grids (50–100 mesh) or, better, single-slot or single-hole grids must be used. To collect ribbons on single-hole grids, see the procedures outlined by Sjöstrand (1967).

For routine grids, pick up section by the following procedure.

1. Use an eyelash glued to the end of a wooden stick to maneuver the ribbon on the water bath. (Incidentally, eyelash hairs are very useful "tools" for the microtomist. In addition to guiding sections, they serve for removing bad sections or dirt particles from the knife edge, for wetting the knife edge, etc.)
2. With one hand, hold a grid with a pair of tweezers and submerge it almost below the ribbon.
3. With the eyelash held in the other hand, guide the ribbon so that *only the first section* touches the grid toward its periphery.
4. Slowly withdraw the grid from the water bath.

5. Eliminate excess water with a piece of filter paper, and store grids in a petri dish with the bottom covered by hairless filter paper.

H. STAINING

The majority of organic molecules contain mostly light atoms unable to scatter electrons enough to confer the necessary contrast upon cellular structures to be viewed and photographed. Osmium, which is primarily used as a fixative, has a fairly high atomic weight, thus acting as an electron "stain" for membranous constituents because of its affinity for lipids. However, osmium alone is not sufficient to induce a satisfactory density of other cell components for proper visualization under the electron beam. Thus, to increase contrast, sectioned tissues must be exposed to a number of heavy-metal ions with an atomic weight approaching or above 200. The most widely used of these ions for staining sections are uranium and lead.

Staining Solutions

These are preparations in which the metal ions are dissolved to be applied as a bulk stain to tissue blocks or to thin sections on the grids.

a. Uranyl. The word "uranyl" indicates complex uranium-containing ions formed in water from uranium salts such as uranyl acetate, uranyl magnesium acetate, and uranyl nitrate.

As uranyl ions bind to several groups, such as carboxyl and phosphate, they produce a general increase in contrast without much specificity (Silva, 1973). Cell membranes, ribosomes, and chromatin are all structures the electron density of which increases with uranyl treatment. Among uranyl salts, uranyl acetate (UA) is preferred as it is more stable and efficient than the other two.

Uranyl acetate can be applied either as a bulk stain, before or during dehydration, or on the sections, or both. When UA is used as a bulk stain before dehydration, tissue samples are soaked in a 0.5–2% aqueous solution for 15 min to several hours. Staining during dehydration is customarily done in 70% ethanol or 70% acetone containing UA from 1% to saturation. The duration of treatment varies again from 15 min to several hours. For staining sections, grids are floated on or immersed in saturated UA solutions in water or 50% ethanol for 1 hr if there was no previous staining or for 10–15 min on 1–2% aqueous or alcoholic UA if tissues were bulk-stained before embedding.

In our laboratory, tissue samples are stained after osmication and thorough rinsing in distilled water, by soaking overnight at 4°C in 0.5% aqueous UA. After dehydration and sectioning, thin sections are poststained for 10 min with a freshly prepared 2% UA solution in 50% ethanol.

Because UA dissolves slowly, solutions must be prepared well in advance of use or by the aid of a mechanical shaker.

b. Lead. Leads salts are also routinely employed for increasing the electron density of biological materials. They are not used for bulk staining, but only for sectioned samples.

Cell structures showing affinity for uranyl and lead atoms are virtually the same. Thus the action of these electron stains is to be regarded as additive rather than differential.

The mechanism whereby lead salts operate is not well understood, mainly because it is not clear what kind of ions are formed in solution. Positively charged lead ions probably attach to polar groups of phospholipids which have acquired a negative charge due to the presence of reduced osmium. Cell membranes, therefore, would not be stained by lead if tissues were not previously osmicated.

In the presence of CO_2, insoluble lead carbonate precipitates will form in lead salt solutions, which will appear as a dark deposit on the specimens. Lead carbonate precipitation, however, can be prevented from chelating lead by tartrate (Millonig, 1961b) or citrate (Reynolds, 1963) or by raising pH considerably (Karnovsky, 1961).

Undoubtedly the most widely used lead-containing formulation is that of Reynolds (1963), with which all lead atoms are chelated by excess citrate so that they can no longer react with CO_2. At any rate, staining is guaranteed by the fact that cell structures have a high affinity for lead cations, which are readily transferred from citrate to tissue binding sites.

Reynolds' (1963) lead citrate is prepared as follows:

1. Pour into a 50-ml volumetric flask 1.33 g of lead nitrate [$Pb(NO_3)_2$] and 1.76 g of sodium citrate [$Na_3(C_6H_5O_7)\cdot 2H_2O$].
2. Add 30 ml of carbon dioxide-free distilled water (CO_2-free water is obtained by boiling water in a flask for 5–10 min, and letting it cool after covering it with a piece of aluminum foil).
3. Shake the mixture at intervals for about 30 min (lead nitrate is converted to lead citrate).
4. Add 8 ml of 1 N NaOH (to prepare this solution weigh rapidly a few solid pellets of NaOH, put them in a stoppered graduated cylinder, and add the required amount of CO_2-free distilled water to make the normal solution. CO_2-free NaOH, 1 N, is also available commercially).
5. Bring to volume (50 ml) with CO_2-free distilled water and mix by inversion until lead citrate dissolves completely and the solution clears up completely.

Lead citrate solutions must be kept well stoppered in the same flask used for preparation. For use, the needed amount is drawn from the flask with a Pasteur pipet, and the first two or three drops are discarded. The solution is never poured from the flask. This simple precaution enables one to use it with negligible waste.

    c. Double Staining with Uranyl Acetate and Lead Citrate. The best way known so far for obtaining a satisfactory contrast of thin-sectioned cells is to expose them to double sequential staining with UA and lead citrate. The following procedure can be used.

1. Fill a suitable small container (such as polypropylene boats used for blood analysis) with a 2% solution of UA in 50% ethanol. Keep the container in a moist petri chamber to prevent evaporation. Only freshly prepared solutions should be used, and care should be taken not to expose them to strong light to avoid photodecomposition.
2. Immerse grids (sections up) in the solution for 10 min. This time is sufficient if tissues were bulk stained before or during dehydration. If not, a longer exposure (up to 1 hr) is needed.
3. Rinse each grid with 20–30 drops of 50% ethanol. Put grids on dry filter paper in a petri dish, taking care not to wet them with residual stain left on tweezers. To avoid contamination put a piece of filter paper between the points of the tweezers and push the grid gently while gradually releasing pressure on the tweezer's arms.
4. Pour several pellets of solid NaOH in a petri dish and place on them a piece of solid hydrophobic material such as a sheet of silicone-treated paper or dental wax.
5. Place on the hydrophobic material as many drops of lead citrate as there are grids that must be stained. Keep the petri dish closed, and avoid breathing on the drops of the staining solution.
6. Float grids (sections down) on the drops for 2 min.
7. Collect grids individually with clean tweezers and rinse with 20–30 drops of warm CO_2-free distilled water. Avoid breathing on the grids.
8. Store the grids in a dry petri dish on filter paper.

When bulk-stained material is used, double staining may be unnecessary, and exposure to lead citrate as short as 30 sec to 1 min may be sufficient (Milne, 1970; Hatta and Matthews, 1974).

I. SEMITHIN SECTIONS

Sections for light microscopy 0.5–2 μm thick can be readily cut with an ultramicrotome using diamond or glass knives. Large sections can be cut dry and collected singly with a toothpick or tweezers. These are placed on a drop of water on a glass slide, which is then warmed up to expand the sections and evaporate the water. Small sections can be floated on the water bath of the knife from which they are collected with a glass strip made by scoring a coverslip with a diamond pencil. The glass strips are held with forceps during collecting operations.

Sections are stained while on the glass strip. The strips are either placed on a microscope slide in a drop of the stain or stuck by capillary attraction to the walls of a small glass container (e.g., a weighing bottle) filled with the staining solution.

For quick examination, staining with toluidine blue (1% in 1% sodium tetraborate) according to Pease (1964) is quite suitable. The sections are heated in a large drop of the stain until it bubbles.

The sections are rinsed in distilled water and ethanol and observed dry or are mounted in Permount after a passage through xylene.

For specific differential staining, other procedures are available as reported in detail by Hoefert (1968) and Roberts and Hutcheson (1975).

Should fluorochromes such as aniline blue for callose detection be applied, embedding plastic is removed by treating sections with sodium methoxide (Peterson et al., 1978).

J. CYTOCHEMISTRY

Visual ultracytochemical methods applied to electron microscopy aim primarily at (1) identifying the chemical nature of normal or novel constituents of the infected cells either by differential enzymatic digestion or by the action of chelating agents; and (2) identifying and localizing enzyme activity by the use of specific marker substrates.

Both approaches have been used in plant virology. Enzyme digestion tests can be carried out on plant specimens before or after embedding. In the former case thin tissue slices or sections 100–200 nm thick are fixed with an aldehyde, exposed to the desired enzyme, postfixed with OsO_4, dehydrated, and embedded. In the second instance, sections of tissues embedded in epoxy resins or GMA are exposed to the enzyme(s) and then stained and viewed.

1. *Differential Enzyme Digestion before Embedding*

This method was successfully applied by Weintraub and Ragetli (1968) to cryostat sections of plant leaf tissues infected with bean yellow mosaic virus (BYMV) for identifying the chemical nature of virus-induced crystalline and cylindrical inclusions. The following procedure is used.

1. Fix tissue pieces 3–5 mm² in 5% glutaraldehyde in phosphate buffer (pH 7.2) at room temperature for 1.5 hr. An additional period of fixation at 4°C for 2 hr may follow.
2. Rinse with the same buffer.
3. With a cryostat microtome, cut sections 60–120 nm thick and immerse

in control or enzyme solutions: 0.25% pepsin in 0.2 N HCl; 1% papain in 0.01 M phosphate buffer (pH 5.5); 0.1% RNase in distilled water (pH 6.8) adjusted with 0.01 N NaOH; 0.05% DNase in distilled water (pH 6.8) adjusted as above. Incubate at 37°C for various times up to 24 hr.
4. Rinse with distilled water or buffer.
5. Fix for 1 hr in Palade's OsO₄, pH 7.2.
6. Rinse, dehydrate, and embed.

A similar approach was used by Martelli and Castellano (1971) to show the proteinaceous nature of inclusion bodies induced by cauliflower mosaic virus. In this case, epidermal strips of infected leaves were used instead of cryostat sections. These were fixed at 4°C for 1–4 hr in Karnovsky (1965) fixative, rinsed with buffer, and incubated at 38°C for up to 6 hr in a 1 mg/ml protease solution in 0.05 M Tris-HCl buffer (pH 7.5). After enzyme treatment, epidermal strips were cut to smaller fragments, osmicated, dehydrated, and embedded as usual.

A method that permits intracellular localization of RNA and determines whether it occurs as single-stranded (ss) or double-stranded (ds) molecules has been developed by Hatta and Francki (1978). This technique exploits the differential ability of pancreatic RNase to digest ssRNA or dsRNA in the presence of low or high salt concentrations. The procedure devised by Hatta and Francki (1978) follows.

1. Fix tissue pieces in Karnovsky (1965) aldehyde fixative, pH 7.5, for 15–16 hr at 4°C.
2. Rinse on 0.06 M phosphate buffer, pH 7.5.
3. Cut tissues into smaller pieces with a razor blade.
4. Wash tissues in four changes of 2× or 0.01× SSC (SSC=0.15 M sodium chloride and 0.015 M sodium citrate) over 6 hr at room temperature.
5. Incubate for 16 hr at 25°C in either 2 × SSC (high-salt medium) or 0.01 × SSC (low-salt medium) without (control) or with 2–50 mg/ml RNase A (type III A, Sigma Chemical Company, St. Louis, MO). Before using, heat RNase solutions for 10 min at 100°C to inactivate any contaminating enzyme.
6. Rinse with phosphate buffer.
7. Fix with 1% OsO₄ in Millonig's phosphate buffer for 3 hr at room temperature; dehydrate and embed.
8. Observe mainly cells cut open with the razor blade, as in these the enzyme has moved without restriction.

Single-stranded RNA is digested regardless of the salt concentration of the incubating medium whereas dsRNA in susceptible to the enzyme only under low salt concentration. The above method has already been success-fully used for the identification of dsRNA in vesicles occurring in vesicu-late-vacuolate inclusions (Hatta and Francki, 1978) or on the tonoplast (Hatta and Francki, 1981a) or in multivesicular bodies (Russo et al., 1983) of cells infected with viruses belonging to different taxonomic groups.

The same technique can be applied to identification of virus particles containing ssRNA, because RNA in ribosomes is digested whereas that of virus particles is not (Hatta and Francki, 1979). However, the RNA of vi-ruses whose structure is stabilized by protein–protein interactions may be removed (Hatta and Francki, 1981b) and, under certain conditions (Russo et al., 1983), also the RNA or virions in which protein–RNA interactions prevail.

2. Differential Enzyme Digestion after Embedding in Epoxy Resins

Sections of epoxy resin-embedded tissues can be treated with proteases, but not with nucleases. Tissues double fixed with glutaraldehyde and os-mium tetroxide or with glutaraldehyde alone can be used, but in the first case sections must be submitted to a preliminary oxidation with H_2O_2 (Monneron and Bernhard, 1966).

a. Double-Fixed Tissues (Monneron and Bernhard, 1966)
1. Fix tissue with aldehydes and osmium and embed in epoxy resins.
2. Collect sections on gold grids.
3. Float grids on 20% H_2O_2 for 20 min at room temperature.
4. Rinse two to three times with distilled water.
5. Float grids on 0.5% subtilisin solution in 0.05 M Tris-HCl buffer (pH 7.8) for up to 180 min at 37°C.
6. Rinse with distilled water and stain with uranyl acetate and lead cit-rate.

In our laboratory, the above method was used satisfactorily for the iden-tification of protein constituents of inclusion bodies induced by broad bean wilt virus (BBWV) (Russo and Martelli, 1975).

b. Tissues Fixed with Glutaraldehyde Alone (Shepard, 1968)
1. Fix tissue for 3 hr with 5% glutaraldehyde in 0.1 M phosphate buffer, pH 7.0.
2. Rinse with buffer, dehydrate, and embed in Araldite.
3. Collect sections on copper grids.
4. Float grids on 0.5% subtilisin solution in 0.04 M Tris-HCl buffer (pH 7.8) for 20 hr at 37°C.
5. Rinse with distilled water and stain with uranyl acetate and lead cit-rate.

A satisfactory application of this technique was found in the study of nuclear and cytoplasmic inclusions induced in tobacco by tobacco etch virus (TEV).

3. Differential Enzyme Digestion after Embedding in GMA

As already mentioned, sections of tissues embedded in water-soluble resins such as GMA can be readily used for enzyme digestion (both proteases and nucleases). The only problem resides in the electron contrast of the material, which is lower than usual because in order to avoid excessive brittleness of tissues, postfixation with osmium is not done.

GMA embedding has been used for enzyme digestion tests of cells infected with several different plant viruses (Weintraub and Ragetli, 1968; Weintraub et al., 1969; Russo and Martelli, 1973).

The procedure, as outlined by Weintraub and co-workers (Weintraub and Ragetli, 1968; Weintraub et al., 1969), follows.

1. Fix tissues for 30–90 min at room temperature in one of the following fixatives prepared in 0.1 M phosphate buffer (pH 7.2): (a) 5% glutaraldehyde; (b) 5% acrolein; (c) a mixture of 5% acrolein and 4% formaldehyde; (d) 5% formaldehyde.
2. Rinse with buffer for 1 hr, or for 15 hr if DNase must be used.
3. Dehydrate and embed in GMA.
4. Mount sections, preferably on silver grids.
5. Incubate grids by floating on enzyme solutions and appropriate controls at 37°C for different lengths of time.
6. Stain with uranyl acetate for 1 hr at 60°C and with lead citrate for 5 min.

4. Differential Staining

An electron staining method based on the use of chelating agents, which gives preferential contrast to certain cell structures, was developed by Bernhard (1969) and used for differentiating ribonucleoprotein from deoxyribonucleoprotein structures in the nuclei of cells infected with bean golden mosaic virus (BGMV), a DNA-containing geminivirus (Kim et al., 1978). With this staining technique (uranyl–EDTA–lead), BGMV-induced nuclear "fibrillar rings" were shown to be composed of deoxyribonucleoproteins, of possible viral origin, whereas the nucleoli, from which fibrillar rings had segregated, were shown to be made up of ribonucleoproteins, as expected.

The procedure (Kim et al., 1978) follows.

1. Fix tissues in 4% glutaraldehyde in 0.05 M cacodylate buffer (pH 7.2) for 2 hr at room temperature.

2. Bulk-stain in aqueous 0.5% UA overnight at 0–4°C.
3. Dehydrate in ethanol and embed in Spurr's medium.
4. Cut sections, stain in 5% aqueous UA for 15 min, and rinse in distilled water.
5. Float sections on 0.5 M aqueous EDTA (pH 7.2) for 15–30 min at room temperature or at 60°C.
6. Rinse sections and stain with lead citrate for 3–5 min.

DNA-containing structures are bleached by the treatment with EDTA, whereas ribonucleoproteins are not.

5. Staining of Enzymes

Although tremendous advances have been made in electron microscopy of enzymes [see the book series edited by Hayat (1973–1975) and Hall (1978)], it has been virtually neglected by plant virologists, with the exception of a few cases referring to the intracellular identification and localization of acid phosphatase (Appiano and Lovisolo, 1979), catalase and peroxidase (De Vecchi and Conti, 1975; De Zoeten *et al.*, 1973; Russo *et al.*, 1983), and glycolate oxidase (Russo *et al.*, 1983).

a. Acid Phosphatase

1. Fix tissue sections 50–100 nm thick for 2–4 hr in 3% glutaraldehyde in 0.1 M cacodylate buffer (pH 7.2).
2. Rinse thoroughly in the same buffer.
3. Incubate for 10–30 min in Gomori medium: sodium β-glycerophosphate, 30 mg; 0.05 M acetate buffer (pH 5), 11 ml; 12% lead nitrate, 0.1 ml. Control samples are incubated in the same medium without β-glycerophosphate.
4. Rinse in 0.05 M acetate buffer, pH 5.
5. Postfix in 1% osmium tetroxide in cacodylate buffer for 1 hr.
6. Dehydrate and embed as usual.
7. Observe sections unstained or stained with UA only.

Using the above procedure Appiano and Lovisolo (1979) identified lysosomal activity in secondary cytoplasmic vacuoles in which apparent digestion of maize rough dwarf virus (MRDV) particles was taking place.

b. Catalase.
The procedure outlined by Frederick and Newcomb (1969) yields satisfactory results for plant tissues.

1. Fix tissue samples in 3% glutaraldehyde in 0.05 M potassium phosphate buffer (pH 6.8) for 1.5–2 hr at room temperature.
2. Rinse in at least two changes of the same buffer.
3. Incubate tissues in standard medium containing 10 mg of 3,3-diaminobenzidine (DAB), 5 ml of 0.05 M 2-amino-2 methyl-1,3-propanediol buffer (pH 10), and 0.1 ml of 3% H_2O_2. The pH is adjusted to

9.0 before adding tissues. Incubation is in stoppered vials or test tubes for 1 hr in a water bath at 37°C.
4. Rinse thoroughly with buffer and postfix in 2% OsO_4 in 0.05 M phosphate buffer at pH 6.8 for 2 hr.
5. Dehydrate and embed as routine.
6. View sections unstained or after double staining with UA and/or lead citrate. Observe mainly cut-open cells.

Controls consist of tissues incubated in the standard medium in the presence of 0.02 M aminotriazole.

This method has been applied to leaf tissues infected with (1) pea enation mosaic virus (PEMV) (De Zoeten *et al.*, 1973), (2) alfalfa mosaic virus (AMV) (De Vecchi and Conti, 1975), and (3) *Cymbidium* ringspot virus (CyRSV) (Russo *et al.*, 1983). In the former two cases it was hoped to find an explanation for the increased peroxidase and catalase activity shown by infected tissues. With CyRSV, catalase detection served as a marker for identifying peroxisomes to demonstrate their relationship with multivesicular bodies.

c. Glycolate Oxidase. Glycolate oxidase, an enzyme typically present in peroxisomes, can be localized in plant cells by the ferricyanide-reduction method (Burke and Trelease, 1975) or by a more recently developed procedure that does not induce formation of background deposits (Thomas and Trelease, 1981). This procedure includes the following steps.

1. Fix tissue samples in ice-cold formaldehyde–glutaraldehyde (4–1%) in 20 mM potassium phosphate buffer (pH 6.9) up to 1 hr at 4%C.
2. Rinse for 10 min in 50 mM potassium phosphate buffer (pH 6.9), then for 20 min in 100 mM buffer (pH 7.5) at room temperature.
3. Preincubate samples for 1 hr in a medium made by 100 mM Trismaleate buffer (pH 7.5) containing 50 mM 3-amino-1,2,4-triazole and 5 mM cerium chloride ($CeCl_3$). Solutions containing $CeCl_3$ should be aerated and filtered (0.45 μm) before use.
4. Incubate samples in the reaction mixture made by adding to the incubation medium 50 mM sodium glycolate. The incubation time may vary from 1–3 hr to 18 hr. Hourly changes with fresh incubation medium are made with short incubation times, one change only for the long incubation time.
5. Rinse tissue samples for 30 min in 100 mM sodium cacodylate (pH 6.0) and for 10 min in the same buffer at pH 7.2.
6. Postfix for 1 hr in 2% osmium tetroxide in 100 mM cacodylate buffer (pH 7.2).
7. Rinse, dehydrate, and embed as usual.
8. Observe sections with or without staining.

Identification of peroxisomes in CyRSV-infected cells by the above reaction was useful for establishing their relationship with multivesicular bodies (Russo *et al.,* 1983).

K. Immunochemistry

Electron microscopic immunochemical methods allow the fine localization in a tissue of an antigen after its reaction with a specific antibody. The reaction product (i.e., the antigen–antibody complex) is not readily detectable in thin sections unless a large number of antibody molecules attach the antigen. This is possible with free virus preparations (Milne and Luisoni, 1977), but not with virus particles or other antigenic molecules inside cells. However, antigen–antibody complexes are revealed if immunoglobulins are conjugated with a marker readily recognizable under the microscope.

Two such markers have so far been used in plant virology: ferritin and ^{125}I. Ferritins are iron-containing globular protein particles about 10 nm in diameter having a molecular weight of 750,000. Because of the iron core, ferritins strongly scatter electrons, making them visible in the electron microscope without further processing. However, sites of deposition of ^{125}I-labeled immunoglobulins can be revealed only by autoradiographic methods.

A third marker, colloidal gold, appears to be promising (Roth *et al.,* 1978; Craig and Millerd, 1981), but it has not yet been used in plant virus studies.

^{125}I-labeling procedures have been used to a limited extent. A detailed description of their application is given by Schlegel and co-workers (Langenberg and Schlegel, 1967; Schlegel and De Lisle, 1971).

Ferritin labeling, on the other hand, has been used more extensively. A description of the method (Shalla and Amici, 1967) follows.

1. Separate immunoglobulin fraction (purified antibodies) from specific antiserum.
2. Conjugate purified antibodies with horse spleen ferritin using toluene 2,4-diisocyanate as a coupling agent (Singer and Schick, 1961).
3. Centrifuge the reaction mixture at 120,000 *g* for 2 hr, discard the supernatant, and resuspend pellets containing conjugated antibodies in 0.05 *M* sodium phosphate buffer (pH 7.2).
4. Fix tissues for 3 hr with 5% glutaraldehyde in neutral 0.1 *M* phosphate buffer at room temperature.
5. Rinse with distilled water.
6. Freeze the tissue in a drop of water at 20°C.
7. With a cryostat shave thin slices from the frozen sample, leaving a piece of tissue with an opened layer of cells. Shaving of frozen tissue can also be done with a razor blade.

8. Thaw tissue pieces, transfer to ferritin–antibody suspension, and incubate overnight at 20°C.
9. Wash with distilled water.
10. Fix with osmium tetroxide, dehydrate, and embed. Sections may be observed unstained to allow ferritin molecules to show up more clearly. Controls consist of ferritin-conjugated normal serum globulins.

Shepard *et al.* (1974) suggested the use of an indirect immunoferritin procedure, in which sheep anti-rabbit globulins are labeled with ferritin. Specific antigens are then localized by first exposing cells to rabbit antivirus globulins (unlabeled) and then to sheep anti-rabbit ferritin-labeled globulins. Advantages of the indirect method are that the ferritin-labeled antibodies can be readily prepared or purchased in large amounts and that a single tagged preparation may serve for investigations of several different antigens. Also, at least theoretically, the intensity of tagging should be higher with the indirect method.

L. Autoradiography

The aim of electron microscope autoradiography is the localization within a tissue section of radioactive precursors incorporated into an insoluble macromolecule, or the site of a serological reaction when antibodies have been coupled with a radioactive substance such as ^{125}I. Tissue sections containing the radioactive substance are coated with a layer of photographic emulsion and stored in the dark.

During this exposure, radiations emitted by the radioactive substance will hit silver halide crystals in the emulsion and form a "latent image," which is transformed into a visible image after photographic development. The visible autoradiographic image is constituted by grains of silver whose position indicates with some approximation (depending on the resolution) the sites of radioactive label.

Plant virologists have been primarily interested in studying incorporation of radioactive materials in different structures of infected cells, such as (1) nucleotides (i.e., [^{3}H]uridine or [^{3}H]thymidine) in viral nucleic acids or inclusion bodies (De Zoeten and Schlegel, 1967b; Laflèche and Bové, 1968, Kamei *et al.*, 1969; Bassi and Favali, 1972; Favali *et al.*, 1973, 1974); (2) amino acids (i.e., [^{3}H]leucine) in inclusion bodies (Hayashi and Matsui, 1967); (3) cell wall precursors (i.e., [^{3}H]glucose or [^{3}H]phenylalanine) in abnormal parietal outgrowths (Bassi *et al.*, 1974). The reader is referred to the cited papers for details on methods for feeding plant tissues with radioactive precursors.

After labeling, specimens are fixed and embedded as usual. Using relatively thick sections (silver to gold interference color) is advisable to avoid an excessively long exposure time because of the low level of radioactive label. Sections thicker than usual also provide higher contrast, and the resolution will remain high enough for autoradiographic purposes considering that resolution of autoradiography is limited by the uncertainty of relating the exact site of the emitting source in the cell to the developed silver grain. The best resolution is about the size of the original silver grain, i.e., 0.1 nm.

If photographic procedures are to be applied after mounting sections on grids, gold grids are preferred, to avoid reaction with developing solutions. Sections can be stained with uranyl acetate and/or lead citrate either before coating with the emulsion or after final photographic processing. Prestaining is generally favored.

Application of Emulsion

The most widely used emulsion for electron microscope autoradiography is Ilford L4, which is stable for at least 4 months if stored at 4°C, is easy to handle, and gives reproducible results from batch to batch.

Two procedures for applying the emulsion will be briefly outlined: the loop method (Caro and Van Tubergen, 1962; Caro, 1969) for sections mounted on grids, and the flat substrate method (Salpeter and Bachmann, 1972).

a. The Loop Method. Grids are attached to one end of a glass slide with double-stick tape. A loop of platinum, nichrome, or other inert metal is made by wrapping a wire around a 150-ml glass beaker (about 4 cm in diameter) and is attached by one end to a bacteriological inoculation loop or a glass rod. Use Ilford L4 emulsion. To avoid irregular distribution of water in the emulsion, before use melt the entire content of the bottle of emulsion by heating in a water bath at 45°C. When the content is melted, mix and allow to gel. The following steps are carried out in the darkroom.

1. With room light on, preweigh a 250-ml plastic beaker; leave it on the scale and add a 15-g weight. Premeasure also 15 ml of distilled water in a cylinder. Have a water bath at 45°C and an ice bath ready. Cover the bath with a Styrofoam sheet with holes to fit the beaker and a thermometer.
2. Turn off the room light, drop pieces of the gel in the beaker, and balance the scale. Add the water, swirl, and place the beaker in the water bath.
3. Allow the emulsion to melt (about 15 min) while stirring gently with a plastic rod.
4. Place the beaker in the ice bath for 1–5 min.
5. Move the beaker to a bath at room temperature until the emulsion is ready to use (in about 10 min). (Alternatively the emulsion can be poured out of the beaker into a plastic petri dish, should the beaker

prove too tall and narrow for the loop to be worked in and our easily.) The emulsion will be usable for about 10 min.

6. Dip the loop in the emulsion and lift slowly.
7. Apply the film by touching the surface of the slide carrying the sections with the loop. The film should be applied after gelling, immediately before it breaks. (To ascertain the right moment of application, make a test film: hold the loop up and count until the film pops. Apply just before breaking occurs.) If the film does not gel rapidly, increase the cooling time in the ice bath. The emulsion can be melted a few times over at 45°C and gelled again, should it become too viscous.
8. Store slides at room temperature in a sealed plastic slide box in a tin can with dry silica gel. Exposure time cannot be predicted, and test grids should be developed at intervals. If sections have been prepared from the same specimen for light microscope autoradiography, use an exposure 5–10 times longer for electron microscope preparation.
9. Develop in Kodak D-19 (2 min), in Microdol-X (5 min), or in a "physical developer" containing 0.1 M sodium sulfite and 0.01 M p-phenylene diamine (1 min). (As the tape dissolves in the developer, grids should be developed individually by transferring them one by one into drops of developer.)
10. Rinse with distilled water and fix with a rapid fixer, such as Kodak's.
11. Rinse by gently swirling the grids in small beakers of distilled water and dry.

Gelatin can be removed (though with some difficulty) or may be left in place. A high accelerating voltage (80 or 100 kV) can be used to overcome problems due to the thickness of specimens.

After staining, before applying the emulsion, it is advisable to cover the grids with a thin layer of evaporated carbon to prevent "chemography," i.e., interaction of the sections with the emulsion and removal of stain by the photographic fluids. Incidentally, this is a point in favor of prestaining, as poststaining would not be possible in the presence of the carbon layer.

b. Flat Substrate Method

1. Clean glass slides carefully with a detergent, rinse, and wipe dry with Kleenex or similar material. Blow slides free of lint.
2. Dip slides in 0.7% collodion dissolved in amyl acetate and dry vertically.
3. Transfer sections with a small loop to a drop of distilled water or 10% on the collodion-coated slide
4. Remove the water with a strip of filter paper. Indicate the location of the sections by marking a circle on the back of the slide.
5. Cover sections with a drop of stain and flush it away with distilled water after staining.

6. Cover the sections with a carbon layer.
7. Apply the emulsion. The loop method is perfectly applicable also to the flat substrate method. Alternatively, slides can be coated by dipping in the emulsion or by applying the emulsion directly on them with a medicine dropper.
8. Store the slides in a sealed box in a dry atmosphere.
9. Develop slides in Kodak D 19, Microdol X, or "physical developer" with phenylenediamine. Sensitivity can be increased with the procedures called "gold latensification," in which gold is deposited on latent images increasing their developability (Salpeter and Bachmann, 1972; Wisse and Totes, 1968).
10. Rinse in distilled water, dip slides in 3% acetic acid for 15 sec, rinse again in distilled water, and fix for 1 min in 20% sodium thiosulfate and 2.5% potassium metabisulfite.
11. Strip the collodion film bearing sections, carbon layer, and emulsion into a water surface. Place grids over the sections (these can be identified because of the interference color), and collect them with a wire gauze or any other suitable tool.

III. Cytological Modifications of Diagnostic Value

When they invade cells, most if not all plant viruses produce cytological changes that are more or less readily detected by examining thin-sectioned material with the electron microscope.

All parts and constituents of a cell of any tissue can be affected by viral infections in a manner that may be either broadly or highly specific, thus representing the "signature" (McWhorter, 1965) that may constitute a lead to identification of the causal agent.

The following section gives some examples of how cytopathological features of infected cells can be used for identifying virus groups or individual viruses. Only those virus groups for which sufficient and characterizing ultrastructural information is available are reviewed, and this section is therefore an addition to, rather than a substitution for, previous papers or books (Esau, 1968; Martelli and Russo, 1977a; Maramorosch, 1977; Rubio-Huertos, 1978; Edwardson and Christie, 1978) in which the same subject matter was treated extensively.

A. Isometric Viruses with ssRNA

There are 15 taxonomic groups (4 monotypic) of ssRNA viruses with isometric or quasi-isometric particles. Of these, 13 groups have virions with a simple structure and chemical composition and a diameter ranging between

25 and 35 nm. Members of four of these groups (tombus-, tymo, como-, and nepoviruses), all with polyhedral particles ~30 nm in diameter, elicit cytological modifications distinctive enough to provide a basis for their identification at the group level.

1. Tombusviruses

Tombusviruses are readily transmitted by inoculation of sap but show a strong tendency to remain localized in infected artificial hosts, in which they induce necrotic lesions. A few hosts, however, are invaded systemically.

Virus particles are sharply defined, very abundant, and readily recognizable, even in necrotic cells. Virions, either scattered or in disorderly accumulations or, more rarely, in crystalline arrays, have been seen in leaf hair cells, epidermis, foliar and stelar parenchymas, and in both differentiating and mature elements of conducting tissues. Except for carnation Italian ringspot virus (CIRV), all definitive members of the group (seven, according to Matthews, 1982) have been thoroughly investigated at the fine structure level (for review, see Martelli et al., 1977; Martelli, 1981). Two cytopathic features are common to tombusvirus infections: (1) cytoplasmic multivesicular bodies (Fig. 1A and B); (2) virus-containing bleblike evaginations of the tonoplast into the vacuole (for review see Martelli and Russo, 1977a; Martelli, 1981).

Other features, such as intranuclear presence of virions (Martelli and Castellano, 1969; Russo and Martelli, 1972) or of membranous inclusions (Martelli and Russo, 1973), cannot be considered to be group specific for their occurrence varies with both the virus and the host.

Multivesicular bodies are membranous cytoplasmic inclusions derived from peroxisomes, which, in infected cells, undergo a progressive vesiculation of the bounding membrane (Fig. 1A) through the possible addition of membranous material from the endoplasmic reticulum. Modified peroxisomes become very plastic, engulfing a portion of the ground cytoplasm either through the invagination of the limiting membrane or through the production of membranous appendages that fold back on the main body (Russo et al., 1983). The end result is a novel structure as large as a small plastid, made up of three major components: (1) a peripheral irregularly thickened membrane sometimes obviously connected to endoplasmic reticulum strands, (2) an electron-dense finely granular or crystalline matrix, and (3) many ovoid to globose vesicles measuring 80–150 nm in diameter and containing finely fibrillar material (Fig. 1A, inset). The intravesicular fibrils represent double-stranded RNA (Russo et al., 1983). Thus, multivesicular bodies may be directly involved in virus multiplication.

Multivesicular bodies occur also in the cytoplasm of cells infected with turnip crinkle and galinsoga mosaic viruses, two putative members of the

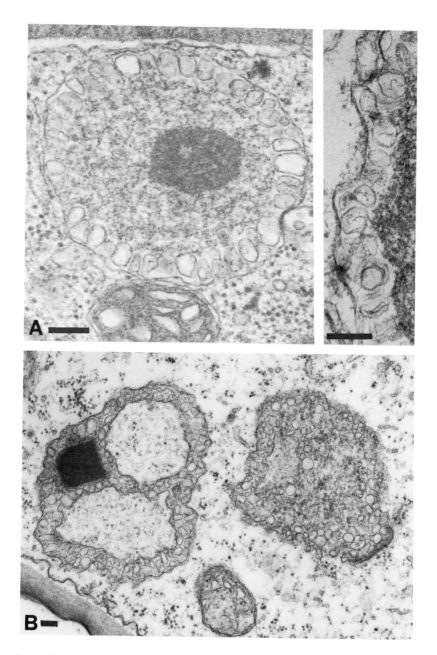

Fig. 1. Multivesicular bodies in tombusvirus infections. (A) A peroxisome in an early stage of transformation. The inset shows a closeup of the peripheral vesicles containing dsRNA strands. (B) Multivesicular bodies originated from profoundly modified peroxisomes. Magnification bars represent 200 nm.

group, but, in this case, the inclusions are made up of vesiculated mitochondria (Russo and Martelli, 1982; Behncken *et al.,* 1982).

The virus-containing blebs protruding from the cytoplasm into vacuole are recurrent structures, the function of which, as originally postulated (Russo *et al.,* 1968), may be that of an active transport of virions beyond cytoplasmic boundaries when the tonoplast is not disrupted by the infection (Russo *et al.,* 1983).

Based on present knowledge, bleblike evaginations of the tonoplast containing virus particles and multivesicular bodies, in particular, are to be regarded as distinctive and diagnostic intracellular features of tombusvirus infections rather than virus crystals, as suggested by Edwardson and Christie (1978). Crystalline structures are frequently found in cells infected by tombusviruses, but their presence is too erratic to be considered useful for diagnosis. Similarly, the presence of virus particles within apparently intact mitochondria is widespread throughout the group but, owing to its frequency and abundance, it may represent a diagnostic character of CyRSV (Martelli and Russo, 1981).

2. *Como- and Nepoviruses*

Viruses belonging in these two groups have morphological and many physicochemical and biological characters in common (Murant, 1981; Stace-Smith, 1981). A major difference lies in the type of natural vectors: comoviruses are transmitted by beetles, nepoviruses by nematodes. Cytopathological features are the same and do not allow for a clear-cut distinction between the two groups.

Several of the 13 members of the comovirus group and of the 27 members of the nepovirus group (Matthews, 1982) have been investigated cytologically (Gerola *et al.,* 1965a, 1969; Walkey and Webb, 1968; Crowley *et al.,* 1969; De Zoeten and Gaard, 1969a; Roberts and Harrison, 1970; Roberts *et al.,* 1970; Stefanac and Liubesic, 1971; Kim and Fulton, 1971, 1972; Hooper *et al.,* 1972; Jones *et al.,* 1973; Kitajima *et al.,* 1974; Saric and Wrischer, 1975; Stace-Smith and Hansen, 1976; Francki and Hatta, 1977; Van Kammen and Mellema, 1977; Hibino *et al.,* 1977; Russo *et al.,* 1978; Savino *et al.,* 1979; Di Franco *et al.,* 1980, 1983; Martelli *et al.,* 1980; Tomenius and Oxelfelt, 1982; Russo *et al.,* 1982).

All these studies indicate that three cytopathological features, i.e., vesiculate-vacuolate cytoplasmic inclusions, tubules containing rows of virus particles, and cell wall outgrowths, are evoked by most, if not all, members of both taxonomic groups, thus qualifying as possible intracellular markers of diagnostic value (Fig. 2). Occasionally, small crystalline aggregates of virus particles are present in the cytoplasm or vacuoles. Virions egress in the vacuole through a mechanism unlike that found with tombusviruses,

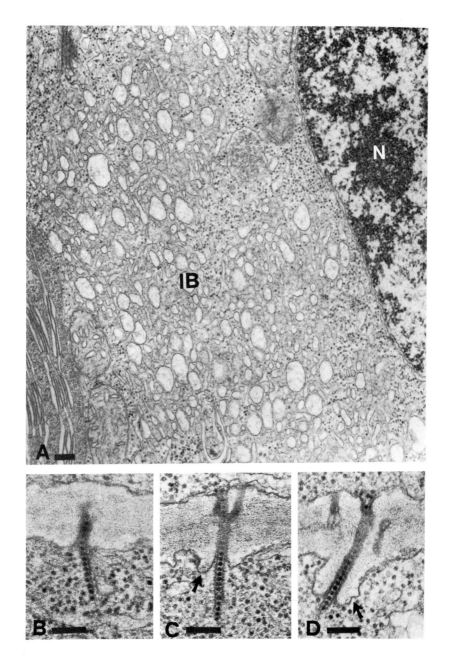

FIG. 2. (A) Vesiculate–vacuolate cytoplasmic inclusion (IB) next to a nucleus (N), typical of como- and nepovirus infections. (B–D). Virus-containing tubular structures associated with plasmodesmata and progressive development of cell wall outgrowths (arrows). Magnification bars represent 200 nm.

but which is apparently able to drive virus particles across the tonoplast without causing it irreversible damage.

Vesiculate-vacuolate inclusions are made up mostly of ribosomes, strands of endoplasmic reticulum, and membranous vesicles with fine fibrils (Fig. 2A). The origin of the fibril-containing vesicles is unclear. They may derive from dictyosomes, endoplasmic reticulum, or the outer lamella of the nuclear envelope. The fibrillar material in broadbean true mosaic virus (BBTMV, a comovirus) infections has been shown to consist of dsRNA (Hatta and Francki, 1978). Moreover, a direct involvement of the inclusion bodies in the replication of cowpea mosaic virus (CpMV, a comovirus) has been demonstrated through autoradiography and cell fractionation studies (reviewed by van Kammen and Mellema, 1977) and by the fact that they contain RF forms of RNA-1 and RNA-2 and the relative RNA polymerase activities (van Kammen, 1982). Hence these inclusions qualify as a most likely site for the replication of viral nucleic acid.

Virus-containing tubules are another salient characteristic of como- and nepoviruses. They are usually single walled, but in the cause of strawberry latent ringspot virus (SLRV, a nepovirus), they are double walled, the external wall being continuous with endoplasmic reticulum strands (Roberts and Harrison, 1970). This feature may be considered to be virus specific for it was found also with SLRV isolates from olive (Pacini and Cresti, 1977) and grapevine (Barabini and Bertaccini, 1982).

The presence of tubule-virus complexes is frequently associated with fingerlike protrusions of the cell wall arising at the level of plasmodesmata, to which tubules are generally connected (Fig. 2,B–D). It has been postulated that formation of wall outgrowths is actually stimulated by the tubules (Martelli, 1980). It is worth mentioning that these outgrowths are not exclusive of como- and nepovirus infections, since they are also present in cells infected by some caulimoviruses, plant reoviruses, and, as a transient feature, pea enation mosaic virus (De Zoeten and Gaard, 1981) and a score of other ungrouped viruses (reviewed by Martelli, 1980).

Broadbean wilt virus (BBWV) is considered a putative comovirus (Matthews, 1982). It has morphology and physicochemical characteristics very similar to those of the true members of the group (Doel, 1975); however, it is transmitted by aphids, not through the seed.

Striking differences exist also in the ultrastructure of infected cells. BBWV produces very complex inclusion bodies that may display, according to the strain, an amorphous and a crystalline area (reviewed by Kishtah *et al.,* 1978). The amorphous inclusions are large and readily visible with the light microscope. They consist usually of two parts, one membranous and one granular. The latter is not present in some strains. The membranous area is made up of convoluted membranes, small vacuoles, and vesicles with fibrils. Osmiophilic droplets are often present, along with accumulations of empty viral capsids. The granular area has a viroplasm-like aspect and consists of accumulations of electron-opaque bodies (~ 50 nm in diameter) with a denser central core.

Cell wall outgrowths have not been reported, and virus-containing tubules are very rare (Hull and Plaskitt, 1973–1974).

Crystalline arrays of virus particles vary according to the strain (Kishtah *et al.*, 1978). They may be in the form of scrolls, cylindrical tubules, quadrangular tubules, or polyhedral or platelike crystals. Thus, ultrastructural characters may assist in distinguishing among different BBWV strains.

3. *Tymoviruses*

Tymoviruses are readily transmitted by sap inoculation and invade all tissues of infected plants. Intracellular virions are abundant and easy to detect in invaded tissue. The group comprises 17 definitive members and one possible member (Matthews, 1982), most of which (15) have been investigated ultrastructurally (Gerola *et al.*, 1966; Chalcroft and Matthews, 1966, Ushiyama and Matthews, 1970; Hatta *et al.*, 1973; Hatta and Matthews, 1974, 1976; Lesemann, 1977; Dale *et al.*, 1975; Shukla *et al.*, 1980). As tymoviruses constitute a rather homogeneous group (Koenig and Lesemann, 1981), great similarities in the cytopathological modifications are to be expected. This has proved to be so for all members so far studied ultrastructurally.

Lesemann (1977) pointed out the existence of group-specific and virus-specific cytological alterations that may assist in the identification of tymoviruses.

Modifications that can be regarded as diagnostic markers of tymovirus infections are essentially two: clumping of altered chloroplasts, and intranuclear accumulations of empty viral capsids.

Chloroplast clumps forming an inclusion, first detected with the light microscope in Chinese cabbage leaves infected with turnip yellow mosaic virus (TYMV) by Rubio-Huertos (1950), are now known to be induced by all tymoviruses. The sequential development of chloroplast abnormalities and clumping has been investigated at the fine-structure level by Hatta and Matthews (1974) for TYMV. In this process, which is likely to occur with all members of the tymovirus group as the available evidence (Lesemann, 1977) suggests, the plastids become progressively swollen (thus acquiring a rounded shape) and vesiculated until they clump together (Fig. 3). A very early and highly significant modification is the appearance of periplastial flask-shaped vesicles bounded by a double membrane and containing finely stranded material. The vesicles originate from localized invaginations of the chloroplast-limiting membrane (Gerola *et al.*, 1966) and have necks point-

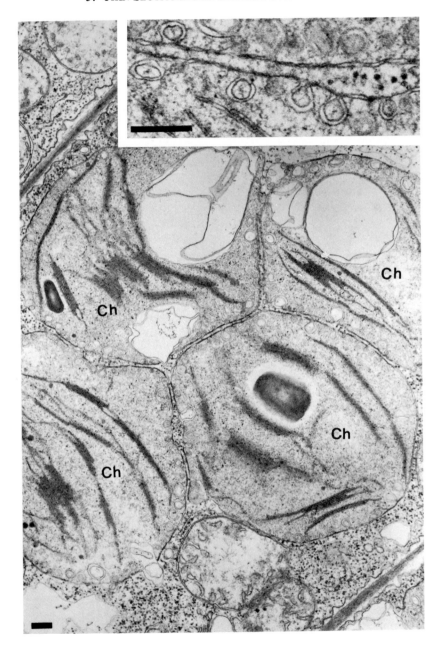

Fig. 3. Clump of swollen and peripherally vesiculated chloroplasts (Ch) typically induced by tymoviruses. Inset shows details of the double-membraned vesicles containing double-stranded viral RNA. Magnification bars represent 200 nm.

ing to and opening at the chloroplast surface. A detailed description of these structures has been given by Matthews (1977). The fibrillar material contained in the vesicles is very likely the replicative form of viral RNA, as biochemical and autoradiographic experiments indicate (Laflèche and Bové, 1968, 1971, Laflèche et al., 1972). Virus assembly takes place at the neck of the vesicles, where coat protein, formed in the cytoplasm, and nucleic acid come in contact (Matthews, 1977). It was suggested that clumping of protoplasts is produced by the gluing action of viral protein that accumulates in the cytoplasm around these organelles (Hatta and Matthews, 1976).

Excess viral protein migrates to the nucleus, where it is assembled into empty shells sometimes forming conspicuous crystalline aggregates (Hatta and Matthews, 1976; Lesemann, 1977). The accumulation in the nuclei of empty particles of okra mosaic virus (OKMV) has been studied biochemically by Marshall and Matthews (1981), who suggested that viral protein enters the nucleic as subunits (monomers or hexamers and pentamers) and is assembled into empty capsids there.

The virus-specific cytological alterations of tymoviruses that Lesemann (1977) listed were primarily connected with modifications of chloroplasts (e.g., lack of vacuoles, sunken regions at the organelle's periphery, presence of empty capsid, tubules, or crystals in the interior), mitochondria (e.g., presence of vacuole-like vesicles, of empty capsids, of doubled-membraned vasicles similar to those of chloroplasts), or cytoplasm (e.g., accumulation of amorphous material, complex virus crystals). Only wild cucumber mosaic virus and two strains of Andean potato latent virus (APLV) showed no additional ultrastructural changes apart from the group-specific ones.

4. Luteoviruses

Viruses belonging in this group are transmissible only by aphids in a persistent circulative manner. All are confined to phloem tissues of infected plants; however, beet western yellows virus (BMYV) can spread from phloem tissues to bundle sheath and mesophyll cells of infected sugarbeet (Esau and Hoefert, 1972a), and tobacco necrotic dwarf virus (TNDV) can infect and replicate in tobacco mesophyll protoplasts (Kubo and Takanami, 1979).

Of the 33 luteoviruses listed by Matthews (1982), only 3 have been studied exhaustively in relation to host cells: BWYV (Esau and Hoefert, 1972a,b), barley yellow dwarf (BYDV) (Gill and Chong, 1975, 1976, 1979a,b), and potato leafroll (PLRV) (Shepardson et al., 1980).

A few other members have been examined though to a lesser extent, in thin-sectioned tissues: soybean dwarf (SDV) (Tamada, 1975a), a virus from alfalfa, possibly pea leafroll virus (Thottapilly *et al.*, 1977), carrot red leaf (CRLV) (Ohki *et al.*, 1979), and subterranean clover red leaf (SCRLV) (Jayasena *et al.*, 1981).

There are no ultrastructural features that can be definitely identified as distinctive for the whole group, although there are cytological modifications that occur rather consistently. These refer to the presence of virus particles and membranous vesicles in the nuclear area and of virions in plasmodesmata.

Virions accumulating around the nucleolus, sometimes in crystalline arrays, as an early event of infection, have been reported for BWYV (Esau and Hoefert, 1972a,b) and some strains of BYDV (Gill and Chong, 1979a,b). The abundance of virus particles and their close association with the nucleolus prompted Esau and Hoefert (1972a) to suggest that nucleoli could be involved in virus multiplication, a contention that was not confirmed by later findings relative to PLRV (Shepardson *et al.*, 1980) and BYDV (Gill and Chong, 1979a).

With PLRV no evidence of intranuclear formation of virus was obtained, although particles were seen in the karyoplasm (Shepardson *et al.*, 1980), whereas no virions could be detected in nuclei of cells infected with some strains of BYDV (Gill and Chong, 1979a).

The vesicular structures first described by Esau and Hoefert (1972a,b) for BWYV have also been found in BYDV (Gill and Chong, 1976, 1979a) and PLRV (Shepardson *et al.*, 1980) infections. These vesicles are bound by a single membrane or, more often, a double one and ordinarily contain a network of fine fibrils resembling nucleic acid (Fig. 4, inset). According to Esau and co-workers, who have studied their development with BWYV and PLRV, these vesicles originate in the parietal cytoplasm in the early stages of infection, move toward the nucleus, where they fuse with its envelope (Fig. 4), and enter the organelle. Their nature and function have not been unequivocally established, but the hypothesis has been repeatedly put foward that these structures may act as carriers of viral nucleic acid, or precursory material into the nucleus for some stages of viral multiplication (Esau and Hoefert, 1972b; Shepardson *et al.*, 1980).

If it is true that fibril-containing vesicles are a common feature of luteovirus-infected cells, striking similar structures are also encountered with pea enation mosaic virus (PEMV) (De Zoeten *et al.*, 1972; Vovlas *et al.*, 1973; Burgess *et al.*, 1974a), cowpea chlorotic mottle (CCMV) (Burgess *et al.*, 1974b), pelargonium zonate spot (PZSV) (Castellano and Martelli, 1981), and velvet tobacco mottle (VToV) (Randles *et al.*, 1981) viruses, none of which belongs to the luteovirus group.

Although the interpretation on the origin of the vesicles is not univocal

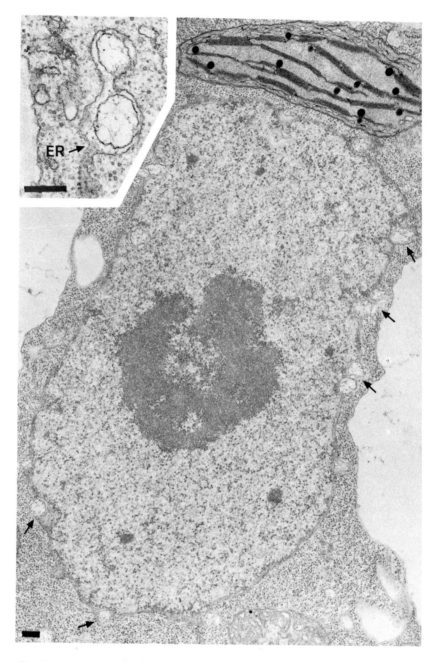

FIG. 4. A nucleus of a luteovirus-infected cell with membranous vesicles (arrows) fused with the nuclear envelope. The inset shows details of the single-membraned vesicles with fibrillar material within a dilated endoplasmic reticulum (ER) cisterna. Magnification bars represent 200 nm. By courtesy of Dr. K. Esau.

because, contrary to the views of Esau and co-workers (Esau and Hoefert, 1972b; Shepardson *et al.,* 1980), in PEMV and CCMV infections they are considered to be the result of a budding process involving both lamellae of the nuclear envelope (De Zoeten *et al.,* 1972; Burgess *et al.,* 1974b), the point remains that the outward appearance of vesiculated nuclei of luteo- and other viruses is much the same. This lowers the level of confidence for a reliable identification of luteoviruses at the group level on this basis.

Apart from these difficulties, ultrastructural differences proved useful for separating BYDV strains into two distinct subgroups (Gill and Chong, 1979a).

5. *Bromoviruses*

The bromovirus group comprises three definitive members and one possible member (Matthews, 1982). These viruses are readily transmitted by inoculation of sap and, in nature, by beetles, and they reach high concentrations in the host plants, regardless of the invaded tissue.

All definitive members have been studied ultrastructurally in infected host tissue, two of them also in protoplasts (De Zoeten and Schlegel, 1967a,b; Hills and Plaskitt, 1968; Paliwal, 1970; Burgess *et al.,* 1974b; Lastra and Schlegel, 1975; Kim, 1977).

All bromoviruses induce formation of inclusion bodies (readily visible also in the light microscope), whose structure and composition have been studied in detail for broadbean mottle virus (BBMV) (De Zoeten and Schlegel, 1967a,b; Lastra and Schlegel, 1975) and cowpea chlorotic mottle virus (CCMV) (Kim, 1977).

Inclusion bodies of BBMV are made up by accumulations of amorphous material, vesicles, and virus particles. They are the site of virus RNA replication and contain viral antigen. This is present also in the nuclei, in high amount in the initial stages of infection, later in the same quantity as in the cytoplasm, thus indicating that virus protein synthesis may take place in nuclei and assemblage of virions in the cytoplasm. However, virus particles can also be found in the nuclei (Hills and Plaskitt, 1968). In the cytoplasm virions are sometimes in crystalline arrays (De Zoeten and Schlegel, 1967a).

The inclusion bodies induced by CCMV consist of a fine granular zone, proliferating endoplasmic reticulum, fibril-containing vesicles, and thin flexuous filaments recalling the filaments form of viral protein found sometimes *in vitro* (Bancroft *et al.,* 1969). CCMV virions are randomly scattered in the cytoplasm without gathering into crystals. The inconsistency with which bromoviruses give rise to crystalline aggregates of particles *in vivo* may be due to faulty fixation schedules (Langenberg, 1979). If so, this adds to the contention that cytoplasmic crystals are too labile to be regarded safely as a diagnostic feature for the group, as was indicated by Edwardson and Christie (1978).

6. *Cucumoviruses*

The type member of this group, cucumber mosaic virus (CMV), is one of the best-studied plant viruses (Kaper and Waterworth, 1981). The group comprises two other definitive members and one possible member (Matthews, 1982). All members are transmitted by aphids, in a nonpersistent manner, and by sap inoculation to a wide range of host species. All tissue types are infected. Besides CMV, only tomato aspermy virus (TAV) has been studied ultrastructurally in infected cells.

Although CMV particles may be distinguished from host ribosomes by their "doughnut" appearance, efforts have been made to devise methods for more reliable identification of CMV virions *in situ*. Thus, Hatta and Francki (1979), using a modification of their enzyme cytochemical method for identifying intracellular single- and double-stranded RNAs (Hatta and Francki, 1978), were able to discriminate CMV particles from ribosomes. The results of these investigations, combined with those of previous studies made either with conventional techniques (Gerola *et al.*, 1965b; Honda and Matsui, 1968) or by first incubating tissue samples in phosphate buffer for destroying ribosomes (Honda and Matsui, 1974), show that the following features characterize CMV infections. (1) Virus particles, which are usually scattered at random in the cytoplasm, tend to aggregate into crystals in the vacuoles (Fig. 5). Ribosomes may inhibit crystallization of CMV particles in the cytoplasm. Crystals, in fact, are not formed *in vitro* when purified virus and ribosomes are mixed together and centrifuged or precipitated with polyethylene glycol, a condition that gives rise to viral crystals when highly purified CMV preparations are used (Ehara *et al.*, 1976). (2) Virus particles occur consistently in the nuclei of many cells (Honda and Matsui, 1974). (3) The site of viral RNA replication may reside in membrane-bound vesicles 50–90 nm in diameter associated with the tonoplast (Fig. 6A) and containing the presumed RF or RI forms of viral nucleic acid (Hatta and Francki, 1981a).

A similar localization of dsRNA was also reported for TAV (Hatta and Francki, 1981a), whose infections in addition show as an outstanding feature the presence of regular arrays of virus particles between dictyosomal cisternae, where assembly of virions may take place (Lawson and Hearon, 1970). It should be noted, however, that this is not a peculiarity unique to TAV-infected cells, for a comparable association of virions and dictyosomes was observed in artichoke yellow ringspot virus (a nepovirus) infections, together with additional visual evidence suggesting that these structures could represent virus assembly sites (Russo *et al.*, 1978).

Fig. 5. An intravacuolar virus crystal made up of cucumber mosaic virus particles. Magnification bar represents 200 nm.

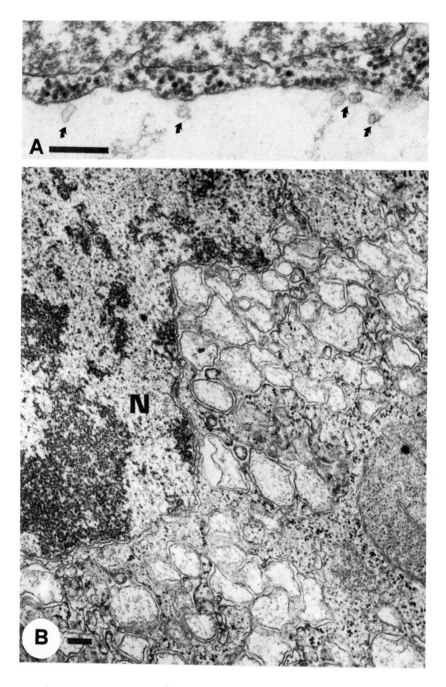

Fig. 6. (A) Tonoplast-associated vesicles (arrows) found in cucumovirus infections. (B) Vesiculated proliferations of the nuclear envelope typically induced by alfalfa mosaic virus. N, nucleus. Magnification bars represent 200 nm.

7. Sobemoviruses

This recently established group comprises two definitive members and four possible members (Matthews, 1982). All are transmitted in nature by beetles and experimentally by sap inoculation to test plants, where they multiply in all types of tissues, reaching fairly high concentrations.

None of the sobemoviruses studied in thin sections appears to elicit any cytopathological feature that may be considered group specific (Milne, 1967; Weintraub and Ragetli, 1970; Bakker, 1975; Chamberlain and Catherall, 1976). Virus particles are found in cytoplasm and nuclei and occasionally they may aggregate into crystals.

Two strains of southern bean mosaic virus (SBMV), the type member of the group, are noteworthy for the differential intracellular behavior: the cowpea strain produces large crystals in the cytoplasm and nuclei, mainly in infected cells near vascular tissues, whereas the bean strain induces formation of crystals very rarely and only in phloem cells (Weintraub and Ragetli, 1970).

The flexuous tubules found in rice yellow mosaic virus (RYMV) infections (Bakker, 1975), and the "pseudonuclei" and network of densely stained material seen in sowbane mosaic virus (SoMV) infections (Milne, 1967) may perhaps be considered specific for the respective viruses, but the consistency of their occurrence and significance warrants further studies.

8. Dianthoviruses

The dianthovirus group has three members (Matthews, 1982); of these, only carnation ringspot virus (CRSV), the type member, has been investigated ultrastructurally (Weintraub et al., 1975). CRSV infections seem to induce unique cytopathological features, to which, however, no diagnostic value at the group level can be assigned at present.

CRSV particles are readily seen in the form of large irregular aggregates in the cytoplasm and nuclei of infected cells and, in the vacuoles, as crystals very reminiscent of those of CMV.

In addition to virus particles, nuclei contain tubules with a diameter slightly greater than that of CRSV particles. Most tubules contain no particles; some contain a single row of virions, but, in this case, tubules seem to be disintegrating. Other virus-containing tubules look intact, and virions appear to be extruded from them. Although these findings suggest tubules as possible sites of viral synthesis and/or assemblage, time sequence studies have failed to clarify the role of tubules in the virus replication cycle.

9. Tobacco Necrosis and Satellite Virus

This small group comprises the type member (tobacco necrosis virus; TNV) and cucumber necrosis virus (CNV) (Matthews, 1982). Both viruses are transmitted in nature by chytrid fungi (*Olpidium*). When sap inoculated, they rarely spread systemically in the plants, but remain localized in lesions that turn rapidly necrotic. Also roots inoculated with viruliferous fungal zoospores tend to respond with local necrotic lesions only.

Frequently TNV is associated with a satellite virus (SV), whose RNA has the coding capacity only for its own protein, but cannot replicate in absence of TNV. The diameter of SV (~ 17 nm) is about half that of TNV (~ 30 nm). No specific cytopathological alterations characterize the group, which has been studied only to a small extent. Kassanis et al. (1970) investigated the intracellular behavior of TNV and associated SV and found a striking difference in the outward aspect of the aggregates of the two viruses. Whereas SV virions formed large stable crystals in situ, TNV particles were never arranged in definite crystalline arrays, even in areas of high concentration. This may be due to an intrinsic instability of TNV crystals, since, after fixing and sectioning, crystals of this virus produced in vitro showed a clear tendency to fall apart.

Although Edwardson et al. (1966) and Gama et al. (1982) reported the occurrence of crystals in cells infected by TNV alone, it seems that these structures are too labile to serve as a possible diagnostic marker.

10. Pea Enation Mosaic Virus

Pea enation mosaic virus (PEMV), the sole member of this group, is transmitted in nature by aphids in a persistent, circulative manner but, unlike luteoviruses, which are similarly transmitted, can be successfully transferred to plants by inoculation of sap. Therefore, PEMV is the only circulative aphid-borne virus that is not phloem limited and is one of the few that are well characterized biochemically and biophysically (Hull, 1981).

Ultrastructurally, PEMV has been studied in infected pea (Shikata et al., 1966; Shikata and Maramorosch, 1966; De Zoeten et al., 1972) and broadbean plants (Vovlas et al., 1973) and in the insect vector (Shikata et al., 1966; De Zoeten et al., 1972; Harris and Bath, 1972; Harris et al., 1975).

As PEMV multiplies only in plants, it is only in plant tissues that specific cytopathological structures related to the replicative process are found. Such structures consist of accumulations of single-membraned vesicles in the perinuclear space of infected cells (Fig. 6B). These vesicles originate in the inner nuclear membrane and then are released into the cytoplasm, acquiring in the process an additional membrane derived from the outer lamella of the nuclear envelope (De Zoeten et al., 1972).

Morphologically, the aggregates of vesicles induced by PEMV resemble those formed by some luteoviruses, but with PEMV the aggregates are much larger, whereas in luteovirus infections vesicles are either single or in groups of very few.

Virus particles of PEMV are found in great quantity in infected pea (De

Zoeten *et al.*, 1972) and broadbean (Vovlas *et al.*, 1973) nuclei and less abundantly in the cytoplasm, where occasionally they may form paracrystalline inclusions in the endoplasmic reticulum (Rassel, 1972).

The nucleus is very likely the site of PEMV replication, as both virus-specific dsRNA and RNA polymerase activities were found in it (De Zoeten *et al.*, 1976; Powell *et al.*, 1977). Isolated nuclei from infected cells were also shown to support at least the initial stages of virus replication (Powell and De Zoeten, 1977).

11. *Alfalfa Mosaic Virus (AMV)*

This monotypic group comprises only the type member. However, as pointed out by van Regenmortel and Pinck (1981), the alfalfa mosaic virus group constitutes a "large conglomerate of strains with different biological properties but unified through a common morphology." In nature, AMV is spread by aphids in a nonpersistent manner and, artificially, is transmitted by sap to a rather large range of plants, most of which are systemically invaded.

The pleomorphism of virus particles greatly facilitates the identification of AMV in thin-sectioned material. It is often possible to recognize a given strain on the basis of the way in which the particles aggregate intracellularly (Hull *et al.*, 1970). AMV strains may or may not form intracellular aggregates, but, when they do, the aggregates are consistently of four types: (1) short rafts of particles arranged in a hexagonal array; (2) long bands of particles aligned side by side, sometimes in a stacked-layer configuration; (3) aggregates made up of a series of whorl-like structures (centers of aggregation) connected to one another by bands of virions to form a three-dimensional irregular network throughout the cell; (4) aggregates consisting of up to four parallel rows of particles packed either in an apparently rhomboid lattice or in a hexagonal one when viewed "end on."

The type of aggregation is constant for a given strain in the same host, but may vary with the host. This was found to be true at least with one strain, which forms aggregates in *Capsicum annuum* and *Vigna cylindrica,* but not in tobacco (De Zoeten and Gaard, 1969b). Some strains forming aggregates in the cytoplasm also form aggregates in the nuclei, and two additional strains elicit intranuclear complexes of long rod-shaped particles, presumably made up of viral protein (Hull *et al.*, 1970).

According to Hull and Plaskitt (1970), in mixed infections it is possible to distinguish cytologically the strains composing the mixture. However, Wilcoxson *et al.* (1974, 1975) found that the type of particle aggregation may vary in different alfalfa organs. Therefore they stressed the necessity of carrying out ultrastructural observations on standard plant organs if such studies are aimed at characterizing AMV strains.

B. Isometric Viruses with dsRNA

Plant Reoviruses

Plant reoviruses comprise two subgroups to which a generic status has been assigned, i.e., *Phytorevirus* (subgroup 1) and *Fijivirus* (subgroup 2), in the family Reoviridae (Matthews, 1982).

Phytoreoviruses (three members) and fijiviruses (nine members) differ from each other morphologically (icosahedral particles with nonspiked core vs spherical particles with spiked core), in genome composition (12 vs 10 segments of dsRNA), and biologically (transmission by leafhoppers vs planthoppers). None of these viruses is transmitted mechanically. In infected hosts, with the exception of rice dwarf virus, they induce tumors or swellings in which the virions are localized.

Three major cytopathic changes are induced by plant reoviruses: (1) accumulations of proteinaceous material (viroplasms) containing dark, spherical bodies about 50 nm in diameter, which consist of immature virus particles (inner cores), (2) crystals made up of mature virus particles, accumulating usually near the viroplasm, and (3) tubules containing a row of virions (Fig. 7).

These features are consistent enough to be distinctive for the group. However, the two subgroups are distinguishable cytologically because of the size and shape of viroplasms, which are small and tendentially spherical in phytoreoviruses and large and elongated in fijivirus (Shikata, 1977, 1981).

Viroplasms induced by maize rough dwarf virus (a fijivirus) were shown to be the site of viral RNA synthesis (Bassi and Favali, 1972; Favali *et al.,* 1974). Likewise, those of rice dwarf virus proved to be the sites of accumulation of viral protein and nucleic acid (Nasu and Mitsubashi, 1968). Thus viroplasm matrices may be regarded as cytoplasmic areas where plant reoviruses replicate (Matthews, 1982).

C. Isometric Viruses with DNA

Two groups of plant viruses contain DNA: (1) caulimoviruses, with particles ~ 50 nm across containing dsDNA, and (2) geminiviruses, with paired spherical particles 18–20 nm in diameter containing ssDNA. Members of both groups induce formation of cytopathological structures that can be used reliably for their identification at the group level.

1. Caulimoviruses

This group includes 10 definitive and possible members, which are transmitted by aphids and some by inoculation of sap (Matthews, 1982).

In all caulimoviruses that have been studied ultrastructurally (see reviews

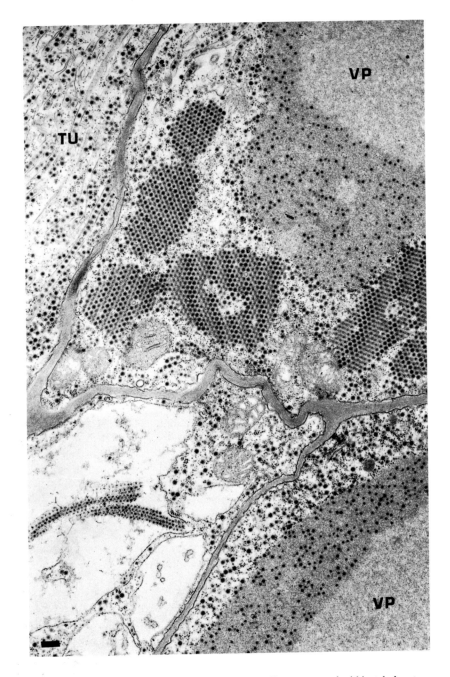

FIG. 7. Viroplasm (VP) and mature virions in crystalline arrays and within tubular structures (TU) in cells infected by a plant reovirus. Magnification bar represents 200 nm. By courtesy of Dr. A. Appiano.

by Martelli and Russo, 1977a; Shepherd, 1977; Edwardson and Christie, 1978; Shepherd and Lawson, 1981), virus particles are readily seen in infected cells. They either occur in parenchyma tissues free in the cytoplasm or, much more commonly and typically, are associated with inclusion bodies. These appear as rounded or elongated structures composed of virus particles embedded in a densely staining granular matrix broken by irregular electron-lucent areas that may contain virions (Fig. 8).

Shalla *et al.* (1980) suggested that in cauliflower mosaic virus (CaMV), the development and final size of the inclusions are controlled by the viral genome and represent a characteristic of individual virus isolates that can be grouped on this basis. These authors, however, could not exclude an influence of the host in determining the inclusion size, thus confirming an idea previously put forward (Martelli and Russo, 1977a).

Although there is a general agreement on the gross structure and constitution of the inclusions, which contain protein, RNA, and DNA (see reviews by Martelli and Russo, 1977a; Shepherd and Lawson, 1981), the interpretation of their function is still controversial. According to some authors (Rubio-Huertos *et al.,* 1968, 1972), they merely represent structures in which virus particles accumulate, a possibility that recent interpretations have reproposed (Shalla *et al.,* 1980).

Other views assumed these inclusions to be viroplasms where virus synthesis and/or assembly takes place (Kitajima *et al.,* 1969; Martelli and Castellano, 1971; Conti *et al.,* 1972; Brunt and Kitajima, 1973; Lawson and Hearon, 1974). Recent investigations have demonstrated that viral DNA replicates in the nuclei (Guilfoyle, 1980; Ansa *et al.,* 1982). Therefore, the inclusions may be regarded as the cytoplastic sites for virus assembly.

There are contrasting reports on the number and molecular weight of polypeptides making up the matrix protein and on whether or not virus coat protein may be derived from the protein pool of the inclusions (Odell and Howell, 1980; Shepherd *et al.,* 1980; Shockey *et al.,* 1980). According to Shepherd *et al.* (1980), CaMV inclusions contain substantial amounts of a polypeptide with a molecular weight of 55,000, a possible nonstructural protein, serologically unrelated to the virus coat. However, the most abundant product obtained in *in vitro* transcription experiments of CaMV DNA was a 62,000 molecular weight polypeptide very similar to the major protein component of inclusion bodies (Covey and Hull, 1981).

In conclusion, inclusion bodies of caulimoviruses constitute a characterizing intracellular feature of the whole group that can be used for diagnosis even with the light microscope. In this connection, it should be noted that Robb (1963) successfully utilized such inclusions as a criterion for selecting healthy dahlia plants.

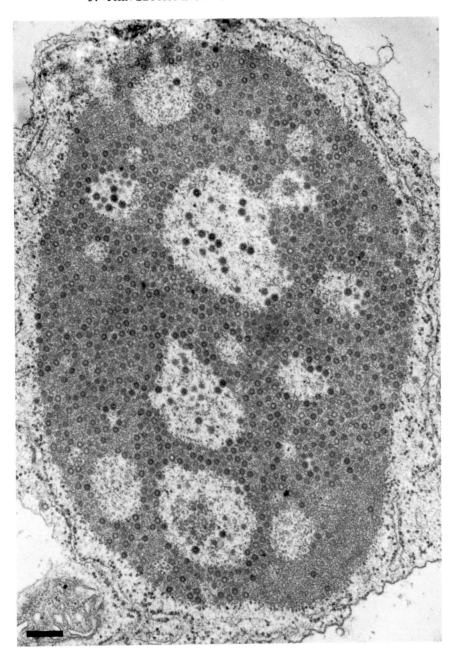

FIG. 8. Cytoplasmic inclusion typical of caulimovirus infections. Magnification bar represents 200 nm.

2. Geminiviruses

Most members of the geminivirus group, which includes five definitive and nine possible members (Matthews, 1982), are limited to phloem tissues. Some are transmitted by leafhoppers, others by whiteflies, and only a few by mechanical inoculation.

Six of these viruses have been investigated ultrastructurally: beet curly top (Esau and Hoefert, 1973; Esau, 1977; Esau and Magyarosy, 1979), bean golden mosaic (Kim et al., 1978), Chloris striate mosaic (Francki et al., 1979), tobacco leaf curl (Osaki and Inouye, 1978; Osaki et al., 1979), Euphorbia mosaic (Kim and Flores, 1979), tomato yellow leaf curl (Russo et al., 1980).

These studies have indicated that there are at least two ultrastructural features that may be regarded as characteristic for geminivirus infections and may have diagnostic value at the group level. Both consist of nuclear modifications: (1) intranuclear accumulations of virus particles accompanied by depletion of chromatin and granular appearance of the karyoplasm; (2) intranuclear ring-shaped inclusions known as "fibrillar rings" (Kim et al., 1978).

Fibrillar rings originate from segregation of granular and fibrillar regions of hypertrophic nucleoli and appear to be composed of electron-dense, very compact fibrils.

One or more such structures can be observed in any one nucleus. Their size is variable. Small rings are usually very compact (Fig. 9), whereas the bigger ones may show electron-clear holes in their matrix.

Fibrillar rings are not host specific. The same virus (e.g., beet curly top) can elicit their formation in different hosts belonging to diverse botanical families (Esau and Hoefert, 1973; Esau, 1977; Esau and Magyarosy, 1979). It is, therefore, plausible to consider these abnormalities as being coded by the viral genome, thus lending support to the suggestion (Kim et al., 1978) that they may be regarded as diagnostic features of geminiviruses.

D. ANISOMETRIC VIRUSES

There are seven taxonomic groups of plant viruses with elongated particles, all containing ssRNA. Particles of these groupings (hordei-, tobra-, and tobamoviruses) are straight and rigid, whereas those of the remaining clusters (potex-, carla-, poty-, and closteroviruses) are filamentous and flexuous. Members of all groups incite cytological modifications of diagnostic significance at the group and/or at the single virus level.

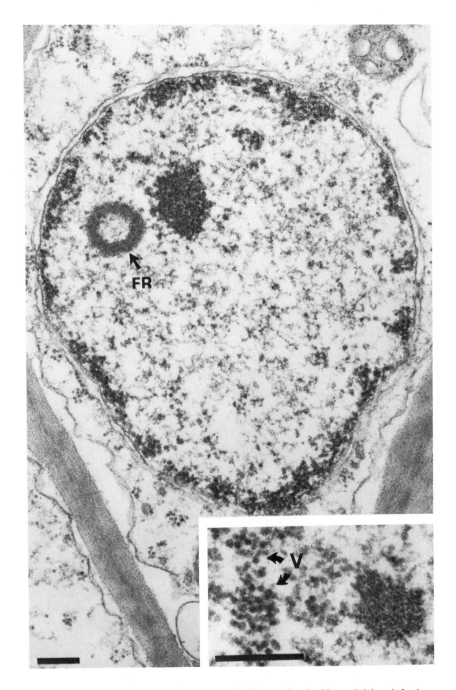

FIG. 9. Nucleus with a fibrillar ring (FR) typically associated with geminivirus infections. Inset shows virus particles (V) in the nucleoplasm. Magnification bars represent 200 nm.

1. *Hordeiviruses*

Of the three viruses included in the hordeivirus group (Matthews, 1982), only the type member, barley stripe mosaic (BSMV), has been studied ultrastructurally in different tissues of the natural host including pollen, ovules, and embryo sac (Shalla, 1966; Gardner, 1967; Carroll, 1968, 1970; Carroll and Mayhew, 1976a,b; McMullen *et al.* 1977, 1978).

A variety of intracellular alterations induced by BSMV have been reported, the most relevant of which are listed.

1. Chloroplast abnormalities consist of severe modifications of the shape, internal structure, and lamellar system, as well as peripheral vesiculation. Plastidial vesicles are small, rounded, or flask-shaped invaginations of the outer lamella of the boundary membrane, containing fine fibrils (Gardner, 1967; Carroll, 1970; McMullen *et al.*, 1978) and recalling those typically elicited by tymoviruses (Matthews, 1977). However, whether these vesicles play a comparable role in viral RNA synthesis is not known.

2. Virion–plastid associations are derived from attachment "head on" of virus particles to the chloroplast envelope, which may mediate plastidial aggregation (Gardner, 1967; Carroll, 1970).

3. Cytoplasmic or intranuclear accumulations of virus particles appear either in disorderly aggregates or in tiers of virions aligned sidewise or in paracrystalline arrays (Gardner, 1967; Carroll, 1970; Carroll and Mayhew, 1976a).

4. Cell wall alterations consist of irregular thickenings of the walls at the level of plasmodesmata filled with very densely staining material and accompanied by extracytoplasmic sacs containing small spherical granules of unknown nature (McMullen *et al.*, 1977). It is worth noting that these cell wall abnormalities are similar to those evoked by beet necrotic yellow vein virus (BNYVV), a tobamovirus (Russo *et al.*, 1981).

2. *Tobraviruses*

Viruses of the tobravirus group (three members) have tubular straight particles of two predominant lengths (Matthews, 1982). In nature they are transmitted by nematodes and, experimentally, by sap inoculation.

All types of tissues can be invaded by tobraviruses, which, depending on the virus and/or virus strain, induce three major kinds of cellular modifications.

1. Mitochondrial aggregates in which individual contiguous organelles are aligned by virions attached with both ends to their surface. Long particles only bridge any two mitochondria, whereas short particles are ran-

domly scattered throughout the cytoplasm. These structures were only observed in cells infected with two Brazilian strains of tobacco rattle virus (TRV) (Harrison and Roberts, 1968; Kitajima and Costa, 1966).

2. Bulky cytoplasmic inclusion bodies made up of mitochrondria (some of which may show peripheral vesiculation), ribosomes, and electron-dense, possibly proteinaceous, material. These are induced both by defective CAM isolates (i.e., naked RNA from long particles) and some ordinary (i.e., nucleoproteins containing both genomic RNAs) forms of different TRV strains such as PRN (Harrison *et al.,* 1970) and peony (Chang *et al.,* 1976).

3. Paracrystalline aggregates of long particles consisting of single or stacked tiers of virions in which the individual elements are packed side by side. Except for the CAM strain of TRV, the above virus aggregates, the size of which may vary with the virus concentration in the cell, have been detected in all tobraviruses infections (De Zoeten, 1966; Chang *et al.,* 1976; M. Russo and A. Di Franco, unpublished information).

3. *Tobamoviruses*

The current formulation of the tobamovirus group is unsatisfactory. The group comprises 10 definitive members and 6 possible members (Matthews, 1982), but, whereas the definitive tobamoviruses (subgroup A) have straight tubular particles 300 nm long, have no known natural vector, are very infectious, and are often transmitted by direct contact, most possible members (subgoup B) have particles of two or three predominant lengths, are soil-borne, and are transmitted by fungi (*Spongospora, Polymyxa*).

Differences are also found at the ultrastructural level.

Crystalline inclusions are typical of viruses of subgroup A. Under the light microscope these appear as hyaline plates with a hexagonal or rounded shape and can be used as a basis for virus identification (see review by Martelli and Russo, 1977a; van Regenmortel, 1981).

Under the electron microscope crystals of ordinary tobacco mosaic (TMV) or tomato mosaic (ToMV) virus isolates appear to be made up of stacked layers of virus particles aligned in a parallel array. Virions in successive rows may be aligned end to end or may be slightly tilted, thus giving rise to a characteristic herringbone pattern. Care must be taken when fixing tissue for thin sectioning, as crystals are usually destroyed by conventional glutaraldehyde–osmic acid fixation. Instead, the native structure of TMV crystals is well preserved by fixing in aqueous or acetone solutions of potassium permanganate (Warmke and Edwardson, 1966a,b), diluted osmic acid solutions (Warmke and Christie, 1967), or chromic acid–formaldehyde (Langenberg and Schroeder, 1972, 1973). When ordinary fixing procedures are used, however, TMV particles are still readily recognizable, even in the absence of a crystalline arrays, as more or less massive aggregates, often bulging into the vacuole, surrounded by the tonoplast (Esau, 1968).

No crystalline inclusions have been detected in cells infected by viruses of subgroup B. Only in the case BNYVV, paracrystalline aggregates of virions similar to the "angle layer" aggregates of subgroup A have been recorded (Tamada, 1975b; Putz and Vuittenez, 1980). "Angled layer" aggregates result from the displacement at an angle (30° or 60°) of one layer of particles over the next and have been found associated with infections by TMV (Warmke, 1967; Herold and Munz, 1967; Shalla, 1968), ToMV (Chen, 1974), and odontoglossum ringspot virus (Corbett, 1974) strains.

Amorphous cytoplasmic inclusion (X bodies) are induced by most ordinary strains of TMV but not by ToMV or other viruses of subgroup A. These structures, which may play a role in virus synthesis, are aggregates of host components such as ribosomes, endoplasmic reticulum and small vacuoles or vesicles, viral protein in granular or tubular form (X tubules), and small pockets of virus particles (Esau, 1968).

Cytoplasmic inclusions are also a common ultrastructural feature of viruses of subgroup B. Their structure and outward appearance, however, differ from that of TMV-induced X bodies. Inclusions of a BNYVV isolate from southern Italy begin as localized proliferations of rough endoplasmic reticulum and enhanced vesiculating activity of dictyosomes. These membranous elements accumulate in the cytoplasm, forming masses that are usually surrounded by mitochondria and become increasingly larger and denser owing to the appearance of virus particles. In fully developed inclusions, virions are plentiful and scattered throughout (Russo et al., 1981).

From the literature it appears that not all BNYVV isolates induce structures comparable to those elicited by the isolate from southern Italy (Tamada, 1975b; Putz and Vuittenez, 1980), which represents a possible way of separating strains.

The above is indeed the case with soil-borne wheat mosaic virus (SBWMV), another member of subgroup B. Inclusion bodies of nine of its isolates could be divided into three groups depending on the composition and relative abundance of their constituents, i.e., strands of endoplasmic reticulum, tubules, vesicles, membranes, and virions (Hibino et al., 1974a,b).

Cucumber green mottle virus (CGMV), a member of subgroup A, is endowed with a remarkable ultrastructural peculiarity that makes it distinguishable from all definitive and possible tobamoviruses. It induces an enlargement and membranous proliferation of mitochondria, which become heavily vesiculated (Hatta and Ushiyama, 1973; Sugimura and Ushiyama, 1975), thus recalling multivesicular bodies of some possible tombusviruses (see preceding section).

Finally, it is worth noting that in several tobramovirus infections tonoplast-associated vesicles containing fibrillar material are found (T. Hatta, unpublished information), which resemble those induced by cucumoviruses.

4. *Potexviruses*

Viruses belonging to this group have modal lengths ranging between 470 and 580 nm and show, when negatively stained with uranyl salts, a clear cross-banding and, sometimes, parallel longitudinal striations (Lesemann and Koenig, 1977). Matthews (1982) lists 18 definitive and 19 possible members. All are mechanically transmissible, invade all types of tissues, and have no known vectors.

The main cytopathological characteristic potexvirus infections is aggregates of virus particles, which may be irregular or fibrous or banded (Iizuka and Iida, 1965; Kitajima and Costa, 1966; Purcifull *et al.,* 1966; Zettler *et al.,* 1968; Turner, 1971; Lawson *et al.,* 1973; Stefanac and Lubešić, 1974; Doraiswamy and Lesemann, 1974; Pennazio and Appiano, 1975; Hanchey *et al.,* 1975; Kitajima *et al.,* 1977).

In fibrous inclusions, virus particles are aligned side by side and aggregated end to end so that the aggregate is often spindle shaped and does not show cross-striations or "banding." Banded inclusions are composed almost entirely of virus particles aligned in horizontal tiers one particle-length wide. Successive layers are separated by thin cytoplasmic strands (Fig. 10) which give the banded appearance to the inclusion.

Two potexviruses, potato virus X (PVX) and clover yellow mosaic virus (ClYMV), induce formation of virus-specific inclusions in addition to the aforementioned aggregates of virus particles. The inclusions characterizing PVX infections ("laminate inclusions") are made up of thin sheets of proteinaceous material sometimes studded with beads about 14 nm in diameter (Shalla and Shepard, 1972). Beads are not ribosomes and, like the proteinaceous sheets, are antigenically different from PVX protein. On the other hand, amorphous inclusions found in the cytoplasm and vacuoles of cells infected with ClYMV are composed of protein that reacts with virus-specific antibodies, and they are thought to represent aggregates of excess viral protein (Schlegel and De Lisle, 1971).

As reviewed by Purcifull and Edwardson (1981), three additional members of the group (dioscorea latent, bamboo mosaic, and papaya mosaic viruses) may be distinguished from each other and the rest because of specific cytoplasmic or nuclear cytopathic structures.

5. *Carlaviruses*

The carlavirus group includes 23 definitive and 12 possible members, all of which are transmissible by sap inoculation and many by aphids in a nonpersistent manner (Matthews, 1982).

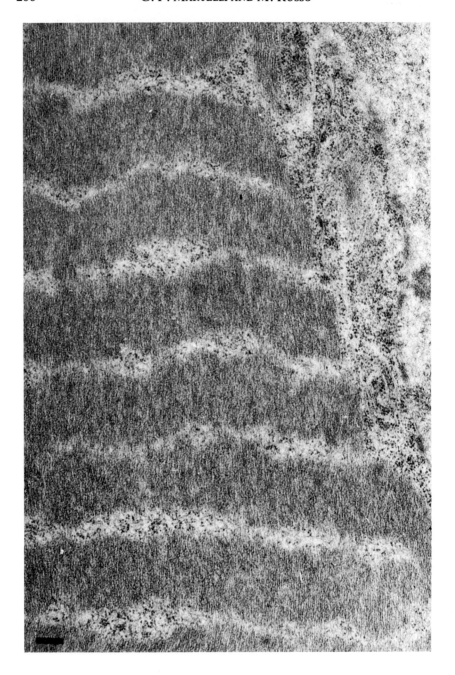

FIG. 10. Banded inclusion made up of potexvirus particles in stacked layers. Magnification bar represents 200 nm. By courtesy of Dr. A. Appiano.

Carlavirus particles have modal lengths ranging between 600 and 700 nm. In negatively stained preparations virions appear as slightly flexuous rods curved to one side, and, when uranyl acetate or uranyl formate is used as a stain, the rods show longitudinal ribbing. Size, shape, and fine structure, then, are the characteristics that distinguish carlavirus particles from those of other filamentous viruses in free preparations (Wetter and Milne, 1981). However, no specific cellular modifications have been found that could be considered distinctive and specific for the group (Lyons and Allen, 1969; Tu and Hiruki, 1970, 1971; de Bokx and Waterreus, 1971; Bos and Rubio-Huertos, 1971, 1972; Hiruki and Tu, 1972; Hiruki and Shukla, 1973; Rubio-Huertos and Bos, 1973; Shukla and Hiruki, 1975; Atkinson and Cooper, 1976; Boccardo and Milne, 1976; Brunt et al., 1976).

The sole recurrent feature of cytological significance consists in the presence of flexuous bundles of virus particles or of banded aggregates similar to, though usually smaller than, those found in potexvirus infections.

Red clover vein mosaic virus (RCVMV) constitutes a remarkable exception, because in addition to the cytoplasmic fascicles of virions, it elicits formation of crystalline inclusions made up by polyhedral particles 10 nm in diameter containing protein and RNA, whose origin and significance are unknown (Rubio-Huertos and Bos, 1973; Khan et al., 1977). These crystals, however, may be regarded as being of diagnostic importance for RCVMV.

6. Potyviruses

With 48 definitive members and 67 possible members, the potyvirus group is by far the largest known grouping of plant viruses (Matthews, 1982). The particles are filamentous and extremely variable in length (range between 275 and 2000 nm). All members (definitive and possible) are mechanically transmissible, but they may have different natural vectors.

The following subgroups can be identified on the basis of particle length and type of vector: subgroup A (104 members), aphid-borne, particles 680–900 nm long; subgroup B (5 members), fungus-borne, particles 275–2000 nm long; subgroup C (1 member), whitefly-borne, particles 850–900 nm long.

Despite these differences, which when better defined may call for a splitting of the group, there is a unifying ultrastructural character (i.e., cylindrical inclusions), that, so far, has been found throughout the group without exceptions.

Cylindrical inclusions were described in detail and so termed by Edwardson (1966a,b). Although later it was shown that they actually have a conical (Andrews and Shalla, 1974) or an elliptic hyperboloid (Mernaugh et al., 1980) shape, the term "cylindrical" is currently in use as an alternative to "pinwheel," which more properly refers to the appearance of the inclusions in cross sections.

Pinwheels are constituted of a central core from which rectangular or triangular curved plates radiate. When viewed in longitudinal section, they appear as a set of parallel lamellae referred to as "bundles." Plates of adjacent pinwheels may fuse to form a series of stacked laminae called "laminated aggregates" or may roll inward to form "scrolls" (in cross section) or "tubes" (in longitudinal section). Laminated aggregates may be long and straight or short and curved. Pinwheels may be associated with laminated aggregates or scrolls or both. These associates give rise to diverse configurations that are consistent enough to be used for subdividing the group into smaller clusters.

Edwardson (1974, 1981) has investigated these aspects thoroughly and has proposed the establishment of four subdivisions (Fig. 11, A–D): Subdivision I: pinwheels and scrolls; Subdivision II: pinwheels and long straight laminated aggregates; Subdivision III: pinwheels, scrolls, and long straight laminated aggregates; Subdivision IV: pinwheels, scrolls, and short curved laminated aggregates.

The nature, composition, formation, and possible role of cylindrical inclusions have been reviewed elsewhere (Martelli and Russo, 1977a; Hollings and Brunt, 1981).

In addition to proteinaceous pinwheel inclusions, infections by many potyviruses are accompanied by the production of complex amorphous cytoplasmic inclusion bodies. Ultrastructurally (see reviews by Edwardson, 1974; Martelli and Russo, 1977a), these appear as aggregates of normal cell constituents such as ribosomes and endoplasmic reticulum strands, membranous vesicles containing finely stranded material, pinwheels, and, often, virus particles in bundles or in paracrystalline arrangement. The significance of these inclusions is not known, although they may be sites of viral nucleic acid synthesis, by analogy with the vesiculate-vacuolate inclusions of como- and nepoviruses, which they resemble. Another characteristic feature of potyvirus infection is secondary vacuolation of the cytoplasm, which gives rise to abnormalities consisting of monolayers of virus particles in parallel array, restrained within narrow cytoplasmic strands delimited by two membranes. These membranous virus-containing structures may bridge secondary vacuoles or project into the central vacuole (see reviews by Martelli and Russo, 1977a; Hollings and Brunt, 1981).

A number of individual potyviruses, in addition to the group-related inclusions, show cytological abnormalities that can be used for specific identification (for review, see Edwardson, 1974; Martelli and Russo, 1977a). A few examples are listed here.

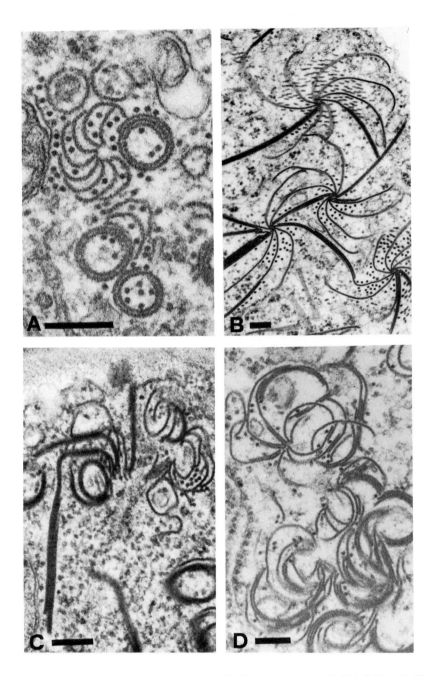

FIG. 11. Cylindrical inclusions of potyiruses. (A) Pinwheels and scrolls (Subdivision I). (B) Pinwheels and laminated aggregates (Subdivision II). (C) Pinwheels, scrolls, and long straight laminated aggregates (Subdivision III). (D) Pinwheels, scrolls, and short curved laminated aggregates (Subdivision IV). Magnification bars represent 100 nm.

1. Bean yellow mosaic virus (BYMV) infections are characterized by intracellular and/or cytoplasmic crystalline formations, appearing as intensely electron-opaque angular or rounded bodies surrounded by spherical ribosome-like particles (Fig. 12A). These inclusions are induced also by BYMV isolate from hosts other than legumes, such as artichoke (Russo and Rana, 1978) and saffron (Russo et al., 1979a).

2. Satellite bodies are most peculiar intracellular inclusions elicited by beet mosaic virus (B˙ 1V). They consist of nucleolus-related amorphous accumulations of prot⸱inaceous material (Fig. 12B), which occur in all hosts so far investigated, regardless of the geographic origin of the virus isolate (Reitberger, 1956; Bos, 1969; Martelli and Russo, 1969; Mérkuri and Russo, 1983).

3. Watermelon mosaic virus 1 (WMV -1) induces cytoplasmic accumulations of electron-dense granular material, sometimes containing thin rod-like structures resembling virus particles (Fig. 12C). These inclusions are made up of proteins and are large enough to be seen also with the light microscope (Martelli and Russo, 1976). Their diagnostic value has been ascertained by comparing several isolates of WMV-1 and WMV-2 from different geographical areas (Russo et al., 1979b).

4. Fimbriate bodies are cytoplasmic or intranuclear accumulations of proteinaceous material typically associated with zucchini yellow fleck virus (ZYFV) infections (Vovlas et al., 1981; Martelli et al., 1981). The inclusions consist of flakes of intensely electron-dense material with a finely granular or fibrous texture that gather together, giving rise to conspicuous structures with a whorl-like organization and an overall fimbriate appearance (Fig. 12D).

5. Tobacco etch virus (TEV) infections are characterized by the presence in the nuclei and cytoplasm of large crystalline bodies readily detectable also with the light microscope (Kassanis, 1939). Nuclear crystals have a three-dimensional morphology resembling that of a truncate four-sided pyramid (Matsui and Yamaguchi, 1964) and are made of protein, which, however, is serologically unrelated to both the viral coat protein and the cylindrical inclusion protein (Shephard et al., 1974).

7. Closteroviruses

The closterovirus group, in the formulation proposed by Bar-Joseph and Murant (1982), includes 16 members (7 definitive and 8 tentative), which are subdivided into three subgroups according to physicochemical properties and particle length. Several of these viruses are transmitted by inoculation of sap and semipersistently by aphids, and others only by aphids. Likewise, some closteroviruses invade and multiply only in the plant phloem, whereas others may spread outside this tissue.

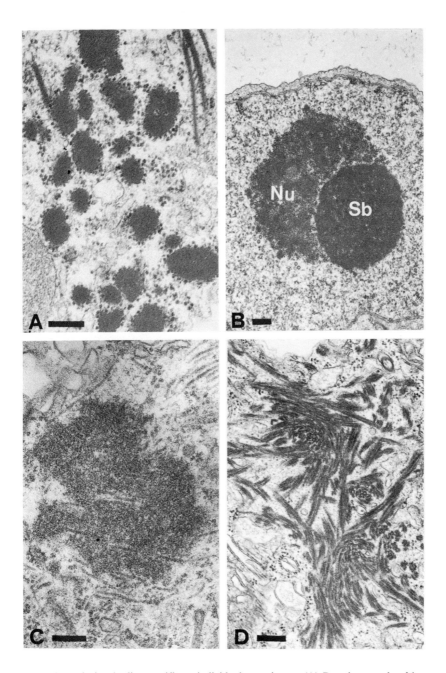

FIG. 12. Inclusion bodies specific to individual potyviruses. (A) Protein crystals of bean yellow mosaic virus. (B) Satellite bodies (Sb) next to a nucleolus (Nu) in a cell infected by beet mosaic virus. (C) Amorphous proteinaceous body induced by watermelon mosaic virus 1. (D) Fimbriate cytoplasmic bodies of zucchini yellow fleck virus. Magnification bars represent 100 nm.

Of the three members of subgroup A (particle length 720–800 nm), two, i.e., a virus from grapevine (Namba *et al.,* 1979; Faoro *et al.,* 1981; Castellano *et al.,* 1983) and heracleum latent virus (Bem and Murant, 1980), have been studied ultrastructurally. No specific cytopathic structures are associated with their infections except for the massive accumulations of virus particles in companion cells and sieve tubes, which are often large enough as to fill up the cell lumen.

Viruses of subgroup B (3 members, particle length 1250–1450 nm) and C (10 members, particle length above 1500 nm) display a different cytopathology. Phloem tissues still represent the elective site of particle accumulation, but virions occur either in cross-banded aggregates made up by stacked tiers of particles packed side by side (Esau, 1968; Hoefert *et al.,* 1970; Esau and Hoefert, 1971b) or in loose wavy paracrystalline aggregates (Esau and Hoefert, 1981), or, more typically, in irregular fascicles intermingled with membranous vesicles containing a network of fine fibrils (Fig. 13). The vesicles are thought to be the possible site of viral RNA replication (Esau and Hoefert, 1971a). Such inclusions may be considered diagnostic for viruses of subgroups B and C, for they were found virtually in all members so far studied at the ultrastructural level (Cronshaw *et al.,* 1966; Price, 1966; Hoefert *et al.,* 1970; Hill and Zettler, 1973; Cadilhac *et al.,* 1975; Ohki *et al.,* 1976; Bar-Joseph *et al.,* 1977; Nakano and Inouye, 1978).

Lilac chlorotic leafspot virus (LCLV) represents a noteworthy exception, for it invades mesophyll and phloem parenchyma cells, inducing cytoplasmic inclusion bodies that bear only a superficial resemblance to those of other closteroviruses. These structures are made up of convoluted and dilated endoplasmic reticulum cisternae, rounded vesicles, and scattered virions (Brunt and Stace-Smith, 1978).

E. Viruses with Enveloped Particles

Only two groups of plant viruses possess virions bounded by a lipoprotein membrane: (1) tomato spotted wilt virus (TSWV), the type member of a monotypic group characterized by quasi-spherical particles ~82 nm in diameter; (2) plant rhabdoviruses, with bacilliform particles up to 380 nm long, included in the family Rhabdoviridae (Matthews, 1982).

1. Tomato Spotted Wilt Virus (TSWV)

TSWV is a cosmopolitan pathogen transmitted in nature by insects (thrips) and, experimentally, by sap inoculation to a wide range of hosts. The virus usually spreads systemically in the host and invades all types of tissues.

In infected cells, TSWV particles appear as large (70–80 nm in diameter),

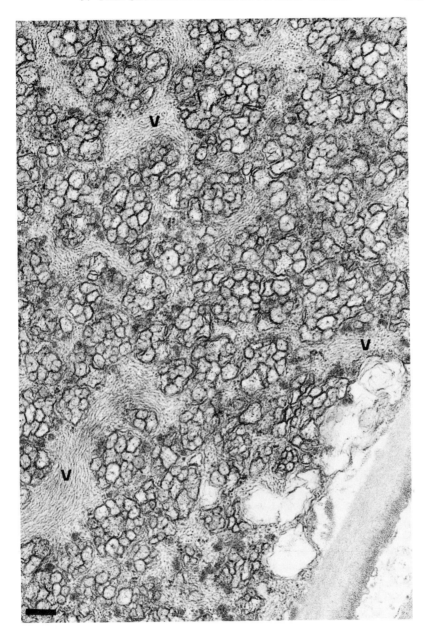

FIG. 13. Typical inclusions elicited by closteroviruses of subgroups B and C, consisting of bundles of virus particles (V) and RNA-containing membranous vesicles. Magnification bar represents 200 nm. By Courtesy of Dr. K. Esau and the copyright holder, University of Wisconsin Press.

rounded, electron-dense bodies gathering in groups within dilated cisternae of the endoplasmic reticulum (Milne, 1970; Ie, 1971) (Fig. 14A). This feature alone is sufficient for reliable identification of TSWV infections. However, there is an additional marker composed of viroplasm-like inclusion bodies. "Viroplasms" are rather large cytoplasmic aggregates of dark-staining, possibly proteinaceous, material (Milne, 1970; Ie; 1971) or of clusters of hollow tubules intermingled with endoplasmic reticulum strands (Francki and Hatta, 1981). The nature and role of these inclusions are unknown, although they may be involved in virus multiplication (Milne, 1970).

2. Plant Rhabdoviruses

Plant rhabdoviruses constitute a large cluster of enveloped viruses with a total of 75 units including definitive, probable, and possible members (Matthews, 1982). In nature, rhabdoviruses are transmitted by insects (aphids, hoppers, and lace bugs) or mites. A few are mechanically transmissible. Based on physicochemical and ultrastructural properties, three subgroups can be identified: viruses of subgroup A, which mature by acquiring the coating membrane from the endoplasmic reticulum and accumulate in the cytoplasm as discrete aggregates within dilated membranous cisternae; viruses of subgroup B, which bud at the inner membrane of the nuclear envelope and accumulate perinuclearly, often in massive paracrystalline aggregates (Fig. 14B); and viruses of the third subgroup, which have nonenveloped particles that associate with the nucleus, at the periphery of which they give rise to highly characteristic structures currently known as "spokewheels" (for reviews see Martelli and Russo, 1977b; Francki et al., 1981).

Additional, though unusual, sites of particle formation and accumulation are the viroplasm-like cytoplasmic bodies found in barley yellow striate mosaic (BYSMV) (Conti and Appiano, 1973) and northern cereal mosaic (MCMV) (Toryiama, 1976) viruses.

Intracellularly, rhabdovirus particles appear as rounded or elongated structures according to whether they were cut transversely or longitudinally (Fig, 14B). In particular, cross-sectioned particles look like multilayered circular bodies (Fig. 14B; top inset). The number of layers in each particle (i.e., two or three) does not reflect intrinsic structural differences between different members of the group. Rather, it depends on the procedures used for fixing, embedding, and staining tissue samples (Russo and Martelli, 1973). Longitudinally sectioned particles may appear either bullet shaped, if immature and in the process of being formed, or bacilliform (Fig. 14B, bottom inset) when mature. Unusually long particles may sometimes be encountered (Di Franco et al., 1979).

Regardless of the outward appearance, the mere presence of recognizable virions is diagnostic for infections by members of this group.

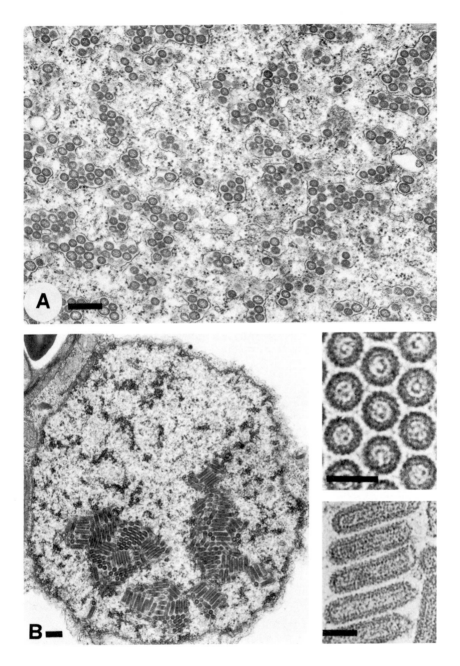

FIG. 14. (A) Tomato spotted wilt virus particles within membranous enclaves in the cyto-plasm of an infected cell. By courtesy of Dr. T. S. Ie. (B) Nucleus with a cluster of rhabdovirus particles. Viral aggregates are in a perinuclear position. Insets show particles in cross and longitudinal section. Magnification bars of (A) and (B) represent 250 nm; bars of insets represent 100 nm.

ACKNOWLEDGMENTS

The authors wish to express their gratitude to Drs. K. Esau and A. Appiano for kindly supplying micrographs and to Dr. R. G. Milne for revising the text and providing unpublished material and information.

REFERENCES

Andrews, J. H., and Shalla, T. A. (1974). *Phytopathology* **64**, 1234.
Ansa, O. A., Bowyer, J. W., and Shepherd, R. J. (1982). *Virology* **121**, 147.
Appiano, A., and Lovisolo, D. (1979). *Microbiologica* **2**, 37.
Appiano, A., Pennazio, S., D'Agostino, G., and Redolfi, P. (1977). *Physiol. Plant Pathol.* **11**, 327.
Atkinson, M. A., and Cooper, J. I. (1976). *Ann. Appl. Biol.* **83**, 395.
Bakker, W. (1975). CMI/AAB Descriptions of Plant Viruses, No. 149.
Bald, J. G. (1966). *Adv. Virus Res.* **12**, 103.
Bancroft, J. B., Bracker, C. E., and Wagner, G. W. (1969). *Virology* **38**, 324.
Barabini, A. R., and Bertaccini, A. (1982). *Phytopathol. Z.* **104**, 304.
Bar-Joseph, M., and Murant, A. F. (1982). CMI/AAB Descriptions of Plant Viruses, No. 260.
Bar-Joseph, M., Josephs, R., and Cohen, J. (1977). *Virology* **81**, 144.
Bassi, M., and Favali, M. A. (1972). *J. Gen. Virol.* **16**, 153.
Bassi, M., Favali, M. A., and Conti G. G. (1974). *Virology* **60**, 353.
Behncken, G. M., Francki, R. I. B., and Gibbs, A. J. (1982). CMI/AAB Descriptions of Plant Viruses, No. 252.
Bem, F., and Murant, A. F. (1980). CMI/AAB Descriptions of Plant Viruses, No. 228.
Bernhard, W. (1969). *J. Ultrastruct. Res.* **27**, 250.
Boccardo, G., and Milne, R. G. (1976). *Phytopathol. Z.* **87**, 120.
Bos, L. (1969). *Neth. J. Plant Pathol.* **75**, 137.
Bos, L., and Rubio-Huertos, M. (1971). *Neth. J. Plant Pathol.* **77**, 145.
Bos, L., and Rubio-Huertos, M. (1972). *Neth. J. Plant Pathol.* **78**, 247.
Brunt, A. A., and E. W. Kitajima (1973). *Phytopathol. Z.* **76**, 263.
Brunt, A. A., and Stace-Smith, R. (1978). *J. Gen. Virol.* **39**, 63.
Brunt, A. A., Stace-Smith, R., and Leung, E. (1976). *Intervirology* **7**, 303.
Burgess, J., Motojoshi, F., and Fleming, E. N. (1974a). *Planta* **119**, 247.
Burgess, J., Motojoshi, F., and Fleming, E. N. (1974b). *Planta* **117**, 133.
Burke, J. J., and Trelease, R. N. (1975). *Plant Physiol.* **56**, 710.
Cadilhac, B., Poupet, A., Cardin, L., and Marais, A. (1975). *C. R. Hebd. Seances Acad. Sci. Sér D* **281**, 639.
Caro, L. G. (1969). *J. Cell Biol.* **41**, 918.
Caro, L. G., and Van Tubergen, R. P. (1962). *J. Cell Biol.* **15**, 173.
Carroll, T. W. (1968). *Virology* **37**, 649.
Carroll, T. W. (1970). *Virology* **42**, 1015.
Carroll, T. W., and Mayhew, D. E. (1976a). *Can. J. Bot.* **54**, 1604.
Carroll, T. W., and Mayhew, D. E. (1976b). *Can. J. Bot.* **54**, 2497.
Castellano, M. A., and Martelli, G. P. (1981). *Phytopathol. Mediterr.* **20**, 64.
Castellano, M. A., Martelli, G. P., and Savino, V. (1983). *Vitis* **22**, 23–39.
Caulfield, J. B. (1957). *J. Biophys. Biochem. Cytol.* **3**, 827.
Chalcroft, J. P., and Matthews, R. E. F. (1966). *Virology* **28**, 555.
Chamberlain, J. A., and Catherall, P. L. (1976). *J. Gen. Virol.* **30**, 41.

Chang, M. U., Doi, Y., and Yora, K. (1976). *Nippon Shokubutsu Byori Gakkaiho* **42,** 325.

Chen, M.-J. (1974). *Plant Protein Bull. (Taiwan)* **16,** 132.

Conti, G. G., Vegetti, G., Bassi, M., and Favali, M. A. (1972). *Virology* **47,** 694.

Conti, M., and Appiano A. (1973). *J. Gen. Virol.* **21,** 315.

Corbett, M. K. (1974). *Int. Symp. Virus Diseases of Ornamental Plants,* 3rd.

Covey, S. N., and Hull, R. (1981). *Virology* **111,** 463.

Craig, S., and Millerd, A. (1981). *Protoplasma* **105,** 333.

Cronshaw, J., Hoefert, L. L., and Esau, K. (1966). *J. Cell Biol.* **31,** 429.

Crowley, N. C., Davison, E. M., Francki, R. I. B., and Owusu, G. K. (1969). *Virology* **39,** 322

Dale, J., Gardiner, J. E., and Gibbs, A. J. (1975). *Newsl. Aust. Plant Pathol. Soc.* **4,** 13.

Dalton, A. J. (1955). *Anat. Rec.* **121,** 281.

de Bokx, J. A., and Waterreus, H. A. J. I. (1971). *Neth. J. Plant Pathol.* **77,** 106.

De Vecchi, L., and Conti, G. G. (1975). *Isr. J. Bot.* **24,** 71.

De Zoeten, G. A. (1966). *Phytopathology* **55,** 744.

De Zoeten, G. A., and Gaard, G. (1969a). *J. Cell Biol.* **40,** 814.

De Zoeten, G. A., and Gaard, G. (1969b). *Virology* **39,** 768.

De Zoeten, G. A., and Gaard, G. (1981). *Int. Congr. Virol. 5th,* Strasbourg, p. 262.

De Zoeten, G. A., and Schlegel, D. E. (1967a). *Virology* **31,** 173.

De Zoeten, G. A., and Schlegel, D. E. (1967b). *Virology* **32,** 416.

De Zoeten, G. A., Gaard, G., and Diez, F. B. (1972). *Virology* **48,** 638.

De Zoeten, G. A., Gaard, G., and Diez, F. B. (1973). *Physiol. Plant Pathol.* **3,** 159.

De Zoeten, G. A., Powell, C. A., Gaard, G., and German, T. L. (1976). *Virology* **70,** 459.

Di Franco, A., Russo, M., and Martelli, G. P. (1979). *Phytopathol. Mediterr.* **18,** 41.

Di Franco, A., Russo, M., and Martelli, G. P. (1980). *Phytopathol. Mediterr.* **19,** 73.

Di Franco, A., Martelli, G. P., and Russo, M. (1983). *J. Submicrosc. Cytol.* **15,** in press.

Doel, T. R. (1975). *J. Gen. Virol.* **26,** 95.

Doraiswamy, S., and Lesemann, D. (1974). *Phytopathol. Z.* **81,** 314.

Edwardson, J. R. (1966a). *Am. J. Bot.* **53,** 359.

Edwardson, J. R. (1966b). *Science* **153,** 883.

Edwardson, J. R. (1974). *Fla. Agric. Exp. Stn. Monogr.* **4,**

Edwardson, J. R. (1981). *Inter. Congr. Virol. 5th,* Strasbourg, p. 244.

Edwardson, J. R., and Christie, R. G. (1978). *Annu. Rev. Phytopathol.* **16,** 31.

Edwardson, J. R., Purcifull, D. E., and Christie, R. G. (1966). *Can. J. Bot.* **44,** 821.

Ehara, Y., Misawa, T., and Nagayama, H. (1976). *Phytopathol. Z.* **87,** 28.

Esau, K. (1968). "Viruses in Plant Hosts. Form Distribution and Pathogenic Effect." Univ. of Wisconsin Press, Madison.

Esau, K. (1977). *J. Ultrastruct. Res.* **61,** 78.

Esau, K., and Hoefert, L. L. (1971a). *Protoplasma* **72,** 255.

Esau, K., and Hoefert, L. L. (1971b). *Protoplasma* **73,** 51.

Esau, K., and Hoefert, L. L. (1972a). *J. Ultrastruct. Res.* **40,** 556.

Esau, K., and Hoefert, L. L. (1972b). *Virology* **48,** 724.

Esau, K., and Hoefert, L. L. (1973). *Virology* **56,** 454.

Esau, K., and Hoefert, L. L. (1981). *J. Ultrastruct. Res.* **75,** 326.

Esau, K., and Magyarosy, A. C. (1979). *J. Ultrastruct. Res.* **66,** 11.

Fahimi, H. D., and Drochmans, P. (1965). *J. Microsc. (Paris)* **4,** 725.

Faoro, F., Tornaghi, R., Fortusini, A., and Belli, G. (1981). *Riv. Pathol. Veg.* **17,** 183.

Favali, M. A., Bassi, M., and Conti, G. G. (1973). *Virology* **53,** 115.

Favali, M. A., Bassi, M., and Appiano, A. (1974). *J. Gen. Virol.* **24,** 563.

Francke, W. W., Krien, S., and Brown, Jr., R. M. (1969). *Histochemie* **19,** 162.

Francki, R. I. B., and Hatta, T. (1977). Nepovirus (tobacco ringspot virus) group. *In* "The

Atlas of Insect and Plant Viruses" (K. Maramorosch, ed.), pp. 222–235. Academic Press, New York.

Francki, R. I. B., and Hatta, T. (1981). Tomato spotted wilt virus. *In* "Handbook of Plant Virus Infections and Comparative Diagnosis" (E. Kurstak, ed.), pp. 491–512. Elsevier/ North-Holland, Amsterdam.

Francki, R. I. B., Hatta, T., Grylls, N. E. and Grivell, C. J. (1979). *Ann. Appl. Biol.* **91,** 51.

Francki, R. I. B., Kitajima, E. W., and Peters, D. (1981). Rhabdoviruses. *In* "Handbook of Plant Virus Infections and Comparative Diagnosis" (E. Kurstak, ed.), pp. 455–489. Elsevier/North-Holland, Amsterdam.

Frederick, S. E., and Newcomb, E. H. (1969). *J. Cell Biol.* **43,** 343.

Gama, M. I. C. S., Kitajima, E. W., and Lin, M. T. (1982). *Phytopathology* **72,** 529.

Gardner, W. S. (1967). *Phytopathology* **57,** 1315.

Gerola, F. M., Bassi, M., and Betto, E. (1965a). *Caryologia* **18,** 353.

Gerola, F. M., Bassi, M., and Belli, G. (1965b). *Caryologia* **18,** 567.

Gerola, F. M., Bassi, M., and Giussani, G. (1966). *Caryologia* **19,** 457.

Gerola, F. M., Bassi, M., and Belli, G. (1969). *G. Bot. Ital.* **103,** 271.

Gill, C. C., and Chong, J. (1975). *Virology* **66,** 440.

Gill, C. C., and Chong, J. (1976). *Virology* **75,** 33.

Gill, C. C., and Chong, J. (1979a). *Virology* **95,** 59.

Gill, C. C., and Chong, J. (1979b). *Phytopathology* **69,** 363.

Glauert, A. M. (1974). *In* "Practical Methods in Electron Microscopy" (A. M. Glauert, ed.), Vol. 3, Part I. North-Holland Publ., Amsterdam.

Glauert, A. M., and Glauert, R. H. (1958). *J. Biophys. Biochem. Cytol.* **4,** 409.

Guilfoyle, T. J. (1980). *Virology* **107,** 71.

Hall, J. L. (1978). "Electron Microscopy and Cytochemistry of Plant Cells." Elsevier/North-Holland, Amsterdam.

Hanchey, P., Livingstone, C. H., and Reeves, F. B. (1975). *Physiol. Plant Pathol.* **6,** 227.

Harris, K. F., and Bath, J. E. (1972). *Virology* **50,** 778.

Harris, K. F., Bath, J. E., Thottappilly, G., and Hooper, G. R. (1975). *Virology* **65,** 148.

Harrison, B. D., and Roberts, I. M. (1968). *J. Gen. Virol.* **3,** 121.

Harrison, B. D., Stefanac, Z., and Roberts, I. M. (1970). *J. Gen. Virol.* **6,** 127.

Harrison, B. D., Murant, A. F., Mayo, M. A., and Roberts, I. M. (1974). *J. Gen. Virol.* **22,** 233.

Hatta, T., and Francki, R. I. B. (1978). *Virology* **88,** 105.

Hatta, T., and Francki, R. I. B. (1979). *Virology* **93,** 265.

Hatta, T., and Francki, R. I. B. (1981a). *J. Gen. Virol.* **53,** 343.

Hatta, T., and Francki, R. I. B. (1981b). *J. Ultrastruct. Res.* **74,** 116.

Hatta, T., and Matthews, R. E. F. (1974). *Virology* **59,** 383.

Hatta, T., and Matthews, R. E. F. (1976). *Virology* **73,** 1.

Hatta, T., and Ushiyama, R. (1973). *J. Gen. Virol.* **21,** 9.

Hatta, T., Bullivant, S., and Matthews, R. E. F. (1973). *J. Gen. Virol.* **20,** 37.

Hayashi, T., and Matsui, C. (1967). *Virology* **33,** 47.

Hayat, M. A. (1968). *Proc. Eur. Reg. Conf. Electron Microsc. 4th,* Rome, Vol. 2, p. 379.

Hayat, M. A. ed. (1970). "Principles and Techniques of Electron Microscopy: Biological Applications," Vol. 1. Van Nostrand-Reinhold, New York.

Hayat, M. A. (1973–1975). "Electron Microscopy of Enzymes," Vols. 1–4. Van Nostrand-Reinhold, New York.

Herold, F., and Munz, K. (1967). *J. Gen. Virol.* **1,** 375.

Hibino, H., Tsuchizaki, T., and Saito, Y. (1974a). *Virology* **57,** 510.

Hibino, H., Tsuchizaki, T., and Saito, Y. (1974b). *Virology* **57,** 522.

Hibino, H., Tsuchizaki, T., Usugi, T., and Saito, Y. (1977). *Nippon Shokubutsu Byori Gakkaiho* **43**, 255.

Hill, H. R., and Zettler, F. W. (1973). *Phytopathology* **63**, 443.

Hills, G. J., and Plaskitt, A. (1968). *J. Ultrastruct. Res.* **25**, 323.

Hiruki, C., and Shukla, P. (1973). *Can. J. Bot.* **51**, 1699.

Hiruki, C., and Tu, J. C. (1972). *Phytopathology* **62**, 77.

Hoefert, L. L. (1968). *Stain Technol.* **43**, 145.

Hoefert, L. L., Duffus, J., and Esau, K. (1970). *Virology,* **42**, 814.

Hollings, M., and Brunt, A. A. (1981). Potyviruses. *In* "Handbook of Plant Virus Infections and Comparative Diagnosis" (E. Kurstak, ed.), pp. 731–807. Elseiver/North-Holland, Amsterdam.

Honda, Y., and Matsui, C. (1968). *Phytopathology* **58**, 1230.

Honda, Y., and Matsui, C. (1974). *Phytopathology* **64**, 534.

Hooper, G. R., Spink, G. C., and Myers, R. L. (1972). *Virology* **47**, 883.

Hull, R. (1981). Pea enation mosaic virus. *In* "Handbook of Plant Virus Infections and Comparative Diagnosis" (E. Kurstak, ed.), pp. 239–256. Elseiver/North-Holland, Amsterdam.

Hull, R., and Plaskitt, A. (1970). *Virology* **42**, 773.

Hull, R., and Plaskitt, A. (1973–1974). *Intervirology* **2**, 352.

Hull, R., Hills, G. J., and Plaskitt, A. (1970). *Virology* **42**, 753.

Ie, T. S. (1971). *Virology* **2**, 468.

Iisuka N., and Iida, W. (1965). *Nippon Shokubutsu Byori Gakkaiho* **30**, 46.

Iwanowski, D. (1903). *Z. Pflanzenkr.* **13**, 1.

Jayasena, K. W., Hatta, T., Francki, R. I. B., and Randles, J. W. (1981). *J. Gen. Virol.* **57**, 205.

Jones, A. T., Kinninmonth, A. M., and Roberts, I. M. (1973). *J. Gen. Virol.* **18**, 61.

Kamei, T., Rubio-Huertos, M., and Matsui, C. (1969). *Virology* **37**, 506.

Kaper, J. M., and Waterworth, H. E. (1981). Cucumoviruses. *In* "Handbook of Plant Virus Infections and Comparative Diagnosis" (E. Kurstak, ed.), pp. 257–332. Elsevier/North-Holland, Amsterdam.

Karnovsky, M. J. (1961). *J. Biophys. Biochem. Cytol.* **11**, 729.

Karnovsky, M. J. (1965). *J. Cell Biol.* **27**, 137 A.

Kassanis, B. (1939). *Ann. Appl. Biol.* **26**, 705.

Kassanis, B., Vince, D. A., and Woods, R. D. (1970). *J. Gen. Virol.* **7**, 143.

Khan, M. Q., Maxwell, D. P., and Maxwell, M. D. (1977). *Virology* **78**, 173.

Kim, K. S. (1977). *J. Gen. Virol.* **35**, 535.

Kim, K. S., and Flores, E. M. (1979). *Phytopathology* **69**, 980.

Kim, K. S., and Fulton, J. P. (1971). *Virology* **43**, 329.

Kim, K. S., and Fulton, J. P. (1972). *Virology* **49**, 112.

Kim, K. S., Shock, T. L., and Goodman, R. M. (1978). *Virology* **89**, 22.

Kishtah, A. A., Russo, M., Tolba, M. A., and Martelli, G. P. (1978). *Phytopathol. Mediterr.* **17**, 157.

Kitajima, E. W., and Costa, A. S. (1966). *Bragantia* **25**, 23.

Kitajima, E. W., Lauritis, J. A., and Swift, H. (1969). *Virology* **39**, 240.

Kitajima, E. W., Tascon, A., Gamez, R., and Galvez, G. E. (1974). *Turrialba* **24**, 393.

Kitajima, E. W., Lin, M. T., Cupertino, F. P., and Costa, C. L. (1977). *Phytopathol. Z.* **90**, 180.

Koenig, R., and Lesemann, D.-E. (1981). Tymoviruses. *In* "Plant Virus Infections and Comparative Diagnosis" (E. Kurstak, ed.), pp. 33–60. Elsevier/North-Holland, Amsterdam.

Kubo, S., and Takanami, Y. (1979). *J. Gen. Virol.* **42**, 387.

Laflèche, D., and Bové, J. M. (1968). *C. R. Hebd. Srancer Acad. Sci. Ser. D* **266**, 1839.

Laflèche, D., and Bové, J. M. (1971). *Physiol. Veg.* **9**, 487.

Laflèche, D., Bové, C., Dupont, G., Mauches, C., Astier, T., Garnier, M., and Bové, J. M. (1972). *Symp. RNA Viruses: Replication Struct. Proc. FEBS Meet.* **8**.

Langenberg, W. G. (1978). *Protoplasma* **94**, 167.

Langenberg, W. G. (1979). *J. Ultrastruct. Res.* **66**, 120.

Langenberg, W. G., and Schlegel, D. E. (1967). *Virology* **32**, 167.

Langenberg, W. G., and Schroeder, H. F. (1972). *J. Ultrastruct. Res.* **40**, 513.

Langenberg, W. G., and Schroeder, H. F. (1973). *Phytopathology* **63**, 1003.

Langenberg, W. G., and Schroeder, H. F. (1974). *Phytopathology* **64**, 750.

Langenberg, W. G., and Schroeder, H. F. (1975). *J. Ultrastruct. Res.* **51**, 166.

Lastra, J. R., and Schlegel, D. E. (1975). *Virology* **65**, 16.

Lawson, R. H., and Hearon, S. S. (1970). *Virology* **41**, 30.

Lawson, R. H., and Hearon S. S. (1973). *J. Ultrastruct. Res.* **48**, 201.

Lawson, R. H., and Hearon, S. S. (1974). *J. Ultrastruct. Res.* **48**, 201.

Lawson, R. H., Hearon, S. S., Smith, F. F., and Kahn, R. P. (1973). *Phytopathology* **63**, 1435.

Leduc, E. H., and Bernhard, W. (1967). *J. Ultrastruct. Res.* **19**, 196.

Lesemann, D.-E., (1977). *Phytopathol. Z.* **90**, 315.

Lesemann, D.-E., and Koenig, R. (1977). Potexvirus (potato virus X) group. *In* "The Atlas of Insect and Plant Viruses" (K. Maramorosch, ed.), pp. 331–345. Academic Press, New York.

Luft, J. H. (1961). *J. Biophys. Cytol.* **9**, 409.

Lyons, A. R., and Allen, T. C. (1969). *J. Ultrastruct. Res.* **27**, 198.

McMullen, C. R., and Gardner, W. S. (1980). *J. Ultrastruct. Res.* **72**, 65.

McMullen, C. R., Gardner, W. S., and Myers, G. A. (1977). *Phytopathology* **67**, 426.

McMullen, C. R., Gardner, W. S., and Myers, G. A. (1978). *Phytopathology* **68**, 317.

McWhorter, F. P. (1965). *Annu. Rev. Phytopathol.* **3**, 1.

Maramorosch, K. (ed.) (1977). "The Atlas of Insect and Plant Viruses." Academic Press, New York.

Marshall, B., and Matthews, R. E. F. (1981). *Virology* **110**, 253.

Martelli, G. P. (1980). *Microbiologica* **3**, 369.

Martelli, G. P. (1981). Tombusviruses. *In* "Plant Virus Infections and Comparative Diagnosis" (E. Kurstak, ed.), pp. 61–90. Elsevier/North-Holland, Amsterdam.

Martelli, G. P., and Castellano, M. A. (1969). *Virology* **39**, 610.

Martelli, G. P., and Castellano, M. A. (1971). *J. Gen. Virol.* **13**, 133.

Martelli, G. P., and Russo, M. (1969). *Annu. Phytopathol.* **1** (hors ser.), 339.

Martelli, G. P., and Russo, M. (1973). *J. Ultrastruct. Res.* **42**, 93.

Martelli, G. P., and Russo, M. (1976). *Virology* **72**, 352.

Martelli, G. P., and Russo, M. (1977a). *Adv. Virus Res.* **21**, 175.

Martelli, G. P., and Russo, M. (1977b). Rhabdoviruses of plants. *In* "The Atlas of Insect and Plant Viruses" (K. Maramorosch, ed.), pp. 181–213. Academic Press, New York.

Martelli, G. P., and Russo, M. (1981). *J. Ultrastruct. Res.* **77**, 93.

Martelli, G. P., Russo, M., and Quacquarelli, A. (1977). Tombusvirus (tomato bushy stunt virus) group. *In* "The Atlas of Insect and Plant Viruses"(K. Maramorosch, ed.), pp. 257–279. Academic Press, New York.

Martelli, G. P., Di Franco, A., and Russo, M. (1980). *Proc. 7th Meet. Int. Council for the Study of Viruses and Virus-like Diseases of the Grapevine,* Niagara Falls, Canada, 1980, p. 217.

Martelli, G. P., Russo, M., and Vovlas, C. (1981). *Phytopathol. Mediterr,* **20**, 193.

Matsui, C., and Yamagucki, A. (1964). *Virology* **22**, 40.

Matthews, R. E. F. (1977). Tymovirus (turnip yellow mosaic virus) group. *In* "The Atlas of Insect and Plant Viruses" (K. Maramorosch, ed.), pp. 347–361. Academic Press, New York.

Matthews, R. E. F. (1982). Classification and nomenclature of viruses. Fourth report of the International Committee on Taxonomy of Viruses. *Intervirology* **17**.

Maupin-Szamier, P., and Pollard, T. D. (1978). *J. Cell Biol.* **77**, 837.

Mérkuri, J., and Russo, M. (1983). *Phytopathol. Mediterr.* **22**, 71–75.

Mernaugh, R. L., Gardner, W. S., and Yocom, K. L. (1980). *Virology* **106**, 273.

Millonig, G. (1961a). *J. Appl. Phys.* **32**, 1637.

Millonig, G. (1961b). *J. Biophys. Biochem. Cytol.* **32**, 1637.

Milne, R. G. (1866a). *Virology* **28**, 520.

Milne, R. G. (1966b). *Virology* **28**, 527.

Milne, R. G. (1967). *Virology* **32**, 589.

Milne, R. G. (1970). *J. Gen. Virol.* **6**, 267.

Milne, R. G. (1972). Electron microscopy of viruses. *In* "Principles and Techniques in Plant Virology" (C. I. Kado and H. O. Agrawal, eds.), pp. 76–128. Van Nostrand-Reinhold, New York.

Milne, R. G., and De Zoeten G. A. (1967). *J. Ultrastruct. Res.* **19**, 398.

Milne, R. G., and Luisoni, E. (1977). *Methods Virol.* **6**, 265–281.

Mohr, W. P., and Cocking, E. C. (1968). *J. Ultrastruct. Res.* **21**, 171.

Mollenhauer, H. H. (1959). *J. Biophys. Biochem. Cytol.* **6**, 431.

Mollenhauer, H. H. (1964). *Stain Technol.* **39**, 111.

Monneron, A., and Bernhard, S. (1966). *J. Microsc. (Paris)* **5**, 697.

Morgan, C., and Rose, H. M. (1967). *Methods Virol.* **3**, 575–616.

Muller, L. L., and Jacks, T. J. (1975). *J. Histochem. Cytochem.* **23**, 107.

Murant, A. F. (1981). Nepoviruses. *In* "Handbook of Plant Virus Infections and Comparative Diagnosis" (E. Kurstak, ed.,) pp. 197–238. Elsevier/North-Holland, Amsterdam.

Nakano, M., and Inouye, T. (1978). *Nippon Shokubutso Byori Gakkaiho* **44**, 96.

Namba, S., Yamashita, S., Doi, Y., Yota, K., Terai, Y., and Yano, T. (1979). *Nippon Shokubutsu Byori Gakkaiho* **45**, 497.

Nasu, S., and Mitsubashi, J. (1968). *Virus (Japan)* **18**, 40.

Odell, J. T., and Howell, S. H. (1980). *Virology* **102**, 349.

Ohki, S. T., Doi, Y., and Yara, K. (1976). *Nippon Shokubutsu Byori Gakkaiho* **42**, 313.

Ohki, S. T., Doi, Y., and Yara, K. (1979). *Nippon Shokubutsu Byori Gakkaiho* **45**, 74.

Osaki, T., and Inoue, T. (1978). *Nippon Shokubutsu Byori Gakkaiho* **44**, 167.

Psaki, T., Kobatake, H., and Inoue, T. (1979). *Nippon Shokubutsu Byori Gakkaiho* **45**, 62.

Pacini, E., and Cresti, M. (1977). *Planta* **137**, 1.

Palade, G. E. (1952). *J. Exp. Med.* **95**, 285.

Paliwal, Y. C. (1970). *J. Ultrastruct. Res.* **30**, 491.

Pease, D. C. (1964). "Histological Techniques for Electron Microscopy," 2nd ed. Academic Press, New York.

Pennazio, S., and Appiano, A. (1975). *Phytopathol. Mediterr.* **14**, 12.

Pennazio, S., Redolfi, P., Cantisani, A., and Vecchiati, M. (1978). *Phytopathol. Z.* **93**, 274.

Peterson, R. L., Hersey, R. E., and Brisson, J. D. (1978). *Stain Technol.* **53**, 1.

Plumb, R. T., and James, M. (1973). *J. Gen. Virol.* **18**, 409.

Powell, C. A., and De Zoeten, G. A. (1977). *Proc. Natl. Acad. Sci. U.S.A.* **74**, 2919.

Powell, C. A., De Zoeten, G. A., and Gaard, G. (1977). *Virology* **78**, 135.

Price, W. C. (1966). Virology **29**, 285.

Purcifull, D. E., and Edwardson, J. R. (1981).

Purcifull, D. E., Edwardson, J. R., and Christie, R. G. (1966). *Virology* **29**, 276.

Putz, C., and Vuittenez, A. (1980). *J. Gen. Virol.* **50**, 201.

Randles, J. W., Davies, C., Hatta, T., Gould, A. R., and Francki, R. I. B. (1981). *Virology* **108**, 111.

Rassel, A. (1972). *C. R. Hebd. Acad. Sci. Scances Ser. D* **274**, 2871.

Reitberger, A. (1956). *Zuechter* **26**, 106.

Reynolds, E. S. (1963). *J. Cell Biol.* **17**, 208.

Ringo, D. L., Brennan, E. F., and Cota-Robles, E. H. (1982). *J. Ultrastruct. Res.* **80**, 280.

Robb, S. M. (1963). *Ann. Appl. Biol.* **52**, 145.

Roberts, D. A., Christie, R. G., and Archer, M. C. (1970). *Virology* **42**, 217.

Roberts, I. M. (1975). *J. Microsc. (Oxford)* **103**, 113.

Roberts, I. M., and Harrison, B. D. (1970). *J. Gen. Virol.* **7**, 47.

Roberts, I. M., and Hutcheson, A. M. (1975). *J. Microsc. (Oxford)* **103**, 121.

Roth, J., Bendayan, M., and Orci, L. (1978). *J. Histochem. Cytochem.* **12**, 1074.

Rubio-Huertos, M. (1950). *Microbiol. Esp.* **3**, 207.

Rubio-Huertos, M. (1978). "Atlas on Ultrastructure of Plant Tissues Infected with Viruses." Consejo Superior de Investigaciones Cientificas, Madrid.

Rubio-Huertos, M., and Bos, L. (1973). *Neth. J. Plant Pathol.* **79**, 94.

Rubio-Huertos, M., Matsui, C., Yamaguchi A., and Kamei, T. (1968). *Phytopathology* **58**, 548.

Rubio-Heurtos, M., Castro, S., Fujisawa, I., and Matsui, C. (1972). *J. Gen. Virol.* **15**, 257.

Russo, M., and Martelli, G. P. (1972). *Virology* **49**, 122.

Russo, M., and Martelli, G. P. (1973). *Virology* **52**, 39.

Russo, M., and Martelli, G. P. (1975). *J. Submicrosc. Cytol.* **7**, 335.

Russo, M., and Martelli, G. P. (1981). *J. Ultrastruct. Res.* **77**, 105.

Russo, M., and Martelli, G. P. (1982). *Virology* **118**, 109.

Russo, M., and Rana, G. L. (1978). *Phytopathol. Mediterr.* **17**, 212.

Russo, M., Martelli, G. P., and Quacquarelli, A. (1968). *Virology* **34**, 679.

Russo, M., Martelli, G. P., Rana G. L., and Kyriakopoulou, P. E. (1978). *Microbiological* **1**, 81.

Russo, M., Cresti, M., Ciampolini, F., and Martelli, G. P. (1979a). *Phytopathol. Mediterr.* **18**, 189.

Russo, M., Martelli, G. P., Vovlas. C., and Ragozzino, A. (1979b). *Phytopathol. Mediterr.* **18**, 94.

Russo, M., Cohen, S., and Martelli, G. P. (1980). *J. Gen. Virol.* **49**, 209.

Russo, M., Martelli, G. P., and Di Franco, S. (1981). *Physiol. Plant Pathol.* **11**, 327.

Russo, M., Castellano, M. A., and Martelli, G. P. (1982). *J. Submicrosc. Cytol.* **14**, 149.

Russo, M., Di Franco, A., and Martelli, G. P. (1983). *J. Ultrastruct. Res.* **82**, 52.

Sabatini, D. D., Bensch, K., and Barnett, R. J. (1963). *J. Cell Biol.* **17**, 19.

Salema, R., and Brandaõ, I. (1973). *J. Submicrosc. Cytol.* **5**, 79.

Salpeter, M. M., and Bachmann, L. (1972). *In* "Principles and Techniques of Electron Microscopy: Biological Applications" (M. A. Hayat, ed.), Vol. 2. Van Nostrand-Reinhold, New York.

Šarić, A., and Wrischer, M. (1975). *Phytopathol. Z.* **84**, 97.

Savino, V., Barba, M., Gallitelli, G., and Martelli, G. P. (1979). *Phytopathol. Mediterr.* **18**, 135.

Schlegel, D. E., and De Lisle, D. E. (1971). *Virology* **45**, 747.

Shalla, T. A. (1966). *In* "Viruses of Plants (A. B. R. Beemster and J. Dijkstra, eds.), pp. 94–97. North-Holland Publ. Amsterdam.

Shalla, T. A. (1968). *Virology* **35**, 194.

Shalla, T. A., and Amici, A. (1967). *Virology* **31**, 78.

Shalla, T. A., and Shepard, J. (1972). *Virology* **49**, 654.

Shalla, T. A., Shepherd, R. J., and Petersen, L. J. (1980). *Virology* **102**, 381.

Shepard, J. F. (1968). *Virology* **36**, 20.

Shepard, J. F., Gaard, G., and Purcifull, D. E. (1974). *Phytopathology* **64**, 418.

Shepardson, S., Esau, K., and McCrum, R. (1980). *Virology* **105**, 379.

Shepherd, R. J. (1977). Cauliflower mosaic virus (DNA virus of higher plants). *In* "The Atlas of Insect and Plant Viruses" (K. Maramorosch, ed.), pp. 159–166. Academic Press, New York.

Shepherd, R. J., and Lawson, R. H. (1981). Caulimoviruses. *In* "Handbook of Plant Virus Infections and Comparative Diagnosis" (E. Kurstak, ed.), pp. 847–878. Elsevier/North-Holland, Amsterdam.

Shepherd, R. J., Richins, R., and Shalla, T. A. (1980). *Virology* **102**, 389.

Shikata, E. (1977). Plant reovirus group. *In* "The Atlas of Insect and Plant Viruses" (K. Maramorosch, ed.), pp. 337–403, Academic Press, New York.

Shikata, E. (1981). Reoviruses. *In* "Handbook of Plant Virus Infections and Comparative Diagnosis" (E. Kurstak, ed.), pp. 423–451. Elsevier/North-Holland, Amsterdam.

Shikata, E., and Maramorosch, K. (1966). *Virology* **30**, 439.

Shikata, E., Maramorosch, K., and Granados, R. R. (1966). *Virology* **29**, 426.

Shockey, M. W., Gardner, C. O., Jr., Melcher, U., and Essenberg, R. C. (1980). *Virology* **105**, 575.

Shukla, D. D., Koenig, R., Gough, K. H., Huth, W., and Lesemann, D.-E. (1980). *Phytopathology* **70**, 382.

Shukla, P., and Hiruki, C. (1975). *Physiol. Plant Pathol.* **7**, 189.

Silva, M. T. (1973). Uranium salts. *In* "Encyclopedia of Microscopy and Microtechnique" (P. Gray, ed.). Van Nostrand-Reinhold, New York.

Singer, S. J., and Schick, A. F. (1961). *J. Biophys. Biochem. Cytol.* **9**, 519.

Sjöstrand, F. S. (1967). "Electron Microscopy of Cells and Tissues." Academic Press, New York.

Spurr, A. R. (1969). *J. Ultrastruct. Res.* **26**, 31.

Stace-Smith, R. (1981). Comoviruses. *In* "Handbook of Plant Virus Infections and Comparative Diagnosis (E. Kurstak, ed.), pp. 171–195. Elsevier/North-Holland, Amsterdam.

Stace-Smith, R., and Hansen, A. J. (1976). CMI/AAB Descriptions of Plant Viruses, No. 159.

Štefanac, Z., and Ljubešić, N. (1971). *J. Gen. Virol.* **13**, 51.

Štefanac, Z., and Ljubešić, N. (1974). *Phytopathol. Z.* **80**, 148.

Sugimura, Y., and Ushiyama, R. (1975). *J. Gen. Virol.* **29**, 93.

Tamada, T. (1975a). *Hokkaidoritsu Nogyo Shikenyo Hokoku* 25.

Tamada, T. (1975b). CMI/AAB Descriptions of Plant Viruses, No. 144.

Thomas J., and Trelease, R. N. (1981). *Protoplasma* **108**, 39.

Thorpe, J. R., and Harvey, D. M. R. (1979). *J. Ultrastruct. Res.* **68**, 186.

Thottapilly, G., Kao, Y.-C., Hooper, G. R., and Bath, J. E. (1977). *Phytopathology* **67**, 1451.

Tomenius, K., and Oxelfelt, P. (1982). *J. Gen. Virol.* **61**, 143.

Toryiama, S. (1976). *Nippon Shokubutsu Byori Gakkaiho* **42**, 563.

Tu, J. C., and Hiruki, C. (1970). *Virology* **42**, 238.

Tu, J. C., and Hiruki, C. (1971). *Phytopathology* **61**, 862.

Turner, R. H. (1971). *J. Gen. Virol* **13**, 177.

Ushiyama, R., and Matthews, R. E. F. (1970). *Virology* **42**, 293.

van Kammen, A. (1982). *Intr. Conf. Comp. Virol. 6th* Banff, Canada, 1982, p. 14.

van Kammen, A., and Mellema, J. E. (1977). Comoviruses. *In* "The Atlas of Insect and Plant Viruses" (K. Maramorosch, ed.), pp. 167–179. Academic Press, New York.

van Regenmortel, M. H. V. (1981). Tobamoviruses. *In* "Handbook of Plant Virus Infections and Comparative Diagnosis" (E. Kurstak, ed.), pp. 541–564. Elsevier/North-Holland, Amsterdam.

van Regenmortel, M. H. V., and Pinck, L. (1981). Alfalfa mosaic virus. *In* "Handbook of Plant Virus Infections and Comparative Diagnosis (E. Kurstak, ed.), pp. 415–421. Elsevier/ North-Holland, Amsterdam.

Vovlas, C., Russo, M., and Martelli, G. P. (1973). *Phytopathol. Mediterr.* **12**, 80.

Vovlas, C., Hiebert, E., and Russo, M. (1981). *Phytopathol. Mediterr.* **20**, 123.

Walkey, D. G. A., and Webb, M. J. W. (1968). *J. Gen. Virol.* **3**, 311.

Warmke, H. E. (1967). *Science* **156**, 262.

Warmke, H. E., and Christie, R. G. (1967). *Virology* **32**, 534.

Warmke, H. E., and Edwardson, J. R. (1966a). *Virology* **28**, 693.

Warmke, H. E., and Edwardson, J. R. (1966b). *Virology* **30**, 45.

Weintraub, M., and Ragetli, H. W. J. (1968). *Virology* **38**, 316.

Weintraub, M., and Ragetli, H. W. J. (1970). *J. Ultrastruct. Res.* **32**, 167.

Weintraub, M., Ragetli, H. W. J., and Veto, M. (1969). *J. Ultrastruct. Res.* **26**, 197.

Weintraub, M., Ragetli, H. W. J., and Leung, E. (1975). *Phytopathology* **25**, 288.

Wetter, C., and Milne, R. G. (1981). Carlaviruses. *In* "Handbook of Plant Virus Infections and Comparative Diagnosis" (E. Kurstak, ed.), pp. 695–730. Elsevier/North-Holland, Amsterdam.

Wilcoxson, R. D., Frosheiser, F. I., and Johnson, L. E. B. (1974). *Can. J. Bot.* **52**, 979.

Wilcoxson, R. D., Johnson, L. E. B., and Frosheiser, F. I. (1975). *Phytopathology* **65**, 1249.

Wisse, E., and Tates, A. D. (eds.) (1968). *Proc. 4th Eur. Reg. Conf. Electron Micros.,* Rome.

Zettler, F. W., Edwardson, J. R., and Purcifull, D. E. (1968). *Phytopathology* **58**, 332.

6 Purification and Immunological Analyses of Plant Viral Inclusion Bodies

Ernest Hiebert, Dan E. Purcifull, and
Richard G. Christie

I. Introduction

Inclusion bodies and their intracellular location are among 49 criteria listed for classifying plant viruses in groups (Harrison *et al.,* 1971). Of the 26 currently recognized plant virus groups, 20 induce inclusions that are considered to be a main characteristic of the group (Matthews, 1982). Studying these inclusions is not only useful for diagnostic (Edwardson and Christie, 1978) and taxonomic (Fenner, 1976; Matthews, 1982) purposes,

225

but also may be invaluable for identifying and characterizing virus-specific, noncapsid proteins and possible sites of viral synthesis.

The structure and composition of plant virus inclusions have been described and reviewed in a number of articles (Christie and Edwardson, 1977; Edwardson and Christie, 1978; Martelli and Russo, 1977; Rubio-Huertos, 1972; McWhorter, 1965). In this presentation basically two types of inclusion bodies are considered. The cylindrical (pinwheel) inclusions induced by potyviruses (Edwardson, 1974) and the nuclear inclusions induced by tobacco etch virus (Knuhtsen *et al.*, 1974) are examples of inclusion bodies consisting of virus-specific, nonstructural proteins. Another type of inclusion has been described as a complex of virus-induced proteins and virions or a complex of host-modified cell constituents and virions. The viroplasms associated with the caulimovirus group (Shepherd *et al.*, 1980) and comovirus group (Assink *et al.*, 1973) are examples of this inclusion type.

In this chapter we describe procedures used to study the two inclusion body types described above and proteinaceous inclusions associated with maize stripe and turnip yellow mosaic viruses. We present procedures for identifying inclusions in infected tissues and assaying inclusions during purification, specific purification procedures, and serological analyses. The characterization studies of plant virus inclusions are still in their infancy and await more research. It is hoped that this presentation of some of the preliminary efforts in plant virus inclusion studies will stimulate new research, especially in the area of the identification and characterization of nonstructural, virus-specific proteins.

II. Inclusion Purification

A. SELECTION OF TISSUE

The choice of host plant for propagating a virus is of critical importance for successful isolation of viral inclusions. For example, tobacco etch virus (TEV) induces nuclear inclusions (NI) in all its hosts, but only in certain hosts, such as *Datura stramonium,* do the NI develop into sufficiently large structures (Fig. 1) so that they can be purified by procedures utilizing selective, low centrifugal forces. Another example illustrating the importance of tissue selection in inclusion purification has been observed with potyviral cylindrical (pinwheel) inclusions (CI). In certain hosts and/or environmental conditions, potyviral infections never develop large masses of CI away from the plasmalemma (Fig. 2). We have been unable to successfully purify CI when they are confined to the plasmalemma. The plant host used, the conditions under which it is grown, and the time at which it is harvested

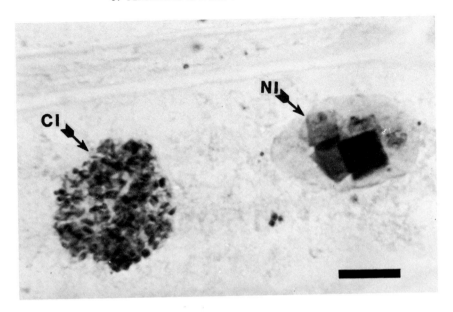

FIG. 1. A light micrograph of an epidermal strip of tobacco etch virus-infected tobacco showing nuclear inclusions (NI) and cylindrical inclusions (CI). Bar marker equals 10 μm.

after inoculation should be chosen to maximize the concentration, distribution, and size of inclusions.

In the selection of tissue for inclusion purification, it should be recognized that symptom expression in many cases cannot be correlated with the presence of inclusions. In cases of severe chlorosis or necrosis, inclusions are often extremely difficult to detect, either because they are too small and undeveloped or because they have disintegrated in the dying cells. In tissues with long-standing infection, although symptoms may be pronounced, the inclusions may have passed their maturity and may begin to disintegrate. This has been observed with the nuclear inclusions associated with tobacco etch virus infections. In many cases the best stages of inclusion development are to be found just prior to symptom development or when symptoms are rather mild. Significantly, well-developed inclusions are often present in hosts that exhibit no apparent symptoms.

Monitoring host tissue for inclusion development and distribution is done most efficiently with the light microscope. While the electron microscope is essential for the study of the inclusion ultrastructural details and inclusion relationships to the host organelles, it is extremely limited in its sampling capacity. Fortunately, most inclusions are large enough to be detected by

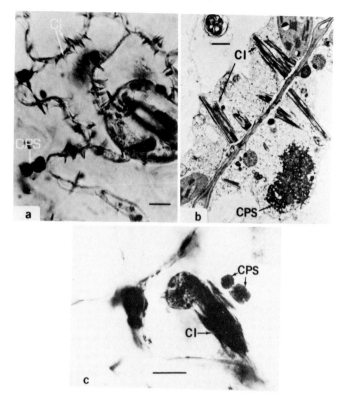

FIG. 2. Cytology of tobacco tissue infected with pepper mottle virus. (a) A light micrograph of an epidermal strip showing the early stages of cylindrical inclusion (CI) development at the cell wall. Amorphous shaped inclusion bodies [also known as cytopathological structures (CPS) and associated with a number of different potyviruses] are seen scattered in the cytoplasm. (b) An ultrathin section of tissue, at a similar infection stage as (a), showing some tangential sections of CI and a section through a CPS. (c) A light micrograph of an epidermal strip showing an advanced infection stage, characterized by the clumping of CI in the cytoplasm. Magnification bar marker equals 10 μm for (a) and (c) and 1 μm for (b).

light microscopy. The light microscope has a large field of view that allows rapid sampling of extensive areas of tissue. Staining procedures have been developed (Christie, 1967; Christie and Edwardson, 1977) that provide a rapid, inexpensive, and simple means for inclusion detection. By the use of selective stains, inclusions can readily be identified and differentiated from host constituents. One selective stain (referred to as the O-G stain) utilizes the textile stains calcomine orange 2RS and Luxol* brilliant green BL and

*Luxol is a trademark of E. I. DuPont de Nemours & Co.

is most useful for the detection of proteinaceous inclusions, especially those associated with potyvirus infections. Another selective stain, azure A, is best suited for detecting inclusions containing nucleoprotein. Most samples such as epidermal strips, whole mounts, and freehand or frozen sections can be readied for study by light microscopy in less than 30 min.

1. *Procedure for the Calcomine Orange–Luxol Brilliant Green BL Technique (Christie and Edwardson, 1977)*

a. Dye Preparation. Luxol brilliant green BL is commercially available from Aldrich Chemical Co., Inc. However, the following simple procedure may be used to prepare small quantities for laboratory staining techniques.

1. Prepare a solution of 1,3-di-*o*-tolylguanidine (DOTG) (Pfaltz and Baur, Inc., 126–04 Northern Blvd., Flushing, NY 11368) by adding 2 g DOTG to 50 ml of 0.7 *M* HCl. Stir the mixture until the DOTG is solubilized. Filter the mixture with a Whatman No. 1 filter paper.
2. Prepare acid green 3-CI 42085 (Pylam Products Co., Inc., 95–10 218th Street, Queens Village, NY 11428; available under different commercial names in 1- to 5-pound lots), by dissolving 3 g of the stain in 50 ml of water.
3. Combine the two solutions while stirring, filter, and wash the precipitate several times with water.
4. Transfer the filter paper with the precipitate to a suitable container and dry under vacuum for 1–2 hr. The yield is approximately 2.7 g of dye.

The color index name for calcomine orange 2RS is direct orange 102. Either this dye or direct orange 26 will work. Both dyes are available under different commercial names from Pylam Products Co., Inc.

b. Stain Solutions. Solutions of the green and the orange dyes are prepared and stored separately. The stock solution of calcomine orange 2RS is prepared by dissolving 1 g of stain in 100 ml of 2-methoxyethanol (stir thoroughly and filter). The stock solution of Luxol brilliant green is prepared by dissolving 1 g in 100 ml of 2-methoxyethanol.

Stock solutions are stable indefinitely at room temperatures. The staining mixture is prepared by mixing 1 part water, 1 part orange stock solution, and 10 parts green stock solution in a small watch glass (~27 mm diameter). The proportions of orange to green can be varied depending on the type of tissue under study, but the water content should be maintained at approximately 10% of the total volume. It is often convenient to prepare small quantities of the O-G stain by mixing aliquots of the stock solutions in the proper proportions just prior to staining.

c. Staining Epidermal Strips

1. Epidermal strips are obtained by inserting the tips of a pair of sharp-pointed tweezers under the epidermis from the lower surface of a leaf and stripping it at an acute angle from the underlying tissues (Fig. 3a). To prevent folding, the epidermal strip should be brought into contact with the surface of the staining solution before it has been completely torn loose from the leaf (Fig. 3b). The torn surface must be in contact with the solution in the watch glass, since the stain will not penetrate the cuticle. The tissue should float in the staining solution and should not be in contact with the glass surface.
2. Stain for about 5–10 min.
3. Remove the staining solution from the watch glass with the aid of a Pasteur pipet. Swirl in several changes of 95% ethanol for about 5–10 sec with each change.
4. Remove the ethanol and then add a small amount of 2-methoxyethyl acetate.
5. After 5 min remove the epidermal strips with the aid of a wooden applicator stick and mount them in a drop of mounting medium (Euparal, Carolina Biological Supply Company, Burlington, NC 27215) on a glass slide. Cover with a No. 1 coverslip and press down on the coverslip to flatten the tissue.
6. Scan with a compound microscope. Magnifications at about $\times 1000$ are needed for good resolution of inclusions, although inclusions may be detected at lower magnifications.

d. Staining Freehand Sections. In plant species for which satisfactory epidermal strips cannot be removed, it is often possible to obtain good samples with freehand sections. Sections can be cut freehand with a razor blade from tissue held in a block of wood pith or Styrofoam. This method is well suited for obtaining samples from stems, roots, and tougher leaf tissues. The sections are completely immersed in one of the staining solutions for an appropriate period of time.

Another method to produce suitable specimens for staining when satisfactory epidermal strips cannot be obtained involves the use of abrasion with an emery cloth (#600 sandpaper) (Ko and Christie, unpublished data). Either the upper or lower surface of a leaf may be removed by selective abrasion, leaving the remaining tissue for staining and examination. The tissue with surface removed by abrasion is cut out and handled like the freehand sections or epidermal strips.

e. Staining Cryostat Sections. Cryostat (frozen) sections are also well suited for the staining techniques presented here. Since the staining solutions are principally nonaqueous, the sections tend to adhere to the surface of the slide and not wash off during the staining process. Adhesives can be

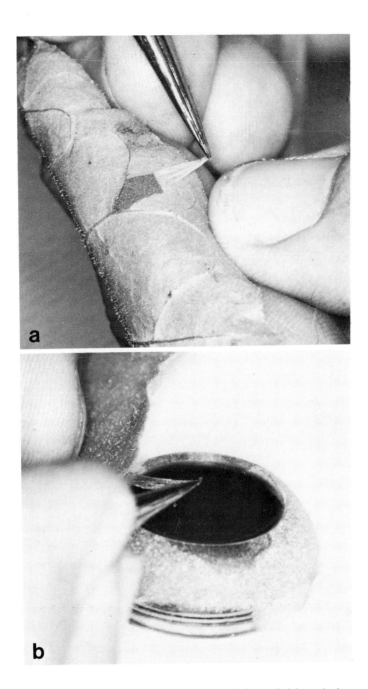

FIG. 3. (a) The photograph shows an epidermal strip being pulled from the lower surface of a tobacco leaf with the aid of sharp-pointed tweezers. (b) The epidermal strip is being brought into contact with the edge of the watch glass and the surface of the staining solution before it is torn loose from the leaf.

used, but slight background staining may result with the O-G stain. The glass slide containing the sections is immersed directly in one of the staining solutions for the appropriate staining period. The slides are then rinsed briefly in 95% ethanol for 15–30 sec and transferred to 2-methoxyethyl acetate for a few minutes. The acetate solution is removed by touching the edge of the slide to a piece of blotting paper. A drop of mounting medium is placed on the slide before the acetate solution has completely dried, and a coverslip is placed over the tissue. Alternatively, the staining can be carried out in a watch glass dish, leaving the sections free. On completion of staining they can be placed directly in the mounting medium and oriented. Cryostat-sectioned material is especially suitable for the study of growing tips, anthers, ovules, stems, and roots.

f. Triton X-100 Treatment. Plastids often obscure small inclusions. The plastids, however, can be dissolved by floating tissues on a 5% solution of Triton X-100 (Sigma Chemical Co., P.O. Box 14508, St. Louis, MO 63178) in a small (27-mm diameter) watch glass for 5 min prior to staining. After treatment, the Triton X-100 solution is removed with the aid of a Pasteur pipet. The tissue is then stained, dehydrated, and mounted as described above. Many inclusions, including the cylindrical inclusions (pinwheels) of the potyvirus group, are stable in Triton X-100 and are easily detected after treatment. Certain viral aggregates, such as the crystalline inclusions of tobamoviruses and the banded inclusions of the potexviruses, are dissociated by treatment with Triton X-100. Therefore, when this procedure is used, nontreated controls should also be studied.

g. Comments. If inclusions are not well differentiated in tissues stained with the O-G combination, the tissue can be removed from the mounting medium, rinsed in absolute ethanol, and restained. If the tissues are overstained green, they should be returned to the O-G stain. If the tissues are overstained orange, they should be returned to the stock solution of the green stain alone (reduce orange concentration in O-G mixture for future work).

The staining procedures described here are primarily designed for fresh tissues. Although fresh material is easy to obtain, it is often difficult to handle, since it has not been fixed and hardened. Fresh tissues are particularly prone to folding and mutilation during the staining process, resulting in erratic staining. Undoubtedly, many failures to detect viral inclusions in fresh tissues result from improper preparation and handling of samples.

Most mounting media are compatible with 2-methoxyethyl acetate. However, those containing xylene should be avoided, since this solvent adversely affects the stains.

Epidermal tissues often contain abundant inclusions. When removing the epidermal strips from the lower surface of the leaf, it is desirable to start the tear at one of the veins so that some subveinal epidermis is also included. These cells contain larger nuclei and inclusions than those of the regular epidermis.

2. *Procedure for the Azure A Technique (Christie and Edwardson, 1977)*

a. Stain Preparation. The source of the stain is Aldrich Chemical Co., Inc.
1. Place 18 drops of 0.05–0.1% azure A (dissolved in 2-methoxyethanol) in a small watch glass (27-mm diameter).
2. Add 2 drops of 0.2 M Na_2HPO_4 (dibasic sodium phosphate) and stir. Prepare fresh stain for each batch of tissue. Do *not* reuse.

b. Staining Epidermal Strips. The procedure is the same as for the O-G stain.

c. Staining Thick Tissue Pieces
1. Cut out leaf pieces from which the epidermis has been removed.
2. Place the tissue in 100% 2-methoxyethanol for 30 min or until chlorophyll is removed.
3. Stain in azure A (immersed with torn surface up). Duration of staining depends on the thickness of the tissue, but 10–15 min has proved to be satisfactory in most cases.
4. Rinse in 95% ETOH, for the same amount of time tissue is stained.
5. Place in 2-methoxyethyl acetate for 15–30 min in a small covered dish. Although the stains are insoluble in this solution, the 2-methoxyethyl acetate solution will remove excess stain dissolved in the 2-methoxyethanol. The tissue should be periodically examined while still in the acetate solution. When the viral inclusions stand out sharply, the tissue can be mounted.

Samples prepared in this manner can be mounted in either the torn surface or the uninjured epidermis next to the coverslip. If the intact epidermis is mounted next to the coverslip, it is possible to study this tissue in an uninjured state. This also gives a good view of the palisade layer in the leaf tissue. This position is advantageous for the study of inclusions, such as the banded inclusions of the potexviruses, which often appear disrupted in epidermal strips. After a tissue sample has been studied in this position, it can be removed and remounted in the other position, i.e., with the torn surface up. In this position the spongy parenchyma and vascular tissue are viewed best. In either position, since the plastids remain unstained, it is possible to detect inclusions that lie several cell layers deep.

d. Freehand or Cryostat (Frozen) Sections. The procedure is the same as described for the O-G stain.

e. Comments. Many viral inclusions that contain nucleic acids are stained by azure A at room temperature. Azure A does not, however, stain the crystalline and paracrystalline inclusions of the tobamovirus group unless heat is applied during the staining period. This is accomplished by placing the tissue in the staining solution and allowing it to remain at room temperature for 5 min. A coverslip is then placed over the tissue, and the slide is gently blotted to remove excess stain. The slide is heated at 60°C for 1–2 min. The tissue is then rinsed in ethanol and mounted. The tobamovirus crystals and paracrystals will now be deeply stained.

The procedure for staining tissue pieces (as described above in subsection 2,c) is useful for detecting inclusions that may be limited to phloem tissue or for inclusions in nonepidermal tissues.

The staining procedures described above stain host cell constituents that are present in both healthy and diseased tissues. The staining reactions are similar for all hosts and are summarized in Table I. The staining reactions of specific inclusions induced by a number of viruses in different plant virus groups have been illustrated by Christie and Edwardson (1977). Some of the data are summarized in Table II.

B. PURIFICATION ASSAYS

During the development of an isolation procedure it is essential to be able to monitor the progress of the isolation. Since there is no known biological assay for the inclusions associated with potyviral infections, we have had to resort to methods that physically detect inclusions. For large inclusions, such as tobacco etch virus (TEV) nuclear inclusions (NI), light microscopy, using a stain such as bromphenol blue (0.01% bromphenol blue in 1% glutaraldehyde and 50 mM citrate buffer, pH 4.0), has been very convenient for detecting inclusions in different fractions. Extracts to be assayed are spread on a glass slide, allowed to dry, and then covered with the stain. A coverslip is placed over the stain, and the slide is viewed at a magnification of at least ×900. The inclusions are visible by their purple staining reaction and by their distinctive morphology. The cylindrical inclusions (CI) of po-

TABLE I
STAINING REACTIONS OF HOST CELL CONSTITUENTS
PRESENT IN BOTH HEALTHY AND VIRUS-DISEASED TISSUE

Stain	Nucleus	Nucleolus	Plastids	Microcrystals	P-protein	Cytoplasm
Azure A	Blue	Red-violet	Faint pink–colorless	Colorless	Colorless	Colorless
O–G stain combination	Orange	Green	Yellow	Green	Olive-green	Pale yellow

TABLE II
INCLUSIONS ASSOCIATED WITH PLANT VIRUSES IN DIFFERENT GROUPS

Group	Recommended stain[a]	Inclusion description	Tissue[b]
Bromovirus	A	Viral aggregates	Epidermis, mesophyll
Carlavirus	A	Cytoplasmic banded bodies (viral aggregates), paracrystals, vacuolate inclusions	Epidermis
Caulimovirus	A	Vacuolate viroplasms	Mesophyll, epidermis
Closterovirus	A	Banded bodies (viral aggregates), paracrystals	Phloem
Comovirus	A	Vacuolate inclusions	Epidermis
		Viral crystals and aggregates	Epidermis, phloem, xylem
Cucumovirus	A	Viral aggregates	Epidermis, mesophyll
Geminivirus	A	Dense nuclear bodies (viral aggregates) and rings	Mesophyll (phloem associated)
Luteovirus	A	Viral aggregates and crystals	Phloem
Nepovirus	A	Vacuolate and viral aggregates	Epidermis, mesophyll
Phytoreovirus	A	Viroplasm, crystalline aggregates	Phloem
Plant rhabdovirus	O–G	Particle aggregates in cytoplasm (subgroup A) and in the perinuclear space (subgroup B)	Epidermis, mesophyll
Potexvirus	A	Cytoplasmic banded bodies (viral aggregates), laminate inclusion components (potato virus X)	Epidermis
Potyvirus	O–G	Cytoplasmic amorphous and cylindrical inclusions	Epidermis
Tobacco etch	O–G	Crystalline nuclear inclusion	Epidermis
Tobacco necrosis	A	Cytoplasmic crystalline inclusions	Epidermis, mesophyll
Tobamovirus	O–G, A	Cytoplasmic crystalline stacked plates, X bodies	Epidermis
Tombusvirus	A	Cytoplasmic crystalline and vacuolate inclusions, nuclear inclusions (viral aggregates)	Epidermis, mesophyll
Tymovirus	A	Viral crystals, clumped chloroplasts	Epidermis, mesophyll

[a] A, Azure A stain; O–G, orange-green combination (see text for details of preparation).
[b] Inclusions are readily observed in the tissues indicated. However the inclusions are not necessarily confined to these tissues.

tyviruses are best monitored by electron microscopy during purification. Extracts to be assayed are mounted on a carbon-coated grid for 1 min, washed with water, and then stained with 2% uranyl acetate. The laminated aggregate plates associated with TEV CI are not stable in phosphotungstate stain. The TEV NI become very electron dense if 2% uranyl acetate is used, but their substructure can be viewed if 0.2% uranyl acetate is used. Therefore, the stability of the inclusions in the negative stains has to be considered if electron microscopy is used in assaying inclusion purification.

The quality of the purified CI of TEV and potato virus Y (PVY) and of the purified NI of TEV has been analyzed by electron microscopic examination of ultrathin sections of purified inclusion pellets (Hiebert *et al.*, 1971; Knuhtsen *et al.*, 1974). The ultrastructures of purified PVY CI, TEV CI, and TEV NI have been compared with those *in situ* using the freeze-etching microscopic technique (McDonald and Hiebert, 1974).

Immunodiffusion techniques (described in Section III,B) have been used to assay for contaminating virions in the CI and TEV NI preparations.

The composition of the purified inclusion preparations can be assayed by ultraviolet spectrophotometry (Hiebert *et al.*, 1971; Knuhtsen *et al.*, 1974). The large size of the inclusions causes considerable light scattering, so they have to be dissociated into subunits prior to analysis. The purified CI and NI are dissociated readily in hot 0.5–1% sodium dodecyl sulfate (SDS) solution. The dissociated inclusions are centrifuged at 10–20 × 10^3 g for 15 min (room temperature) to remove insoluble material. The spectra of purified inclusions are typical of protein with maximum absorbance around 277–280 nm, a minimum around 250 nm, and a tryptophan shoulder at 290 nm. The extinction coefficient of TEV CI and pepper mottle virus (PeMV) CI is 1 A_{280}, i.e., 1 mg of protein as determined by a colorimetric protein assay (E. Hiebert, 1982, unpublished data).

Inclusion purification can also be monitored by analysis in SDS–polyacrylamide gel electrophoresis (PAGE). Aliquots from the preparations to be assayed are dissociated in electrophoresis sample buffer, centrifuged if there is considerable insoluble material, and then subjected to SDS–PAGE. The CI of potyviruses have a subunit size around 67–72 × 10^3 daltons. The inclusion band can be easily detected by using suitable protein standards. The TEV NI consist of equimolar amounts of two protein subunits with sizes of 49 × 10^3 and 54 × 10^3 daltons (Section II,D,2). There are a large number of host proteins around 45–60 × 10^3 daltons, so using SDS–PAGE at early stages of TEV NI purification gives confusing results. However, SDS–PAGE is useful in analyzing the final preparations of TEV NI, especially for detecting virus and CI protein contamination.

C. PURIFICATION OF POTYVIRAL CYLINDRICAL INCLUSIONS

Various factors have been considered in developing a purification procedure for the cylindrical inclusions (CI). Most important of these is the stability of the cylindrical inclusions to shear forces and to clarifying agents, such as the Triton X-100 detergent, and to organic solvents, such as chloroform and carbon tetrachloride. The large sizes of the inclusions (up to several micrometers in length) have been utilized in developing selective centrifugation techniques during isolation. However, the inclusions are not uniform in size, and thus the selective centrifugation techniques can lead to considerable losses of inclusions during these procedures. A number of modifications and variations in the inclusion purification have been used. Some of these are discussed below.

1. *General Procedure*

1. Systemically infected tissue (mature infections, 3–8 weeks after inoculation) is homogenized in a Waring blender for 2 min with a cold solution containing 2 ml of 500 mM potassium phosphate buffer (PB) (pH 7.5), 5 mg sodium sulfite, 0.5 ml chloroform, and 0.5 ml carbon tetrachloride per gram of tissue.

2. The homogenized material is centrifuged at 1020 g_{max} for 5 min.

3. The pellet, containing organic solvents, is reextracted with 0.5 ml of PB per gram of initial tissue and recentrifuged at 1020 g_{max} for 5 min.

4. The supernatants from steps 2 and 3 are combined and centrifuged at 13,200 g for 15–20 min.

5. The pellets are resuspended, with the aid of a Sorvall Omnimixer, in 0.5 ml of 50 mM PB (pH 8.2) containing 0.1% of 2-mercaptoethanol (2-ME) per gram of initial tissue. The supernatant may be saved for virus purification (Lima *et al.,* 1979).

6. Triton X-100 is added to the resuspended material to a final concentration of 5%. The mixture is stirred for 1 hr at 4°C.

7. The mixture is centrifuged at 27,000 g for 15 min.

8. The pellet is resuspended in 20 mM PB (pH 8.2) containing 0.1% 2-ME (5 ml of PB, pH 8.2, per 100 g of initial tissue). The resuspended material is homogenized for 30 sec.

9. The homogenized material is layered (5 ml per centrifuge tube) onto a sucrose step gradient made of 10 ml of 80%, 7 ml of 60%, and 7 ml of 50% (w/v) sucrose in 20 mM PB (pH 8.2). The gradient is centrifuged for 1 hr at 70,000 g_{max} in a swinging-bucket rotor.

10. The inclusions that appear at the 60 and 80% sucrose interface are collected dropwise from a puncture in the bottom of the centrifuge tube.

11. The inclusion fraction is diluted 4-fold with 20 mM PB (pH 8.2) and centrifuged at 27,000 g for 15 min.

12. The pellet is resuspended in a small volume (1–2 ml per 100 g of starting tissue) of 20 mM Tris-HCl (pH 8.2). Yields up to 24 mg per

100 g of tissue have been obtained. A purified preparation of potato virus Y CI is illustrated in Fig. 4.

2. Comments and Variations

Characteristics of the final pellet usually indicate the quality of the preparation. The pellet should be white with a bluish tint and be readily soluble in buffer. The resuspended pellet should be opalescent and show swirling when agitated. A pure white appearance suggests starch contamination, and insolubility suggests the presence of either starch or other host contaminants.

The different cylindrical inclusions purified to date are listed in Table III. We have been unsuccessful in purifying CI from wheat streak mosaic virus, zucchini yellow fleck virus, and araujia mosaic virus, although the presence of CI in the tissues was established by light microscopy. The lack

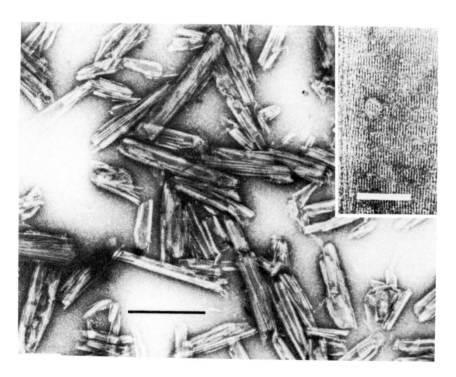

Fig. 4. Electron micrograph of a preparation of purified potato virus Y cylindrical inclusions, negatively stained. Bar marker equals 1 μm. *Inset:* Enlargement of an inclusion showing striations on the surface of the inclusion. Bar marker equals 0.1 μm. From Hiebert *et al.* (1971).

TABLE III

LIST OF CYLINDRICAL INCLUSIONS PURIFIED FROM DIFFERENT POTYVIRUS INFECTIONS

Potyvirus	Inclusion morphology[a]	Reference
Tobacco etch	Laminated aggregates (LA)	Hiebert et al. (1971); Hiebert and McDonald (1973); Dougherty and Hiebert (1980b)
Bidens mottle	LA	Hiebert and McDonald (1973)
Bean yellow mosaic	LA	Nagel et al. (1983)
Clover yellow vein	LA	Nagel et al. (1983)
Cardamom mosaic virus	LA	S.-D. Yeh and D. Gonsalves (1982, personal communication)
Turnip mosaic	LA and scrolls	Hiebert and McDonald (1973); McDonald and Hiebert (1975)
Soybean mosaic	LA and scrolls	Hiebert (1982, unpublished data)
Dasheen mosaic	LA and scrolls	Zettler et al. (1978)
Watermelon mosaic-2	LA and scrolls	Hiebert (1982, unpublished data); Baum (1980)
Potato virus Y	Scrolls	Hiebert et al. (1971); Hiebert and McDonald (1973)
Pepper mottle	Scrolls	Hiebert and McDonald (1973)
Watermelon mosaic Moroccan	Scrolls	Baum (1980)
Papaya ringspot virus (PRSV)	Scrolls	Yeh and D. Gonsalves (1982, personal communication)
Watermelon mosaic-1 (strain of PRSV)	Scrolls	E. Hiebert (1982, unpublished data); Baum (1980)
Tobacco vein mottling	Scrolls	Hellmann et al. (1983)
Blackeye cowpea mosaic	Scrolls	Lima et al. (1979)
Zucchini yellow mosaic	Scrolls	E. Hiebert and D. Purcifull (unpublished data)

[a] Taken from Edwardson (1974).

of success for the first two appears to be due to instability of the CI under the buffer conditions outlined above. The failure with araujia mosaic virus appears to be due to insufficient concentration of inclusions in the tissues used in the purification attempts.

Thorough homogenizations of the resuspended pellets in steps 5 and 8 are crucial for subsequent steps. Clarification of tobacco tissue extracts with organic solvents is more efficient at pH 7.5 than at pH 8.2. However, potyviruses, such as TEV, appear to be more soluble at pH 8.2 than at pH 7.5, and therefore pH 8.2 is used after the initial extraction in order to reduce virus contamination.

The use of urea up to concentrations of 0.8 M in the extraction buffer (Hiebert and McDonald, 1973) may reduce virus contamination in the final preparation and eliminates nuclear inclusion (NI) contamination in the TEV CI preparations. Starch contamination (Hiebert et al., 1971) may be reduced by exposing infected plants to a dark period before harvesting the tissue. At any rate most of the starch is eliminated by the sucrose gradient centrifugation or by a short centrifugation at 500 g_{max} for 5 min after step 8, as indicated in the modification outlined below.

Variable results may occur at step 10 because the inclusions are not uniform in size and thus sediment at different rates. Therefore the 60 and 80% sucrose zones and sucrose gradient pellet should be monitored for the presence of significant amounts of inclusions. One modification to avoid the loss of inclusions during this step involves omitting the sucrose centrifugation. This modification has been used for tobacco vein mottling virus (TVMV) CI purification (Hellmann *et al.*, 1983), for papaya ringspot virus CI, watermelon mosaic virus-1 CI, and cardamom mosaic virus CI purification (S.-D. Yeh and D. Gonsalves, personal communication), and for bean yellow mosaic virus CI and clover yellow vein virus CI purification (Nagel *et al.*, 1983).

A modification of the procedure outlined above (Section II,C,1) follows.

1. After step 8, the homogenized material is centrifuged at 500 g for 5 min.
2. The supernatant is centrifuged at 27,000 g for 15 min.
3. The pellet is resuspended in a small volume of 20 mM Tris-HCl (pH 8.2).

A comparison of this modification with the previous procedure outlined is illustrated in Fig. 5. This modification will reduce starch and large cellular debris contaminations.

3. *Purification of Cylindrical Inclusion Protein*

After step 8 if sucrose gradient centrifugation is avoided, or after step 12, the CI protein can be further purified by preparatory gel electrophoresis as outlined in steps 1–8 below.

1. Purified CI (6–10 mg) is dissociated in 1 ml of a solution containing 100 mM Tris-HCl (pH 6.8), 2.5% SDS, 5% 2-ME, and 5% sucrose. This is heated for 1–2 min in boiling water.
2. In order to detect the inclusion band by fluorescence, an aliquot of purified CI (0.5 mg in 100 μl) is dissociated with 50 μl 10% SDS, 10 μl 1 M Tris-HCl (pH 8.8), and 10 μl 10% dansyl chloride (Talbot and Yphantis, 1971). The solution is heated in boiling water for 2 min. Forty microliters of 2-ME is added, and the solution is heated for another 2 min. The solution prepared in step 1 is combined with the dansyl chloride-treated solution and centrifuged at 5000 g for 10 min at room temperature.
3. The supernatant is layered onto a 1.5-cm stacking gel formed on top

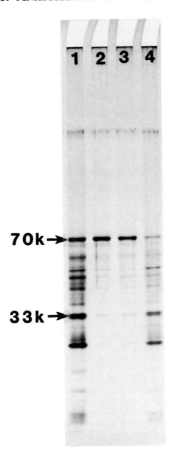

FIG. 5. Analysis of various purification fractions of soybean mosaic virus cylindrical inclusions by SDS-PAGE. Lane 1, the 500 g pellet fraction, and lane 2, the 27,000 g pellet fraction, both from the modified procedure after step 8 in the general procedure; lane 3, the fraction from the 60–80% sucrose interface; lane 4, the sucrose pellet after gradient centrifugation. Fraction 4 comes from step 10 of the general cylindrical inclusion purification procedure. The proteins were detected with a silver stain. Cylindrical inclusion protein is indicated by the 70,000 (70K) mulecular weight marker on the figure.

of a linear 7.5–15% polyacrylamide gel gradient using the Laemmli (1970) discontinuous buffer system. Slab gel apparatus dimensions are 3 × 140 × 150 mm. Electrophoresis is at a constant 70 V for 15–16 hr.

4. The inclusion protein band is visualized by placing the gel on a UV (302 nm) light box.

5. The fluorescing band about 60–70 mm from the top of the stacking gel is sliced out of the separating gel.
6. For elution of the protein, the gel slice is crushed with a mortar and pestle. The gel pieces are washed into a beaker by the addition of 10 ml of H_2O and followed by a rinse of 5 ml of H_2O. The gel pieces in H_2O are frozen and then thawed and incubated at room temperature for 24 hr.
7. Gel pieces are removed by Miracloth filtration. The gel pieces on the Miracloth filter are rinsed with 5 ml of H_2O. If all traces of acrylamide pieces are to be removed, the eluate is filtered through a (Millipore) filter (type HA 0.45 μm). The water is removed from the filtrate by freeze-drying.
8. The residue is resuspended in 3–4 ml of H_2O and dialyzed for 24 hr against 1 liter of H_2O (changed several times). The dialyzed inclusion protein (not completely soluble at this stage) is freeze-dried again and then stored at $-20°C$. Yields up to 6 mg of protein have been recovered from a preparatory electrophoresis run.

D. Purification of Tobacco Etch Virus Nuclear Inclusions

The nuclear inclusions (NI) induced by TEV are not as stable as most of the cylindrical inclusions associated with potyviruses. The TEV NI are subject to shear forces, denaturation of substructure in the absence of the sodium sulfite (Knuhtsen et al., 1974), and dissociation by urea at concentrations normally used to reduce viral aggregation. The TEV NI, present in infected Datura stramonium L., are so large that they will readily sediment in a few minutes at 1000 g or sediment through a dense and viscous sucrose gradient at low centrifugal force. This property was used in developing a purification procedure.

1. Procedure Steps

1. Datura leaf tissue (300 g) at 3–4 weeks after inoculation is homogenized in a Waring blender for 1.5 min at 4°C in 900 ml of 100 mM potassium phosphate buffer (pH 7.5) and 0.2% sodium sulfite.
2. The homogenate is expressed through cheesecloth and Miracloth.
3. Triton X-100 is added to the filtrate to a final concentration of 5%. The mixture is stirred for 1 hr at 4°C.
4. The mixture is centrifuged at 1000 g_{max} for 10 min.
5. The pellets are resuspended in 400 ml of 20 mM potassium phosphate buffer (PB) (pH 8.2) containing 0.5% sodium sulfite.
6. The resuspended material is centrifuged at 1000 g_{max} for 10 min.

7. The pellets are resuspended in 30 ml of 20 mM PB containing 40% (w/v) sucrose and 0.5% sodium sulfite. The material is homogenized in a Sorvall Omnimixer (~3000 rpm for 2 min).

8. The homogenized material is layered (5 ml per tube) onto a sucrose discontinuous gradient made up of 8 ml of 50%, 6 ml of 60%, and 6 ml of 80% (w/v) sucrose in 20 mM PB containing 0.5% sodium sulfite. The gradients are centrifuged in a swinging-bucket rotor at 32,000 g_{max} for 20 min (4°C).

9. The entire 80% sucrose zone is collected and diluted with 3 volumes of 20 mM PB containing 0.5% sodium sulfite. The inclusions are pelleted by centrifugation at 1000 g for 10 min.

10. The pellets are resuspended in 100 ml of 20 mM PB containing 0.5% sodium sulfite and 5% Triton X-100. The solution is stirred for 1 hr at 4°C, and the inclusions are centrifuged again at 1000 g for 10 min.

11. The pellets are resuspended in buffer and homogenized as in step 7. The homogenized material is centrifuged as in step 8.

12. The inclusions layered at the 60 and 80% sucrose interface are collected in droplets after the bottom of the tube has been punctured. The collected material is diluted with 3 volumes of 20 mM PB containing 0.5% sodium sulfite.

13. The inclusions from the collected fraction are pelleted (1000 g for 10 min). The inclusion pellet is resuspended in a small volume of 20 mM Tris-HCl (pH 8.2) containing 0.1% sodium sulfite and stored in the freeze-dried state at −20°C. Yields up to 27 A_{280} units of protein per 100 g of tissue have been obtained.

2. Separation of the TEV Nuclear Inclusion Protein Subunits

The TEV nuclear inclusions (NI) consist of equimolar amounts of two different protein subunits of 49,000 (49K) and 54K daltons. These proteins can be separated (see steps 1–5 below) by preparatory SDS–PAGE using the conditions described for cylindrical inclusion protein preparatory electrophoresis (Section II,C,3).

1. Six to eight milligrams of purified TEV NI are dissociated in 1 ml of a solution containing 100 mM Tris-HCl (pH 6.8), 2.5% SDS, 5% 2-ME, and 5% sucrose. This is heated for 1–2 min in boiling water. An aliquot (0.5 mg) of TEV NI is treated with dansyl chloride as described for cylindrical inclusion protein (Section II,C,3, step 2).

2. The dissociated NI plus the dansyl chloride-treated NI are layered onto a gel gradient as described in Section II,C,3, step 3.

3. After electrophoresis, the 49K and 54K protein bands are visualized by placing the gel on a UV (302 nm) light box.

4. The fluorescing bands (82–87 mm for 54K and 88–91 mm for 49K, measured from the top of the gel) are sliced out of the gel (Fig. 6).
5. The proteins are eluted from the gel strips as before. Approximately 3–4 A_{280} units have been recovered for each protein band.

3. Comments

A number of other potyviruses also form nuclear inclusions (Edwardson, 1974; Christie and Edwardson, 1977), but we have not attempted to isolate them using the procedure for TEV NI. The procedure outlined above was developed for the NI induced by the American Type Culture Collection PV-69 strain. The procedure has also been used to isolate the NI of two other strains of TEV that form morphologically distinct NI (Christie *et al.*, 1982; W. G. Dougherty, 1982, unpublished data). The PV-69 strain forms NI described as thin, truncated pyramids with square faces (Knuhtsen *et al.*, 1974), whereas an "Oxnard" isolate forms bipyramidal-NI (Christie and Edwardson, 1977) and a "Madison" isolate forms diamond-shaped (octahedral) NI (Edwardson and Christie, 1979).

Thorough homogenization of the resuspended pellets in steps 7 and 11 is essential for the subsequent treatments.

E. PURIFICATION OF POTYVIRAL AMORPHOUS INCLUSIONS

Amorphous cytoplasmic inclusions (AI) (also known as cytopathological structures, Fig. 2; or irregular inclusions, Christie and Edwardson, 1977) have been reported for a number of potyviruses. The AI associated with papaya ringspot virus (PRSV) type W (previously known as watermelon mosaic-1, Purcifull *et al.*, 1984) and PeMV have been purified and partially characterized (de Mejia et al., 1984). Antisera prepared to the constituent AI proteins, which have a size of 51K, react with the respective major *in vitro* translation products of PRSV type W and PeMV RNAs. The constituent AI protein of PRSV type W is serologically related to the helper component protein (size 53K) (de Mejia *et al.*, 1984) associated with tobacco vein mottling virus (Thornbury and Pirone, 1983). Thus, the constituent protein associated with potyviral AI represents another (different from those described in Section II,C and D) virus-coded nonstructural protein and may be a pool or reservoir of helper component protein.

1. Procedure

1. Systemically infected tissue (mature infections, 3–8 weeks after inoculation) is homogenized in a Waring blender for 2 min with a cold extraction buffer (EB), consisting of 0.1 M Tris-HCl (pH 7.5) and 0.5% sodium sulfite, in the proportion of 1 ml EB per gram of tissue.

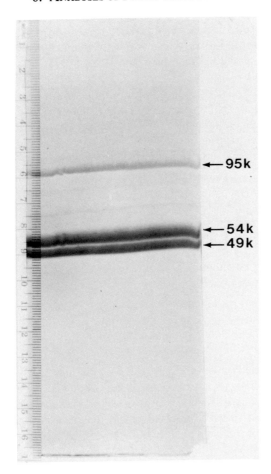

FIG. 6. Preparatory gel electrophoresis of the purified tobacco etch virus nuclear inclusions (NI). This shows the separation of the two NI proteins. The molecular weights of the two proteins are given to the right of the gel. A protein band around 95K may be a dimer of the two protein subunits. The gel was fixed in methanol–acetic acid. The proteins were stained with Coomassie brilliant blue R250. Photograph by courtesy of W. G. Dougherty.

2. The homogenate is filtered through two layers of cheesecloth.
3. The filtrate is layered over a 30-ml cushion of 20% sucrose in EB and centrifuged at 4000 g for 5 min (Sorval GSA rotor).
4. The pellets are resuspended in EB, Triton X-100 is added to a final concentration of 5%, and the resulting mixture is stirred for 1 hr at 4°C.
5. The mixture is layered over a 15-ml cushion (Sorvall SS 34 rotor) at 40% sucrose in EB and centrifuged at 4000 g for 5 min.

6. The pellets are resuspended, Triton X-100 is added (final concentration 5%), and the mixture is stirred for 15 min.
7. The mixture is centrifuged as before (step 5). This is repeated two more times.
8. The final pellets are resuspended in a small volume of 0.1 M Tris-HCl (pH 7.5) and stored at $-20°C$ or further purified by preparatory gel electrophoresis as described for CI (Section II,C, Step 3). Yields up to 4 A_{280} units of protein (size 51K) have been recovered from a preparatory electrophoresis run. Up to 35 A_{280} units per kg tissue have been obtained.

2. Comments

The AI purification can be readily monitored by light microscopy. Extracts to be assayed are spread on a glass slide, allowed to dry, and then covered with the azure A or O-G stain or 0.15% aqueous phloxine B (Christie, 1967). A coverslip is placed over the slide and the slide is viewed at a magnification of $\times 500$ to $\times 1000$.

F. PURIFICATION OF VIROPLASMS

1. Cauliflower Mosaic Virus

Cauliflower mosaic virus (CaMV), the type member of the caulimovirus group, induces cytoplasmic inclusions consisting of a dense matrix embedding virions in host cells (Shepherd *et al.*, 1980). These inclusions are called viroplasms (Fig. 7a and b). The viroplasm matrix is proteinaceous. The major viroplasm protein is considered to be encoded by the CaMV DNA coding region VI (Xiong *et al.*, 1982b). The apparent molecular weight estimated by SDS–PAGE for this protein ranges from 61K to 66K, but it is calculated to be 58K from the coding nucleotide sequence (Xiong *et al.*, 1982b). This protein may accumulate to a concentration estimated to be 10% of the extractable protein in infected plants (Odell and Howell, 1980). Shepherd *et al.* (1980) reported that the major protein species of the viroplasm matrix has an estimated size of 55K. Based on the fact that the 55K protein's N-terminal amino acid sequence does not appear in the complete CaMV DNA sequence, Daubert *et al.* (1981) have concluded that this is a host plant protein. Therefore the question whether the CaMV viroplasms consist of virus-specified protein or host protein produced in response to virus-infection has not yet been resolved in the scientific literature.

   a. Purification Procedure of CaMV Viroplasms according to Al Ani et al. (1980)
      1. Twenty-five grams of CaMV-infected (21–28 days after inoculation) turnip (*Brassica rapa*) leaves without midribs are homogenized in a

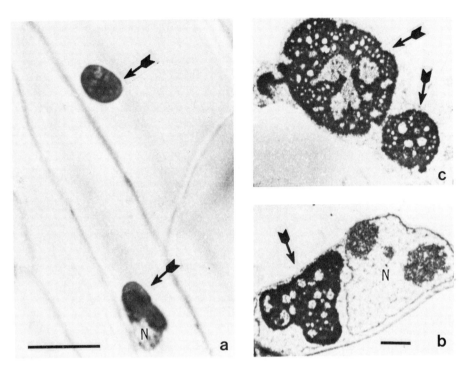

FIG. 7. Cytology of caulimovirus-infected tissues. (a) An epidermal strip of cauliflower mosaic virus-infected mustard tissue, stained with azure A stain and showing viroplasm (marked with arrows). (b) An ultrathin section through a viroplasm (marked with an arrow) in tissue as in (a). (c) An ultrathin section through two viroplasms (marked with arrows) in *Dianthus* sp. infected with carnation etched ring virus. Magnification bar equals 10 μm for (a) and 1 μm for (b) and (c).

VirTis homogenizer with 100 ml of a pH 6.5 buffer consisting of 5 mM 2-(N-morpholino)ethanesulfonic acid (MES), 0.25 M sucrose, 1 mM CaCl$_2$, and 5 mM 2-ME.

2. The homogenate is filtered through nylon screens having 25-μm and 10-μm pore diameters.

3. The filtrate is treated with 1% (v/v) Triton X-100 for 10 min.

4. The detergent-treated material is centrifuged at 800 g for 15 min through a 1.2 M sucrose cushion.

5. The pellets are resuspended in the pH 6.5 buffer and treated again with 1% Triton X-100.

6. The treated material is centrifuged at 800 g for 15 min through a 1.2 M sucrose cushion.

7. The pellets are resuspended in 2 ml of buffer as above, but without sucrose.

b. *Further Purification of Viroplasmic Protein according to Xiong et al. (1982b).* The semipurified viroplasms are subjected to electrophoresis on a 10% polyacrylamide–SDS preparative slab gel with 124 mM boric acid and 91 mM Tris (Laemmli, 1970) as a buffer system. The protein band is located by chilling the gel to 4°C in the presence of 0.5 M KCl. After 10 min, the opaque band is cut out of the gel, rinsed three times with 0.5 M NaCl at room temperature, and stored at −20°C. The protein is eluted from the gel slice by electrophoresis. The eluted protein is dialyzed against deionized H$_2$O, concentrated by lyophilization, and stored at −20°C.

The relative amounts of the major protein (66K) present in the partially purified viroplasms fractions from two isolates of CaMV were in good agreement with relative amounts of viroplasms seen in infected tissues (Xiong *et al.*, 1982a).

c. *Purification of CaMV Viroplasms (Vacuolate Inclusions) according to Ansa et al. (1982)*

1. Four hundred grams of leaf tissue is homogenized in a Waring blender for 1 min with 600 ml of pH 7.5 buffer consisting of 25 mM Tris-HCl, 50% saturated (NH$_4$)$_2$SO$_4$, and 7.5% (v/v) glycerol. Forty milliliters of 25% (v/v) Triton X-100 are added, and blending is continued for 30 sec.
2. The homogenate is filtered through four layers of cheesecloth and then four layers of Miracloth.
3. The filtrate is centrifuged at 7500 g for 30 min.
4. The pellet is resuspended in 20 ml of resuspension buffer (RB) consisting of 50 mM Tris-HCl (pH 7.5) and 15% (v/v) glycerol.
5. Two milliliters of 25% (v/v) Triton X-100 are added, and the mixture is stirred for 5 min.
6. The mixture is centrifuged at 4000 g for 10 min.
7. The pellet is resuspended in 15 ml of RB.
8. The resuspended material is underlaid with 10 ml of a 40% (w/w) CsCl and centrifuged at 16,000 g for 10 min.
9. The band at the CsCl–buffer interface is collected, diluted to 20 ml with RB, and then centrifuged at 400 g for 10 min. The final pellets are resuspended in a small volume.

All the steps in the above procedure are carried out at 4°C. The high salt concentrations in the extracting buffer are used to disintegrate the contaminating nuclei. Contaminating chloroplasts sediment much slower than the viroplasms and thus are eliminated by selective low-speed centrifugations. Glycerol in all the buffers reduces viroplasm aggregation and the tendency for viroplasm to stick to walls of lusteroid tubes and hydrophobic surfaces (Shepherd *et al.,* 1980).

   d. Purification of CaMV Viroplasms according to Covey and Hull (1981)
   1. Turnip leaves are ground in a mortar in 50 m*M* Tris-HCl (pH 7.6), 60 m*M* KCl, and 6 m*M* 2-ME (SSB).
   2. Tissue extract is squeezed through four layers of muslin.
   3. The filtrate is centrifuged at 2000 *g* for 10 min at 10°C.
   4. The pellet is resuspended in SSB containing Triton X-100 (1%) and EDTA (5 m*M*).
   5. Resuspended material is centrifuged at 2000 *g* for 10 min.
   6. Steps 4 and 5 are repeated.
   7. The pellet is washed by resuspending in SSB, centrifuged at 2000 *g* for 10 min, and finally resuspended in a small volume of SSB.

Electron microscopic examination indicated that the preparation was composed largely of viroplasms, the main contaminants being starch grains and chromatin fragments. A 62K-molecular weight polypeptide was the major protein in the viroplasm preparation.

   e. Purification of CaMV Viroplasms according to Shockey et al. (1980)
   1. Surface-sterilized leaves are cut into 1-cm strips and homogenized in 5 volumes of 50 m*M* Tris-HCl, 10 m*M* Na$_2$SO$_3$, 8.5% sucrose, 19 m*M* MgCl$_2$, 25 m*M* KCl, pH 7.3 (homogenizing buffer) in an Omnimixer for 60 sec maintained at 4°C.
   2. The homogenate is strained through four layers of cheesecloth.
   3. Fifteen milliliters of the filtrate are layered onto 15 ml of saturated sucrose in a 50-ml centrifuge tube.
   4. The tubes are centrifuged in a swinging-bucket rotor at 1200 *g* for 30 min.
   5. The top of the sucrose cushion is collected, mixed with an equal volume of 10% Triton X-100, and stirred at 4°C for 1 hr.
   6. The mixture is layered on top of the sucrose cushion as in step 3 and centrifuged as in step 4.
   7. The top of the sucrose cushion is collected, diluted with 1 volume of homogenizing buffer, and centrifuged at 3000 *g* for 20 min.
   8. The partially purified viroplasm pellets are resuspended in a small amount of homogenizing buffer.

The purification of the viroplasms was monitored by light microscopy using one drop of azure A stain (Christie and Edwardson, 1977) (see Section II,A) per drop of extract. Centrifugations which pellet viroplasms produce aggregates that are difficult to disrupt. This aggregation was reduced by sedimenting viroplasms into a saturated sucrose cushion. The structure of the isolated viroplasms was indistinguishable from those observed in leaf thin sections when compared by electron microscopy (Shockey *et al.*, 1980). Two major polypeptides of 61K and 43K (virion protein) were detected in the viroplasm preparations.

2. *Carnation Etched Ring Virus*

Purification of Carnation Etched Ring Virus (CERV) Viroplasms (Fig. 7c) according to Lawson and Hearon (1980)

1. Tissue is harvested (13–15 days after inoculation) from mechanically inoculated *Saponaria vaccaria* grown in a growth chamber at 27°C with 21,520-lux fluorescent illumination on a 16-hr light period.
2. Thirty grams of tissue including inoculated leaves and the first pair of systemically infected leaves are homogenized in 4 or 5 volumes of distilled water or 50 mM Tris buffer (pH 7.2).
3. Leaf homogenates are stirred for 2 hr at 4°C after the addition of Triton X-100 to a final concentration of 5% (v/v).
4. The extract is sieved successively through 417-, 149-, 105-, 74-, and 45-μm (mesh) screens.
5. The sieved extract is centrifuged at 3000 g for 10 min.
6. Pellets are resuspended in one-half the volume of the medium used to homogenize the tissue.
7. The resuspended material is centrifuged at 3000 g for 10 min.
8. Steps 6 and 7 are repeated twice.
9. The final pellet is resuspended in 8 ml of the original extraction medium.
10. The resuspended material (1.5–2 ml/tube) is layered on 50–80% discontinuous sucrose density gradient (5, 5, 8, and 10 ml, respectively). Centrifugation is at 1000 g for 5 min.
11. Each concentration of sucrose in the gradient is collected in separate tubes.
12. The viroplasms are pelleted from the 60% sucrose fraction by centrifugation at 5000 g for 15 min.

About 60% of the viroplasms layered on the gradient were collected from the 60% sucrose fraction. The heterogeneity of the viroplasms resulted in their distribution throughout the gradient during centrifugation.

3. *Cowpea Mosaic Virus*

Cowpea mosaic virus (CpMV) infection is accompanied by the appearance of cytopathic structures that consist of amorphous electron-dense material and a large number of vesicles (Fig. 8) (Assink *et al.*, 1973; de Zoeten *et al.*, 1974; Rezelman *et al.*, 1982). Double-stranded CpMV RNA (Assink *et al.*, 1973, viruslike particles (Hibi *et al.*, 1975), replicase activity (de Zoe-

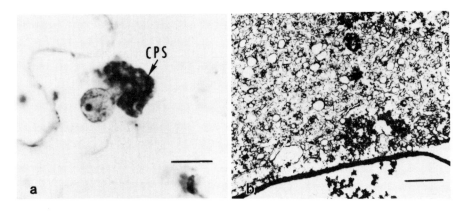

FIG. 8. (a) A light micrograph of an epidermal strip of cowpea infected with cowpea mosaic virus, stained with azure A. The cytopathological structure (CPS) is identified in the micrograph. (b) An ultrathin section of a cowpea mosaic virus-induced CPS. Magnification bar equals 10 μm in (a) and 1 μm in (b).

ten *et al.*, 1974), and virus-specific proteins (Rottier, 1980) have been associated with these structures.

Purification Procedure according to Assink et al. (1973) and de Zoeten et al. (1974)

1. Forty-five grams of deribbed cowpea leaves (4–5 days after inoculation) are chopped with a razor blade at 0°C in 45 ml of Honda medium [0.25 M sucrose, 2.5% Ficoll, 5% Dextrose T4, 2.5 mM Tris (pH 7.8), 1 mM magnesium acetate, 4 mM 2-ME].

2. The extract is filtered through two layers of Miracloth.

3. The filtrate is centrifuged at 1000 g for 15 min.

4. The pellet is gently resuspended in 5–6 ml of Honda medium.

5. The suspension is layered in 1.5–2-ml portions onto a discontinuous gradient made of 10 ml of 60%, 10 ml of 45%, and 5 ml of 20% sucrose (w/w) in 10 mM Tris buffer (pH 8.2), containing 1 mM MgCl₂. The gradients are centrifuged at 58,000 g for 2 hr at 4°C.

6. The chloroplast fraction at the interface between the 20 and 45% sucrose layers is collected and diluted with gradient buffer (10 mM Tris pH 8.2, and 1 mM MgCl₂).

7. The diluted fraction is centrifuged at 18,000 g for 30 min.

8. The pellet is resuspended in 6 ml of gradient buffer and layered (2 ml per tube) on a second gradient made of 5 ml each of 37, 39, 41, 43, and 45% sucrose (w/w) in gradient buffer. The gradients are centrifuged at 58,000 g for 15 min.

9. The material from the top of the gradient down to the middle of the 39% sucrose zone is collected and diluted to a final volume of 10 ml.
10. The diluted fraction is centrifuged at 12,000 g for 10 min.
11. The pellets are either fixed for electron microscopy or resuspended in buffer for nucleic acid and protein analyses.

Approximately 70% of the total cytopathic structures layered on the gradient (step 8) were recovered from the top fraction. Hybridization experiments performed with labeled CpMV RNA and nucleic acid extracted from this fraction showed that it contained 60–70% of the total hybridizable material (de Zoeten et al., 1974).

G. PURIFICATION OF MAIZE STRIPE VIRUS INCLUSIONS

Maize stripe virus (MStpV) is being considered as a member of a new virus class (Gingery et al., 1981; Toriyama, 1982). This virus has a novel particle morphology, described as a filamentous nucleoprotein about 3 nm in diameter. Associated with MStpV infections are large amounts of needle-shaped crystals (Gingery et al., 1981; Falk and Tsai, 1983) (Fig. 9). These crystals are composed of a single protein with a subunit molecular weight of 16,500. These crystals reach a concentration up to 2 mg/g fresh weight of tissue, about 10–20 times the concentration of MStpV. The crystals have been purified and used for antiserum production. The protein crystals are serologically unrelated to MStpV and are considered to be noncapsid, viral protein.

The purification procedure for MStpV crystals is according to Gingery et al. (1981).

1. MStpV-infected tissue is ground in a phosphate–citrate buffer, pH 5.5 (1 g of tissue per 3 ml of buffer). Phosphate–citrate buffer is prepared by combining 0.2 M K_2HPO_4 and 0.1 M citric acid.
2. The extract is centrifuged at 12,000 g for 5 min.
3. The crystals in the pellet are dissolved in phosphate–citrate, pH 7.0 (1 ml/g tissue).
4. The suspension is centrifuged at 12,000 g for 2 min.
5. The pH of the supernatant is lowered by adding an equal volume of phosphate–citrate, pH 3.0.
6. The pH-adjusted solution is immediately centrifuged at 180,000 g for 30 min at 10°C.
7. The supernatant is incubated at 4°C overnight to allow recrystallization.
8. Crystals are collected by centrifugation at 12,000 g for 15 min and washed twice with phosphate–citrate, pH 5.5

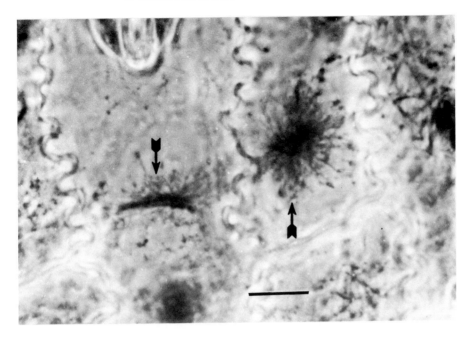

FIG. 9. A light micrograph of an epidermal strip from corn infected with maize stripe virus, stained with O-G stain. The needle-shaped inclusions are marked with arrows in the micrograph. Magnification bar equals 10 μm. Photograph by courtesy of M. A. Overman.

9. CsCl is added to the crystal suspension to a final density of 1.27 g/ml. The material is centrifuged at 140,000 g overnight at 20°C.
10. The isopycnically banded crystals are removed with a syringe.

Recrystallized inclusion proteins have a needle-shaped morphology when viewed by phase-contrast light microscopy. Falk and Tsai (1983) have further purified the crystal protein by preparatory SDS-polyacrylamide gel electrophoresis.

H. PURIFICATION OF STRUCTURAL PROTEIN INCLUSIONS

A necrotic strain of turnip yellow mosaic virus (tymovirus) has been shown to produce tubes as well as virus particles in infected tissue (Hitchborn and Hills, 1968). These tubes appeared to occur in coaxial groups with diameters between 110 and 250 nm and up to several micrometers in length. These tube-shaped inclusions have been purified and shown to consist primarily of protein that is serologically related to the virus.

The purification procedure is according to Hitchborn and Hills (1968).

1. Infected Chinese cabbage tissue (midribs removed) is homogenized in a blender with 500 ml of 10 mM MgSO$_4$ per 100 g of tissue.
2. The homogenate is expressed through muslin.
3. A suspension of magnesium bentonite (Dunn and Hitchborn, 1965) consisting of 10 mg/ml in 10 mM MgSO$_4$ (1/20 volume of extract volume) is added slowly to the extract with rapid stirring.
4. The mixture is centrifuged at 300 g for 5 min in a swinging-bucket rotor.
5. The supernatant is removed and treated again with bentonite as in steps 3 and 4. This cycle is repeated 2–4 times until the supernatant is no longer green.
6. Residual bentonite is removed by centrifugation at 5900 g for 20 min.
7. The supernatant is centrifuged at 27,000 g for 30 min.
8. The pellet is resuspended in 10 mM phosphate buffer (pH 7.4) or water.
9. Centrifugation as in step 7 and resuspension as in step 8 are repeated twice.
10. The final pellet is resuspended in a small volume and stored at 8°C.

The addition of more bentonite than necessary to remove the green material in steps 3–5 resulted in partial or total loss of the tubular inclusions. The tubular inclusions appeared to disintegrate upon storage overnight (Hitchborn and Hills, 1968).

III. Serological Analyses

A. PREPARATION OF ANTISERA

Antisera have been prepared to inclusion bodies induced by potyviruses and caulimoviruses (Table IV). Adult New Zealand white rabbits are injected intramuscularly, subcutaneously, or into toe pads with antigen emulsified with Freund's adjuvant. Usually, complete adjuvant is used for the first injection, and incomplete adjuvant for subsequent injections. Intravenous injection with nonemulsified antigen preparations also have been utilized (Table IV).

1. General Procedures

Following are some general procedures that have been useful for preparing antisera for potyviral inclusions and antisera to other antigens.

a. Processing of Serum. Normal (preimmune) serum is collected from the rabbit prior to immunization. Blood is collected by nicking the marginal ear veins or by cardiac puncture and placed in glass centrifuge tubes (30 ml). The blood is allowed to clot in a water bath at 37°C for 30–45 min and centrifuged at 525 g for 10–15 min. The serum is decanted into 15-ml conical, plastic tubes and centrifuged at 2100 g for 15 min. The supernatant, which consists of the clarified serum, is removed by decanting or pipetting and placed in appropriate vials. The serum is stored at -20°C or lyophilized in 1–5 ml lots and stored at 2–4°C or at -20°C. Lyophilized serum is reconstituted before use by adding an appropriate amount of deionized or distilled water to restore the original volume.

b. Emulsification of Inclusion Bodies and Proteins. For initial injections, proteinaceous inclusion bodies or isolated inclusion body proteins at concentrations of 1–3 A_{280} units in a volume of 1 ml are emulsified with an equal volume in Freund's complete adjuvant. The emulsification can be accomplished directly in the syringe to be used for injection. Remove the plunger from the syringe, cap the tip with a rubber stopper or other suitable means, partially submerge the syringe in an ice bath, add the required amount of adjuvant, and then add the antigen dropwise while stirring with a mechanical stirrer at 5000–7000 rpm. A "Phillips"-head screwdriver (with handle removed) inserted into the chuck of the stirrer serves as suitable stirring device to facilitate emulsification.

c. Injection Procedure. Emulsified immunogen is injected initially into a toe pad of a hind foot (0.15–0.2 ml per toe pad injection) using a 26-gauge needle, and the remainder is injected intramuscularly, using a 21-gauge needle. Subcutaneous injections in the shoulder or back also may be administered. One to four subsequent injections, with immunogen emulsified with incomplete Freund's adjuvant (both the complete and incomplete forms are available from Difco Laboratories, Detroit, MI), are administered at intervals of 1–2 weeks, and booster shots can be given later, as required.

d. Evaluation of Sera. Beginning about 3 weeks after the first injection, sera are collected weekly for evaluation. The course of immunogen administration will be influenced by the qualitative and quantitative evaluation of sera obtained, and by the availability of purified immunogen.

2. *Antisera to Purified, Nontreated Inclusions*

Purified preparations of virus-induced inclusions have been used to produce antisera that react in immunodiffusion tests. Included are cylindrical (pinwheel) inclusions, such as those induced by tobacco etch, potato Y, turnip mosaic, pepper mottle, and Bidens mottle viruses (Hiebert *et al.,* 1971; Purcifull *et al.,* 1973: McDonald and Hiebert, 1975); tobacco etch crystalline nuclear inclusions (Knuhtsen *et al.,* 1974); amorphous inclusions (de Mejia, 1984; de Mejia *et al.,* 1984); and carnation etched ring virus viroplasms (Lawson and Hearon, 1980).

TABLE IV

Examples of Immunogen Treatments and Immunization Schedules Used in Preparing Antisera to Plant Virus-Induced Inclusions

Virus[a]	Type of inclusion	Immunogen treatment	Route and schedule for immunization[b]	Reference[c]
Tobacco etch (potyvirus)	Cylindrical	Purified inclusion preparation	Intramuscular injection, followed 10 days later by an intravenous injection and a second intramuscular injection	3
Potato Y (potyvirus)	Cylindrical	Purified inclusion preparation	Intramuscular injection, followed by intravenous and intramuscular injections 1 and 5 months later, respectively	7
Bidens mottle (potyvirus)	Cylindrical	Purified inclusion preparation	Intramuscular and toe pad injections given initially, followed 1 week and 8 weeks later with similar injections	11
Pepper mottle (PeMV) (potyvirus)	Cylindrical	Sodium dodecyl sulfate (SDS)-treated inclusion preparation	Intramuscular and subcutaneous injections initially, followed by an intramuscular injection 1 month later	1, 6
Watermelon mosaic-1 (strain of papaya ring-spot WMV-1 (potyvirus)	Cylindrical	Inclusion body protein purified by SDS-gel electrophoresis	Intramuscular and toe pad injections performed four times at weekly intervals	10
Tobacco etch (potyvirus)	Nuclear	Purified inclusion preparation	Two intramuscular injections were administered 7 days apart. Two booster injections were injected 60 and 67 days, respectively, after the initial injection	4
Tobacco etch (potyvirus)	Nuclear	Nuclear inclusion protein subunits purified by SDS–poly-	Antigen was emulsified and injected in a toe pad and the remainder was administered intramuscularly. The	2

Virus[a]	Form	Preparation[c]	Injection procedure	Ref.[b,c]
		acrylamide gel electrophoresis	procedure was repeated thrice more at weekly intervals	
PeMV and WMV-1	Amorphous	Amorphous inclusion protein purified by SDS-polyacrylamide gel electrophoresis	Antigen was emulsified and injected in a toe pad and the remainder was administered intramuscularly. The procedure was repeated twice more at biweekly intervals	12
Carnation etched ring virus (caulimovirus)	Viroplasm	Purified inclusion preparation	Two intravenous injections given at intervals of 9 days, followed 14 and 25 days later by intramuscular injections	5
Cauliflower mosaic (caulimovirus)	Viroplasm	Isolated protein from guanidine-HCl-treated, purified inclusion preparations	Subcutaneous injections given at intervals of 2 weeks, followed by a booster injection 3 months later	8
Cauliflower mosaic (caulimovirus)	Viroplasm	Viroplasm protein purified by SDS–gel electrophoresis	Four intramuscular injections given at intervals of 2 weeks	9
Maize stripe	Nonstructural protein	Protein purified by preparatory gel electrophoresis	Three intramuscular injections of 100 µg in 1 ml emulsified with Freund's incomplete adjuvant and given at weekly intervals	13

[a] Group name of virus is given in parentheses.
[b] All antisera produced in rabbits.
[c] 1, Batchelor (1974); 2, Dougherty and Hiebert (1980b); 3, Hiebert et al. (1971); 4, Knuhtsen et al. (1974); 5, Lawson and Hearon (1980); 6, Purcifull and Batchelor (1977); 7, Purcifull et al. (1973); 8, Shepherd et al. (1980); 9, Xiong et al. (1982b); 10, S.-D. Yeh and D. Gonsalves (personal communication); 11, Zurawski (1979); 12, de Mejia (1984); 13, Falk and Tsai (1983).

3. *Antisera to Sodium Dodecyl Sulfate (SDS)-Dissociated Inclusion Body Proteins*

Treatment of plant virus-induced inclusions with SDS, an anionic detergent, has been used effectively to obtain dissociated proteins that stimulate the production of antibodies in rabbits. Two basic approaches have been used: (1) treatment of inclusions and subsequent removal of excess SDS and reconcentration of the protein, and (2) treatment with SDS followed by further purification of dissociated proteins by SDS–polyacrylamide gel electrophoresis (SDS–PAGE).

a. Chromatographic Procedure. The following method has been used for potyviral-induced nuclear and cylindrical inclusions (Batchelor, 1974).

1. To 1 ml of an inclusion preparation containing 3–5 A_{280} units/ml, add SDS and 2-mercaptoethanol (2-ME) to give final concentrations of 3% (w/v and v/v, respectively).
2. Place samples in a boiling water bath for 1–2 min.
3. Centrifuge for 15 min at 2700 g to remove insoluble material.
4. Layer 1 ml of the preparation on a K 15/30 Sephadex column packed with G-50-150. Elute by gravity flow using 20 mM sodium borate (pH 8.0) at a flow rate of 1–2 ml per minute.
5. Monitor chromatographic events by ultraviolet absorbancy. The inclusion body proteins appear in the exclusion volume.
6. Reconcentrate the preparation recovered from the column. Place the preparation in a 12-ml syringe that has been fitted at the tip with a 400-mesh nylon net. Seal the tip with Parafilm, add Sephadex G-25 and allow to swell for 10–15 min. Remove the Parafilm from the tip, and centrifuge the syringe at 300 g for 15 min. Collect the liquid containing the sample and repeat as necessary to obtain the desired volume. About 4 concentrations are required to reduce the volume from 10 to 1 ml.
7. Use directly as immunogen or freeze-dry sample for future use in preparing antiserum.

This procedure has been used to prepare highly reactive antisera to several potyviruses and their virus-induced inclusion proteins, for use in SDS–immunodiffusion and immunoprecipitation analyses. The protein peak is separated from the SDS and 2-ME, which elute in a second peak. As much as 90% of the immunogen preparation applied to the column was recovered in the final concentrated preparation, based on UV absorbancy. After removal of free SDS and 2-ME the denatured proteins became progressively less soluble, and it was necessary to work as rapidly as possible. Amorphous materials but no intact inclusions, were observed in electron micrographs of SDS-denatured tobacco etch virus-induced cylindrical inclusion proteins.

b. Gel Electrophoretic Procedure. Examples of procedures utilizing SDS–PAGE to purify potyviral cylindrical inclusion proteins and tobacco etch virus-induced nuclear inclusion proteins for use as immunogens were described in Sections II,C and II,D. In addition, S.-D Yeh and D. Gonsalves (personal communication, 1982) and Yeh (1984) developed a scheme involving SDS–PAGE for cylindrical inclusions of papaya ringspot and watermelon mosaic-1 viruses. They detected inclusion-containing bands in gels by soaking the gels in 0.2 M KCl for 5–10 min at 4°C. Yields of protein ranged from 1 to 5 mg per 100 g of tissue.they resuspended the final inclusion subunit fractions in 0.12 M guanidine-HCl, 20 mM Tris-HCl (pH 8.2) and stored samples frozen. Such preparations were used as immunogens.

Xiong *et al.* (1982b) prepared antiserum to cauliflower mosaic virus viroplasm protein purified by SDS–PAGE. Partially purified inclusions (Al Ani *et al.,* 1980) were subjected to preparatory SDS-slab gel electrophoresis. The gels were chilled at 4°C in 0.5 M KCl to facilitate detection of protein, and the bands were cut out, rinsed 3 times with 0.5 M NaCl at room temperature, and stored at −20°C. Gel slices were electrophoresed to elute the protein into dialysis tubing. The samples were dialyzed against deionized water, lyophilized, and stored at −20°C until used as immunogen.

B. IMMUNODIFFUSION TECHNIQUES

Double radial immunodiffusion (Ouchterlony) tests are very suitable for analyses with dissociated inclusion body components. Sodium dodecyl sulfate (SDS) has been very effective as a dissociating agent against potyviral inclusions and numerous plant viruses (e.g., Purcifull and Batchelor, 1977), and its use is considered here in some detail. Other treatments (e.g., alkali and phosphotungstate) also are mentioned.

1. *Sodium Dodecyl Sulfate-Treated Antigens*

a. Antiserum. Production of an antiserum that will react with the SDS-dissociated inclusion body proteins is essential. Methods for obtaining such antisera are described in Section III,A.

b. Immunodiffusion Medium. The medium developed by Gooding and Bing (1970) for dissociating potyvirus particles into diffusible antigens has been adapted for degrading potyviral cylindrical inclusions (Hiebert *et al.,* 1971; Purcifull *et al.,* 1973; Purcifull and Batchelor, 1977) and tobacco etch virus-induced nuclear inclusions (Knuhtsen *et al.,* 1974). The medium consists of 0.8% agar, 0.5% SDS, and 1.0% sodium azide (NaN$_3$), and it can be prepared as follows. Add 4 g of Noble agar (Difco Laboratories, Detroit,

MI) to about 300 ml of distilled water and autoclave for 5 min to melt the agar. Add 5 g of sodium azide and mix well, maintaining the agar in a molten state; dissolve 2.5 g of SDS (Sigma Chemical Co., St. Louis, MO) in about 150 ml of hot water (70–95°C), and add to the agar–azide solution; bring mixture to a final volume of 500 ml with hot distilled water, and mix thoroughly. Place petri dishes on a level surface, and add 12 ml per dish (round, plastic, disposable dishes, 90 mm in diameter from American Scientific Products). This makes about 40 plates. After agar solidifies, place dishes in plastic bags or plastic boxes to prevent desiccation and store at 2–4°C until use (within 1–2 months). (Notes: SDS plates become cloudy when chilled, but will clear when warmed to room temperature. Also, sodium azide is toxic to humans and should be handled accordingly.)

A modified medium used for certain potyviruses (Tolin and Roane, 1975) also is useful for potyvirus-induced inclusions (F. Morales, F. W. Zettler, and M. Abo-El Nil, unpublished data, 1977; Zurawski, 1979). This medium contains 0.2% SDS, 0.7% NaCl, 0.8% Noble agar, and 0.1% NaN_3 in deionized water.

c. *Preparation of Inclusion Antigens for Testing.* Purified inclusions at concentrations of 1.0 A_{280} unit/ml in water or in dilute Tris buffer have given excellent results as test antigens. In some cases it may be helpful to accelerate dissociation by adding 0.1–0.5% SDS to the preparations. The lower limit of viral antigen detectability in the SDS–immunodiffusion systems is in the range of 1–10 $\mu g/ml$ (Purcifull and Batchelor, 1977; Garnsey et al., 1979).

Analyses of protein bands recovered from gels after SDS–PAGE of cylindrical inclusions (Purcifull et al., 1973) and tobacco etch nuclear inclusions (Batchelor, 1974) have provided useful information about the antigenic properties of the inclusions.

Appropriately treated extracts from potyvirus-infected plants have given positive results in immunodiffusion in many instances (Table V; Fig. 10). For each gram of tissue containing virus-induced inclusions, triturate in 1 ml of water with a mortar and pestle; then add 1 ml of 3% SDS, triturate briefly, and express through cheesecloth. Such extracts may be used directly (Purcifull et al., 1973) or boiled for 1–2 min (Batchelor, 1974). Antigens may be stored at −20°C or lyophilized (McDonald and Hiebert, 1975).

d. *Preparation of Viral Capsid Protein Antigens.* In determining the antigenic specificity of inclusion body proteins, it is advisable to make appropriate comparisons with the capsid proteins of the viruses that induce them. For potyviruses, procedures similar to those used for inclusion body proteins also are suitable for preparing capsid proteins (Hiebert et al., 1971; Purcifull and Batchelor, 1977) (Fig. 11).

TABLE V

DETECTION OF CYLINDRICAL (PINWHEEL) INCLUSION BODY ANTIGENS
BY SDS–IMMUNODIFFUSION TESTS WITH EXTRACTS FROM POTYVIRUS-INFECTED PLANTS[a]

Inclusion induced by	Host	Reference[b]
Bean yellow mosaic virus	*Pisum sativum*	6
Bidens mottle virus	*Fittonia* sp.	12
	Helianthus annuus	9
	Lactuca sativa	7
	Nicotiana × *edwardsonii*	9, 12
	Zinnia elegans	9
Blackeye cowpea mosaic virus	*Vigna unguiculata*	3
Dasheen mosaic virus	*Philodendron selloum*	5
Papaya ringspot virus	*Cucumis metuliferous*	11
Pepper mottle virus	*Capsicum annuum*	9, 10
	N. × *edwardsonii*	9
	Nicotiana tabacum	1, 9
Potato virus Y	*Datura metel*	9
	N. × *edwardsonii*	9
	N. tabacum	8, 9
	Lycopersicon esculentum	9
	Petunia hybrida	8, 9
Soybean mosaic virus	*Glycine max*	3
	Nicotiana benthamiana	7
Tobacco etch virus	*D. metel*	9
	Datura stramonium	9
	L. esculentum	9
	N. × *edwardsonii*	9
	N. tabacum	1, 8, 9
	P. hybrida	8, 9
Turnip mosaic virus	*Brassica perviridis*	4, 9
	N. × *edwardsonii*	9
Watermelon mosaic virus-1 (strain of papaya ringspot)	*Cucurbita pepo*	2, 11, 13
Watermelon mosaic virus-2	*C. pepo*	2
Watermelon mosaic virus (Moroccan isolate)	*C. pepo*	2
Zucchini yellow mosaic virus	*C. pepo*	14

[a] Antigens were prepared by triturating plant tissue in water and adding SDS before expressing the triturate through cheesecloth (1 ml of water: 1 ml of 3% SDS: 1 gm of tissue). Tests were conducted in medium consisting of 0.8% Noble agar, 0.5% SDS, and 1.0% sodium azide, with undiluted whole serum.

[b] 1, Batchelor (1974); 2, Baum (1980); 3, Lima and Purcifull (1980); 4, McDonald and Hiebert (1975); 5, Zettler *et al.* (1978); 6, Nagel *et al.* (1983); 7, D. E. Purcifull (unpublished data); 8, Purcifull and Batchelor (1977); 9, Purcifull *et al.* (1973); 10, Purcifull *et al.* (1975); 11, S.-D. Yeh and D. Gonsalves (personal communication); 12, Zurawski (1979); 13, Purcifull *et al.* (1984); 14, D. Purcifull and E. Hiebert (unpublished data).

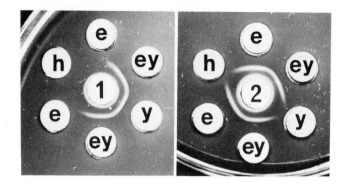

FIG. 10. Immunochemical specificity of antisera to cylindrical inclusions of two different potyviruses in SDS immunodiffusion tests. The center wells contain (1) antiserum specific for potato virus Y cylindrical inclusions; (2) antiserum specific for tobacco etch virus cylindrical inclusions. The peripheral wells contain SDS-treated extracts from *Nicotiana* × *edwardsonii:* (e) tobacco etch virus-infected plants; (y) potato virus Y-infected plants; (ey) plants infected with both potato Y and tobacco etch viruses; (h) healthy, noninoculated plants.

e. Host Proteins. Host proteins are used in immunochemical studies with inclusion body proteins to test the antigenic nature of inclusions, to determine the purity of preparations, and to absorb antihost antibodies. Sap from healthy plants treated with SDS as described in Section III,B,1,c is used as a routine control when testing crude sap from infected plants as an antigen. Host plant antigens concentrated by the following method are useful as controls or absorbing antigens (Purcifull *et al.*, 1973; McDonald and Hiebert, 1975; Zurawski, 1979).

1. Leaf tissue from healthy plants (30 g) is homogenized in 60 ml of 0.1 *M* potassium phosphate (pH 7.4) containing 1% sodium sulfite.
2. The homogenate is frozen for one to several days, thawed, and centrifuged at 27,000 *g* for 10 min. The pellet is discarded.
3. The supernatant from step 2 is centrifuged at 237,000 *g* for 3 hr, and the pellet is resuspended in 2–4 ml of 20 m*M* Tris-HCl (pH 7.4). This resuspended material is used as test antigen or absorbing antigen.

f. Preparation of Agar Plates for Double-Diffusion Tests. Antigens and antisera are placed in circular wells cut in the agar with cork borers or a gel cutting device (e.g., an adjustable cutter, such as manufactured by Grafar Corp., Detroit, MI, is useful). A convenient pattern consists of a central well (7–10 mm in diameter) and two to six peripheral wells (7 mm in diameter), with a distance of 4–5 mm between wells. A single plate (90 mm in diameter) can accommodate 1–4 such patterns, depending on the nature of the tests and on the well spacing.

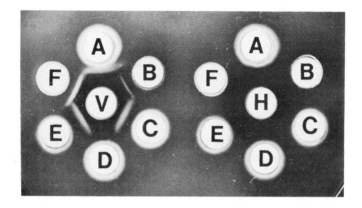

FIG. 11. Detectability of tobacco etch virus (TEV) capsid protein, nuclear inclusion proteins, and cylindrical inclusion protein by sodium dodecyl sulfate (SDS) immunodiffusion tests. The central wells contain boiled, SDS-treated sap from *Datura stramonium* leaves taken from TEV-infected plants (V) and from noninoculated, healthy plants (H). The peripheral wells contain undiluted sera as follows: (A) antiserum to untreated, purified TEV; (B) antiserum to SDS-treated, purified TEV-induced cylindrical inclusions; (C) antiserum to the 49,000-molecular weight TEV-induced nuclear inclusion protein; (D) normal (control) serum; (E) antiserum to the 54,000-molecular weight TEV-induced nuclear inclusion protein; (F) antiserum to SDS-treated, purified TEV.

The agar plugs created by the cork borers are removed by aspiration with a Pasteur pipet or cannula attached to a vacuum line. The gel patterns should be punched just prior to use.

g. Addition of Reactants, Incubation, and Recording of Results. Usually the sera are added to center wells and antigens to peripheral wells, especially when the degrees of relationship between antigens are being assessed. A Pasteur pipet or a pipettor with disposable plastic tips is used to add the solutions. Plates are incubated at 25°C for 3–4 days, and results are noted at least every day, beginning not later than 24 hr after adding the reactants. The appearance of precipitin lines in SDS–immunodiffusion tests is ephemeral, and the lines usually disappear 4–5 days after the reactants are added. The lines can be partially stabilized by emptying all wells after 44–48 hr of incubation and filling them with a slurry of Norit A charcoal (15%, w/v, in water; Fisher Scientific).

The precipitin lines can be visualized by viewing the plates with darkfield illumination. Results can be recorded photographically and by drawings.

h. Intragel Absorption Tests. Intragel absorption tests (van Regenmortel, 1967) are useful for studying inclusion body relationships (Purcifull *et al.*, 1973) and for removing antibodies to host proteins from a serum. Inclusions at a concentration of 1 A_{280} unit/ml or concentrated host proteins are added to the central wells of seven-well patterns. After incubation for 16–20 hr, the central well is emptied, then refilled with appropriate serum. Test antigens are added to the peripheral wells. The immunoprecipitate resulting from the absorption occurs as a white ring around the serum well.

i. Comments. The SDS–immunodiffusion system has been used successfully to detect the inclusion body proteins of 14 potyviruses and for detecting the capsid proteins of at least 66 viruses in 16 groups (Purcifull, unpublished compilation from the literature, personal communications, and unpublished results), and for detecting viral aggregate inclusions of citrus tristeza virus (Lee *et al.,* 1982). Because of its wide applicability and special suitability for analysis of proteins isolated by SDS–PAGE, the SDS–immunodiffusion method should continue to be useful for analysis of virus-induced inclusion body proteins.

2. Immunodiffusional Analyses of Antigens Prepared by Other Methods

Tobacco etch virus cylindrical inclusions are disrupted in the presence of potassium phosphotungstate at pH 6.9 (Purcifull *et al.,* 1970) to give antigenic components that react against antisera to cylindrical inclusions but not with virus antiserum (Hiebert *et al.,* 1971). Those immunodiffusion tests were conducted in media composed of 0.85% Noble agar, 0.85% sodium chloride, and 0.1% sodium azide. This procedure, however, was unsuccessful with potato Y virus.

Cauliflower mosaic inclusion matrix protein was serologically distinct from viral coat protein (Shepherd *et al.,* 1980). Shepherd *et al.* used protein suspended in a buffer consisting of 25 mM Tris-HCl (pH 8.2), 50 mM NaCl, 1 mM EDTA, 1 mM 2-mercaptoethanol, and 0.5% Triton X-100. The immunodiffusion medium consisted of 0.6% agarose containing 0.5% Triton X-100, 50 mM Tris-HCl (pH 8.2), and 50 mM NaCl. Virus and matrix protein were both tested at 0.2 mg/ml.

Alkaline-treated carnation etched ring virus inclusions yielded a rapidly diffusing antigen that was not detected in extracts from healthy plants. The antigen was detected under a variety of preparative conditions (Lawson and Hearon, 1980). Immunodiffusion media consisted of 1% agarose and 0.1% sodium azide with or without 0.85% NaCl. Antivirus sera were tested against inclusions in the same media.

Ethanolamine and pyrrolidine have been used effectively to degrade potyviruses into diffusible, antigenic fragments (Purcifull and Shepherd, 1964; Purcifull and Gooding, 1970; Shepard *et al.,* 1974a; Batchelor, 1974). Although weak reactions have been obtained with cylindrical inclusion antisira and pyrrolidine- or ethanolamine-treated sap from infected plants, the reactions were unsatisfactory by comparison with the SDS–immunodiffusion system (Purcifull, 1982 unpublished data; Batchelor, 1974). For example, Batchelor (1974) tested various materials for effectiveness against TEV, TEV cylindrical inclusions, and TEV nuclear inclusions in a gel medium consisting of 0.6% Noble agar, 0.03% sodium azide, and 50 mM borate (pH 8.0), and found that 3% pyrrolidine and 3% ethanolamine treatments resulted in very strong reactions with virus, weak reactions with cylindrical inclusions, and no reactions with nuclear inclusions.

C. Immunoprecipitation of in Vitro Translation Products

Antibodies prepared to potyviral inclusion proteins have been very useful in analyzing the products of the *in vitro* translations of potyviral RNAs. The experimental approach for these studies utilized antibodies prepared to potyviral cylindrical inclusion and TEV nuclear inclusions and *Staphylococcus aureus* strain Cowan 1 as an immunoabsorbent (Dougherty and Hiebert, 1980b). The immunoprecipitated, *in vitro* translation products are then analyzed by SDS–PAGE. *In vitro* translation products have been identified as virus-specific proteins, premature terminations, adjacent gene readthroughs, and "polyproteins" on the basis of size, serological reactions, and peptide mapping (Dougherty and Hiebert, 1980b,c; Hiebert, 1981). The immunological analyses of the *in vitro* translation products can provide significant information regarding viral genome structure, expression, and relationships (Dougherty and Hiebert, 1980b; Hiebert, 1981; Hiebert, 1982, unpublished data).

1. *Immunoprecipitation Procedure according to Dougherty and Hiebert (1980a,b)*

1. *In vitro* translation products are dissociated by the addition of 3 volumes of solution containing 2% (w/v) SDS, 5% (v/v) 2-ME, and 10% (v/v) glycerol in 62.5 mM Tris-HCl (pH 6.8) and heating at 90°C for 3–5 min.
2. Forty microliters of dissociated products are added to 80 μl of NET buffer [150 mM NaCl, 5 mM EDTA, 50 mM Tris-HCl (pH 7.4), and 0.02% sodium azide], 0.05% Nonidet P-40 (NP-40; Sigma N 6507), ovalbumin (1 mg/ml), and 2 mM methionine. Twenty microliters of antiserum are added, and the mixture is incubated for 1 hr at room temperature.
3. The antibodies are immunoprecipitated by the addition of 20 μl (10% w/v) of *Staphylococcus aureus* strain Cowan 1 and incubation for 30 min at room temperature.

4. Immunoabsorbent is collected by centrifugation at 2000 g for 5 min.
5. The pellet is resuspended in 250 μl of 0.05% NP-40 in NET buffer.
6. Immunoabsorbent is washed three times by repeating steps 4 and 5.
7. The final pellet is resuspended in 40 μl of solution containing 62.5 mM Tris-HCl (pH 6.8), 2% SDS, 5% 2-ME, and 10% glycerol, and heated at 90°C for 3-5 min.
8. Immunoprecipitated products, 10–40 μl (depending on the radioactivity in the sample) per well, are analyzed by SDS–PAGE.
9. Products are detected in dried gels by fluorography (Bonner and Laskey, 1974) (Fig. 12).

The immunoprecipitation procedure outlined above is a modification of the procedure described by Kessler (1975). The major modification involves the use of SDS to reduce or eliminate nonspecific serological reactions during immunoprecipitation. SDS concentration of less than 0.4% in the final incubation mixture resulted in nonspecific immunoprecipitation of products using control antisera such as preimmune serum (W. G. Dougherty, 1982, unpublished data). Specific reactions were detected even at 1% SDS concentrations.

In order to optimize the efficiency of the immunoprecipitation using *Staphylococcus aureus* strain Cowan 1, the bacterial cells should be washed within 24 hr prior to use (Kessler, 1975). A 1-ml aliquot of a bacterial suspension (10% w/v) is centrifuged at 2000 g for 5 min. The pellet is resuspended in 1 ml of 0.5% NP-40 NET buffer and incubated at 24°C for 15 min. The bacteria are centrifuged at 2000 g for 5 min and resuspended in 1 ml of 0.05% NP-40 NET buffer. They are centrifuged again and resuspended in 1 ml of 0.05% NP-40 NET buffer containing 1 mg/ml ovalbumin, and 2 mM of the amino acid analog used in labeling of the translation products. The treatments outlined above should eliminate free protein A or fragmented cells that would reduce the efficiency of the immunoprecipitations.

2. Immunoprecipitation according to Hellmann et al. (1983)

The *in vitro* translational products of tobacco vein mottling virus (TVMV) have also been immunologically analyzed using antisera to TVMV cylindrical inclusions, TVMV helper factor, and TEV nuclear inclusions (Hellmann *et al.*, 1983). The immunoprecipitation procedure described here does not use SDS. Aliquots of translation mixtures are incubated with 2 μl of antiserum in 175-μl reaction mixtures containing 1.5% Triton X-100, 1.5% sodium deoxycholate, and the appropriate nonradioactive amino acid at 20 mM. After 30 min of incubation at 25°C with shaking, 15 μl of a 90 mg/ml solution of *S. aureus* strain Cowan 1 (IgGsorb, The Enzyme Center) in physiological saline is added and incubation is continued for 30 min. Im-

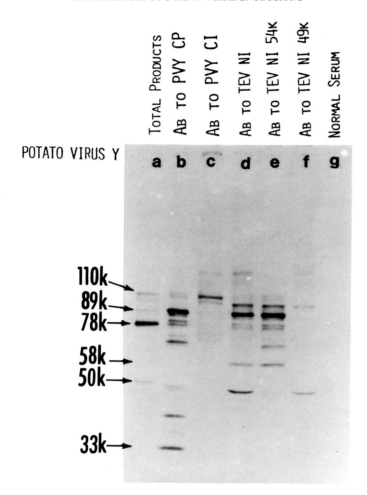

FIG. 12. Analysis of the *in vitro* translation products of potato virus Y (PVY) RNA by immunoprecipitation and SDS–PAGE. Lanes: Total products of PVY RNA translation (a); products immunoprecipitated by PVY coat protein antiserum (b), by PVY cylindrical inclusion antiserum (c), by tobacco etch virus (TEV) nuclear inclusion (NI) antiserum (d), by antiserum to 54K TEV NI protein (e), by antiserum to 49K TEV NI protein (f), and by normal serum (g). The [³⁵S]-methionine-labeled products were detected by fluorography. The molecular weights of some major translation proteins are given on the left.

munoprecipitates are collected by sedimentation through sucrose (Rhoads *et al.,* 1973).

3. Immunoprecipitation Procedure according to Xiong et al. (1982b)

Antibodies to the CaMV major viroplasm protein have been used to identify *in vitro* translation products of CaMV mRNA (Xiong *et al.,* 1982b).

1. The translation mixture is centrifuged at 40,000 g for 1 hr. The pellet is discarded.
2. Ten microliters of antiserum are added to 35 μl of supernatant.
3. The mixture is diluted by the addition of 35 μl of 2.3-fold concentrated phosphate-buffered saline (PBS) (pH 7.4).
4. After incubation for 1 hr at room temperature, 15 μl of protein A-Sepharose CL-4B (Sigma) (concentration 30 mg/ml) is added and incubation at room temperature is continued for 3 hr.
5. The immunoprecipitate is collected by centrifugation.
6. The pellets are washed with PBS containing 1% Triton X-100 and 0.1% 2-ME until no radioactivity can be detected in the supernatant.
7. The final pellet is dissociated in 2% SDS and 2% 2-ME at 100°C for 2 min and electrophoresed in SDS–PAGE.

D. Immunofluorescence Microscopy

Antibodies prepared to the potyviral cylindrical inclusions, amorphous inclusions, TEV nuclear inclusions, TVMV helper factor (Hellmann *et al.,* 1983; Thornbury and Pirone, 1983), and capsid protein represent probes for more than 80% of the total genetic information in the potyviral genome. These serological probes may be used to analyze the development and distribution of the virus-specific proteins in infected cells and tissues. By incubating infected cells and tissues with antibody conjugated to a fluorescent dye, one may visualize the distribution and localization of antigens in these tissues and cells directly under the microscope.

The principles of immunofluorescence staining and the methods involved with animal viruses have been given in this series by Emmons and Riggs (1977).

1. *Fractionation of Serum*

The immunoglobulin G (IgG) is purified by the use of protein A–Sepharose CL-4B (Pharmacia) as a specific-affinity absorbent using the procedure outlined by Miller and Stone (1978).

1. Antiserum (2–3 ml) is centrifuged at 27,000 g for 20 min.
2. The supernatant is layered on top of a 5-ml protein A–Sepharose CL-4B gel column (Bio-Rad 1.5/15 Econo-column) equilibrated with buffer containing 16 mM boric acid, 12 mM NaCl, 2.5 mM NaOH, 0.1 mM phenylmethylsulfonyl fluoride (PMSF), and 0.02% NaN$_3$ (pH 8.0). The column is washed with 30 ml of buffer as above at a flow rate of 10–15 ml/hr. Chromatography is carried out at 4°C.
3. The bound IgG is eluted from the column with buffer containing 0.1 M glycine-HCl (pH 3.0), 0.1 mM PMSF, and 0.02% NaN$_3$, at a flow

rate of 15–20 ml/hr. Protein fractions eluting from the column are measured for absorbance at 280 nm. Fractions containing absorbance values greater than 1 are combined and dialyzed against three 1-liter changes of the same buffer as in step 2, but without PMSF and NaN₃.

2. Conjugation

Purified IgG is conjugated with fluorescein isothiocyanate (FITC) (Research Organics, Inc., Cleveland, OH) or with tetramethylrhodamine isothiocyanate (TRITC) (BBL Microbiology Systems, Becton Dickinson and Co.) by dialysis according to Research Organics Inc. (Cleveland, OH 44125, Bull. 95).

1. Purified IgG (2–4 mg/ml), 1.2 ml, is mixed with 0.8 ml of buffer containing 1 M NaCl and 0.1 M borate (pH 9.3). The mixture is added to a dialysis bag.
2. The protein solution is dialyzed for 12–15 hr at 4°C against 100 ml of 50 mM borate buffer (pH 9.3) containing 3 mg of TRITC or 3 mg of FITC.
3. After dialysis, the unconjugated fluorochromes are separated from the IgG conjugates by chromatography on a Sephadex G-50 column equilibrated with buffer containing 20 mM sodium phosphate (pH 7.4) and 0.85% NaCl.
4. The IgG conjugate (~4 ml) is filtered through a (Millipore) 0.1-μm filter and stored at 4°C after the addition of 0.02% NaN₃.

3. Immunofluorescent Staining of Infected Tissues

a. Procedure. Our preliminary attempts to study the distribution of virus-specific antigens in infected tissue have been limited to use of fresh, unfixed tissue. Epidermal tissues from virus-infected plants often contain abundant inclusions. Satisfactory epidermal strips can be removed from many plant species by inserting the tips of a pair of sharp-pointed tweezers under the epidermis and stripping it from the underlying tissues (see Section II,A). The epidermal strips are floated onto a drop of conjugated antibody with the torn surface in contact with the antibody. The drop of conjugated antibody [27 μl antibody, 3 μl 10% dimethyl sulfoxide (DMSO) in saline] is placed in a 50 × 9-mm style petri dish with a tight lid (Falcon 1006). The DMSO is added to the antibody to enhance immunocytochemical staining and antibody penetration (Herbert et al., 1982). The epidermal strips in contact with the antibody are incubated in a moist chamber at 30°C for 1.5 hr. After incubation the epidermal strips are blotted, washed three times with saline, and incubated for 1 hr on top of a large drop (100–200 μl) of saline. The epidermal strips are picked up with the aid of a wooden applicator stick, blotted, and transferred to an aqueous nonfluorescing medium

(Aqua-mount, Cat. No. M-800, Lerner Laboratories) on a glass slide. A coverslip is placed over the mounting medium. Observations are made with epifluorescence microscopy using the appropriate filters for FITC or TRITC.

b. *Comments.* Examples of the specificity of the TRITC- conjugated antibodies are illustrated in Fig. 13. Unfortunately the TEV NI (Fig. 1) and the cytopathological structures associated with pepper mottle virus infections (Fig. 2) show autofluorescence at the epi-illumination wavelength used for FITC studies. We have not tested counterstains (Schenck and Churukian, 1974) that may reduce or eliminate this autofluorescence in the presence of blue-light excitation.

Nonspecific binding of antibodies may be reduced or minimized by a 30-min preincubation of antisera (27 μl) with healthy tissue extract (27 μl of tissue extracted in 10-fold (weight) saline buffer (Ko and Hiebert, unpublished). After this preincubation the serum is used for immunofluorescence as described in Section III,D,3,a.

With the availability of protein A labeled with either FITC or rhodamine (TRITC) (i.e., LKB Instruments Ltd.) there is no need to conjugate specific antisera with fluorescent compounds. In this approach the epidermal strips are floated onto a drop of nonlabeled antiserum, incubated as described in Section III,D,3,a, and washed with saline and incubated as before. Then the epidermal strips are floated onto a drop of labeled protein A (diluted 1/50 in saline) and incubated for 1 hr. The epidermal strips are washed several times with saline and incubated on top of a large drop of saline for 30 min. The epidermal strips are mounted and viewed as described in Section III,D,3,a.

Penetration of the conjugated antibody to the antigen sites in the fresh tissue is a problem. We have not tested our conjugated sera against tissue slices to see whether antibody penetration is more efficient under these conditions. Immunohistofluorescence studies with plant tissue slices have been reported (e.g., Hapner and Hapner, 1978; Jeffree and Yeoman, 1981; Weidemann, 1981).

The use of acetone to fix the tissue prior to incubation with conjugated antibody may also aid in the penetration of antibody to the antigen sites. Fujisawa *et al.* (1974) used FITC-conjugated cauliflower mosaic virus (CaMV) antiserum to study the serological reaction of dahlia mosaic virus-induced inclusions (viroplasms) in acetone-fixed tissue. The epidermal layer is stripped from infected dahlia leaves and fixed with acetone for 2 hr at room temperature. After fixation the specimen is washed four times with phosphate-buffered saline (PBS). The specimen is immersed in CaMV-conjugated antiserum and incubated for 24 hr at 36°C. Then the specimen is washed four times in PBS. The stained specimen is immersed in PBS containing 10% glycerin and examined under a fluorescence microscope.

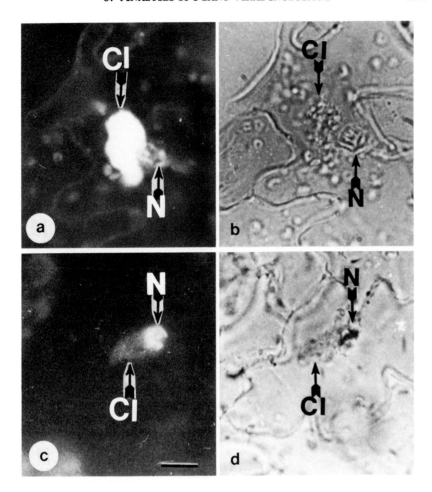

FIG. 13. Immunofluorescence of the nuclear inclusions and cylindrical inclusions (CI) in tobacco infected with tobacco etch virus (TEV). (a) An epidermal strip stained with TRITC-labeled TEV CI antiserum. (b) A nonfluorescing view of the same cell as in (a) for orientation. (c) An epidermal strip stained with TRITC-labeled TEV nuclear inclusion antiserum. (d) A nonfluorescing view of the same cells as in (c). N, nucleus containing nuclear inclusions. Magnification bar equals 10 μm for the micrographs. No specific fluorescence was observed when preimmune serum labeled with TRITC was tested on TEV-infected tissues (data not shown).

Enzymatic digestion of cell walls prior to incubation with conjugated antibody has been used to facilitate antibody penetration of plant cells. Nishiguchi *et al.* (1980) followed the spread of tobacco mosaic virus in tomato leaf epidermis by staining the isolated epidermis with fluorescent antibody after enzymatic digestion of the upper cell wall. The lower epidermis is isolated from leaf tissue with forceps and placed on the surface of a glass slide which has been smeared with Mayer's albumin. The epidermis is dried for 30 min with cool air using a hair dryer, and fixed in acetone for 10 min at room temperature. The epidermis is covered with a drop of an enzyme solution consisting of 0.5% (w/v) Macerozyme R-10, 1% (w/v) Cellulase Onozuka R-10, 0.1% (w/v) potassium dextran sulfate, 0.2 M KCl, and 0.2 M mannitol (pH 7.0). The slide is then placed on a moist filter paper in a petri dish and maintained at 36°C for 30 min. Next the glass slide is immersed in a solution of 0.2 M KCl and 0.2 M mannitol, then washed in PBS for 30 min. The excess PBS is blotted from the slide with filter paper and then the specimen on the slide is covered with fluorescent antibody solution in PBS. The slide is incubated with moist filter in a petri dish sealed with Parafilm for 2 hr at 36°C or overnight at 4°C in the dark. The specimen is washed again with PBS for 1 hr and mounted with PBS including 10% (v/v) glycerol for examination under a fluorescence microscope.

E. Ferritin–Antibody Electron Microscopy

The conjugation of electron-dense ferritin to an antibody combines the specificity of antibodies for homologous antigen with high-resolution electron microscopy. Details regarding techniques of ferritin-tagged antibodies are given in this series (Breese and Hsu, 1971), so the discussion here will be limited to examples involved in plant viral inclusion studies.

1. Procedures

Shepard and Shalla (1969) used the method of direct ferritin tagging to study the serological reaction of cylindrical inclusions in TEV-infected tobacco tissue. Glutaraldehyde-fixed tissue pieces are frozen in a drop of water and shaved paradermally with a razor blade. In order to make these shaved pieces more permeable to the conjugated antiserum, they are exposed to ultrasonication for 20–40 sec immediately prior to placement into conjugated antiserum. The shaved and sonicated slices are soaked in the ferritin-antibody solution for 2 hr, rinsed in 0.1 M neutral phosphate buffer, postfixed with buffered 2% osmium tetroxide, and embedded in Araldite epoxy resin. The material is then sectioned and stained for electron microscopy.

Shalla and Shepard (1972) used direct ferritin-labeled antibodies to determine whether the potato virus X (PVX) cellular (laminate) inclusions were antigenically related to PVX or to its antigenically distinct depolymerized viral structural protein. In order to ensure adequate antibody penetration to the site of inclusions in the virus-infected tissues, two methods were used. In one set of experiments, tissue was treated as described for the TEV cylindrical inclusion study (Shepard and Shalla, 1969). The second approach involved the labeling of subcellular fractions. Infected leaves are cut into 1-cm² pieces, infiltrated with 2% buffered glutaraldehyde under a mild vacuum for 10 min, and then disrupted with a Waring blender for 30 sec. Cell-free inclusions of PVX are collected by centrifugation at 1000 g for 10 min. Inclusion components in the pellet are washed and then incubated with the ferritin–antibody preparation. After rinsing to remove unbound ferritin, the inclusion components are treated with osmium tetroxide, dehydrated, and embedded for thin sectioning. The material is then sectioned and stained for electron microscopy.

In another study Shepard et al. (1974b) used indirect immunoferritin experiments to study serological reactions of the TEV-induced nuclear and cylindrical inclusions in fractured plant cells. In order to ensure adequate penetration of the ferritin–antibody conjugate to the antigen sites, a method was devised to fracture cells, nuclei, and cytoplasmic regions. Tissue strips 2–4 mm wide are cut with a razor blade from TEV-infected and uninfected leaves and fixed for 2–3 hr in a 3% glutaraldehyde solution buffered to pH 7.0 with 0.1 M phosphate. The strips are rinsed 1–2 hr in several changes of phosphate buffer and then pulverized with a mortar and pestle in the presence of liquid nitrogen. The powder with a small volume of liquid nitrogen is poured into a clinical centrifuge tube. The tube is filled with phosphate buffer. The large tissue pieces are allowed to settle for 5 min, and then the buffer with smaller fragments is decanted into a second centrifuge tube. Cell fragments are collected by centrifugation at 100 g for 10 min. The pellet is resuspended in 1 ml of 0.1 M phosphate buffer, pH 7.0. One milliliter of rabbit antibody is added to the resuspension and gently agitated with a magnetic stirrer for 1 hr. The cell fragments are separated from unreacted antibody by centrifugation at 100 g and then rinsed and recentrifuged three or four times over the course of 2 hr. After the final rinse the pellet is dispersed in 2 ml of phosphate buffer, and then 0.5 ml of ferritin-tagged sheep anti-rabbit-IgG antibody is added. Incubation is for 1 hr with gentle stirring. Unreacted antibody is removed by centrifugation and four washes during a 2-hr period as described above. The final pellet is covered with 1% solution of osmium tetroxide and postfixed in the cold overnight. The material is dehydrated with acetone, stained with uranyl acetate in acetone, and embedded. The material is then sectioned and stained for electron microscopy.

2. Comments

Limited penetration of the conjugated antibodies into cytoplasm and nuclei in plant tissue is a problem in the immunoferritin procedures. Methods that aid in the penetration of the ferritin conjugate to the reaction sites must be balanced with the need to preserve cell structure integrity in order to identify the site of reaction.

Another problem in this technique is the high levels of nonspecifically bound ferritin. Shepard *et al.* (1974b) found that treating cell fragments with 2 ml of sheep immunoglobulin (not tagged with ferritin) for 30–60 min prior to adding rabbit antibody greatly reduced the levels of nonspecific ferritin binding.

F. Liquid Precipitin Tests

Conventional liquid precipitin or microprecipitin tests (Ball, 1974) have been useful for detecting the reactivity of virus-induced inclusion antisera to virions of carnation etched ring virus (Lawson and Hearon, 1980). Microprecipitin tests also were used to detect contaminating antibodies to viral capsid proteins in antisera prepared to partially purified tobacco etch virus-induced inclusions (Hiebert *et al.*, 1971).

Hitchborn and Hills (1968) used liquid precipitin tests to determine that the unusual tubular inclusions (Section II,G) associated with infection of plants by a strain of turnip yellow mosaic virus (a tymovirus) consisted of aggregates of viral capsid protein.

G. Western Blotting and Radioimmunoassay

Xiong *et al.* (1982b) used radioimmunoassays (Muller *et al.*, 1982) and a protein blotting procedure combined with autoradiography (Towbin *et al.*, 1979) to evaluate the reactivity and specificity of antiserum to the cauliflower mosaic virus-induced viroplasm protein.

1. *Procedures*

For Western blotting, the partially purified viroplasms are subjected to SDS–gel electrophoresis on 10% polyacrylamide slab gels. The protein samples are electrophoretically transferred to nitrocellulose sheets (0.45 μm pore size), additional binding sites are saturated with 3% bovine serum albumin, and the sheets are washed in phosphate-buffered saline containing Tween 20 and EDTA. The sheets are incubated independently in sera (cauliflower mosaic virus antiserum, cauliflower mosaic virus viroplasm protein antiserum, and normal serum, each diluted 1:100), and the reaction mixtures are shaken at room temperature for 9 hr. The sheets are washed and treated with ^{125}I-labeled protein A, incubated for 45 min at 37°C, washed to remove unbound ^{125}I-labeled protein A, air dried, and exposed for autoradiography at −70°C. Details of the protein blotting and autoradiographic procedures are described by Xiong *et al.* (1982b), Towbin *et al.* (1979), and de Mejia (1984).

Solid-phase radioimmunoassays of cauliflower mosaic virus viroplasm protein were conducted by Xiong *et al.* (1982b) according to the general procedures of Muller *et al.* (1982). Test antigens are prepared by treatment of partially purified viroplasm preparations with 2% SDS and 2% 2-mercaptoethanol at 100°C for 2 min. The samples are centrifuged to remove insoluble material, and the supernatants are used as test antigens. The antigens are diluted in 0.05 M sodium carbonate buffer (pH 9.6), and 250 μl of the diluted samples is added to the V-shaped wells of flexible, polyvinyl chloride microtiter plates (Dynatech Laboratories, Alexandria, VA). After incubation overnight at 37°C, nonabsorbed antigen is removed by washing the plates with PBS-T-E (phosphate-buffered saline, pH 7.4, containing 0.05% Tween 20 and 1 mM EDTA). Antisera diluted in PBS-T-E are added to the wells (250 μl/well) and the plates are incubated for 3 hr at 37°C. The plates are washed twice with BBS-T (borate-buffered saline, pH 8.5, and containing 0.1% Tween-20) and three times with PBS-T-E. Then 250 μl of ^{125}I-labeled protein A (4 $\times$ 10^4 cpm/ml, obtained from Amersham International, Amersham, England) is added, and the plates are incubated for 1 hr at 37°C. The plates are washed extensively with BBS-T, the individual wells are cut out, and bound radioactivity is measured with a gamma counter.

2. *Comments*

The solid-phase radioimmunoassays employed enabled Xiong *et al.* (1982b) to detect concentrations of viroplasm protein estimated to be as low as 2.5 ng/ml. The radioimmunoassays also showed the antigenic specificity of viroplasm protein as being distinct from that of viral capsid protein. The autoradiographic results also demonstrated the specificity of the antiviroplasm protein antiserum.

H. ENZYME-LINKED IMMUNOSORBENT ASSAYS (ELISA)

The ELISA tests have become widely used since 1977 for the detection of plant viral antigens (e.g., Clark and Adams, 1977; Koenig, 1981; van Regenmortel, 1982). Falk and Tsai (1983) prepared antiserum to the nonstructural protein (Section II,F) associated with infection of corn by maize stripe virus and used indirect ELISA to detect the protein in corn sap. Yeh (1984) used indirect ELISA to detect papaya ringspot virus cylindrical inclusion protein.

1. *Indirect ELISA for Maize Stripe Nonstructural Protein (Falk and Tsai, 1983)*

1. Polystyrene microtiter plates are coated with sap from infected corn plants diluted from 10^{-1} to 10^{-5} in coating buffer (0.05 M sodium

carbonate, pH 9.6) overnight at 6°C. Sap from healthy corn plants is treated similarly for use as a control.

2. The plates are washed with PBS–Tween (0.02 M phosphate, 0.15 M NaCl, 0.05% Tween-20, pH 7.4).

3. The IgG fraction of antiserum to the maize stripe virus-induced non-structural protein is purified according to Clark and Adams (1977). The IgG fraction is diluted in PBS–Tween to a concentration of 1 $\mu g/$ml and added to the plates for 3 hr at 25°C.

4. The plates are washed with PBS–Tween.

5. Goat anti-rabbit antiserum (Miles Laboratories) is conjugated with alkaline phosphatase (Sigma) according to the procedures outlined by Clark and Adams (1977). The conjugate is diluted to 10^{-3} in PBS–Tween containing 2% polyvinylpyrrolidone (PVP-40; Sigma) and added to the plates for 2 hr at 25°C.

6. The plates are washed with PBS–Tween.

7. P-Nitrophenyl phosphate (Sigma) is prepared at a concentration of 0.6 mg/ml in diethanolamine buffer (pH 9.8) and added to the plates. After incubation at 25°C for 30 min to 2 hr, the reactions are stopped by adding 3 M NaOH (Clark and Adams, 1977), and the results are assessed photometrically at 405 nm.

2. Indirect ELISA for Papaya Ringspot Virus (Potyvirus) Cylindrical Inclusion Protein (Yeh, 1984)

1. Using a stock solution of purified cylindrical inclusion protein at a concentration of 1 mg/ml in 0.125 M Tris-HCl (pH 6.8) containing 0.5% SDS, make dilutions in coating buffer (0.05 M sodium carbonate, pH 9.6, containing 0.01% sodium azide). (Crude plant extracts diluted in coating buffer also were used.) Add diluted antigens to wells of polystyrene microliter plates and incubate at 6°C for 15–18 hr or at 30°C for 4–6 hr.

2. Rinse plates three times with PBS–Tween.

3. Using IgG purified from cylindrical inclusion protein antiserum by the method of Clark and Adams (1977), dilute IgG to a concentration of 1 $\mu g/$ml in enzyme buffer [PBS–Tween, 2% polyvinylpyrrolidone (molecular weight, 40,000) and 0.2% ovalbumin]. Add 200 μl of diluted IgG per well and incubate at 30°C for 4 hr.

4. Rinse three times with PBS–Tween, and add 200 μl of goat anti-rabbit IgG–alkaline phosphatase conjugate (Sigma Chemical Co., No. A-8025) at a dilution of 1/1000 in enzyme buffer. Incubate 4 hr at 30°C.

5. Rinse plates with PBS–Tween, and add 220 μl of p-nitrophenyl phosphate at a concentration of 1 mg/ml in 0.1 M diethanolamine buffer (pH 9.6) per well, and incubate at room temperature. Stop reactions by adding 50 μl of 3.0 M NaOH per well, if necessary.

3. Comments

For maize stripe mosaic virus-induced nonstructural protein, the indirect ELISA gave strong reactions with sap from infected plants when diluted to 10^{-6} to 10^{-7}, and the method was much more sensitive than the double-antibody sandwich procedure. With papaya ringspot virus, the indirect ELISA detected cylindrical inclusion protein at concentrations as low as 1.6 ng/ml and at sap dilutions as low as 3.2×10^{-5}. The double-antibody sandwich method gave nonspecific reactions in tests to detect cylindrical inclusion protein.

IV. Discussion

The value of studying plant virus-induced inclusions is illustrated in the research reported for the potyviral cylindrical inclusions, TEV nuclear inclusions, and caulimovirus viroplasm protein. The cylindrical inclusion protein (Purcifull et al., 1973; Dougherty and Hiebert, 1980b), the TEV nuclear inclusion proteins (Knuhtsen et al., 1974; Dougherty and Hiebert, 1980b), the amorphous inclusion protein (de Mejia, 1984; de Mejia et al., 1984), and the cauliflower mosaic virus viroplasm protein (Xiong et al., 1982b) have been demonstrated to be virus-specific proteins distinct from the respective virion capsid protein. Furthermore, the antisera prepared to the potyviral cylindrical inclusions and TEV nuclear inclusions have been invaluable in analyzing the in vitro translations of potyviral RNAs and in the development of a genetic map for the potyviral genome (Dougherty and Hiebert, 1980c; Hiebert, 1981).

The cylindrical inclusion protein, the amorphous inclusion protein, and the TEV nuclear inclusion proteins represent ~70% of the total protein coding capacity of the potyviral genome. Antisera to these inclusion proteins are potentially useful in viral diagnosis and in studying potyviral relationships. Serological comparisons limited to antisera for the virion capsid protein probe only ~10% of potyviral protein coding capacity. Thus potyviral inclusion research has led to the development of new serological probes for analyzing a much greater proportion of the potyviral genome.

The techniques described here, such as the detection of inclusions by light microscopy, purification by selective low speed centrifugations, and immunogen preparation for antisera production, should be useful in studying other plant virus inclusion bodies. Cytological studies of infected tissue show that most, if not all, plant viruses induce recognizable inclusions (Christie and Edwardson, 1977). Many of these inclusions very likely are aggregated virions, based on staining reactions and morphological and ultrastructural studies. Studying these inclusions may not lead to the detection of nonstructural virus-specific proteins but could be valuable in providing a novel way for obtaining purified virions. An example of such an approach has been reported for citrus tristeza virus (Lee *et al.,* 1982). The laminate inclusion component associated with potato virus X (Shalla and Shepard, 1972; Christie and Edwardson, 1977) and X bodies associated with certain tobamoviruses (Christie and Edwardson, 1977) are examples of virus-induced inclusions that may contain nonstructural, virus-specific proteins, but the information is lacking. In addition to containing nonstructural, virus-specific proteins, some of the virus-induced inclusions could be sites of viral synthesis.

Information is needed regarding the possible function(s) of the cylindrical and nuclear inclusion proteins induced by potyviruses and the viroplasm protein induced by caulimoviruses. Since a significant proportion of the limited genetic content of members in these two virus groups is devoted to inclusion protein, it is reasonable to assume that the inclusion proteins have essential functions. The elucidation of the functions of the virus-specific proteins may be important in understanding how viruses act as pathogens.

It has been demonstrated that potyviruses and caulimoviruses produce very significant amounts of nonstructural proteins *in vitro* (Hiebert and McDonald, 1973; Odell and Howell, 1980). There is no reason not to believe that members of other plant virus groups may also accumulate significant amounts of nonstructural proteins *in vivo,* either in specific inclusions or in ill-defined cytopathological structures and localized sites of viral synthesis. It is our opinion that studies of nonstructural, virus-specific proteins provide interesting opportunities in plant virus research.

Acknowledgments

We wish to thank W. G. Dougherty, D. Gonsalves, and S.-D. Yeh for allowing us to include some of their unpublished data in this chapter. We also thank E. Balàzs for making manuscripts available to us before publication. We thank S. R. Christie for advice and assistance at various stages in our inclusion body research. We are indebted to W. E. Crawford for technical assistance. It is a pleasure to thank J. R. Edwardson, whose dedication to the study of plant viral inclusions is a continual inspiration, for his support during the writing of this chapter and for his critical review of it.

Our research has been supported, in part, by the National Science Foundation (Grants GB-32093, BMS75-14014, and PCM-7825524).

References

Al Ani, R., Pfeiffer, P., Whitechurch, D., Lesot, A., Lebeurier, G., and Hirth, L. (1980). *Ann. Virol. (Institut Pasteur)* **131E,** 33–55.

Ansa, O. A., Boyer, J. W., and Shepherd, R. J. (1982). *Virology* **121,** 141–156.

Assink, A. M., Swaans, H., and van Kammen, A. (1973). *Virology* **53,** 384–391.

Ball, E. M. (1974). "Serological Tests for the Identification of Plant Viruses." American Phytopathological Society, St. Paul, MN.

Batchelor, D. L. (1974). Ph.D. Dissertation, University of Florida.

Baum, R. H. (1980). Ph.D. Dissertation, University of Florida.

Bonner, W. M., and Laskey, R. A. (1974). *Eur. J. Biochem.* **46,** 83–88.

Breese, S. S., Jr., and Hsu, K. C. (1971). *Methods Virol.* **5,** 399–422.

Christie, R. G. (1967). *Virology* **31,** 268–271.

Christie, R. G., and Edwardson, J. R. (1977). *Fla. Agr. Exp. Stn. Monogr. Ser.* No. 9.

Christie, R. G., Hiebert, E., and Dougherty, W. G. (1982). *Phytopathology* **72,** 938.

Clark, M. F., and Adams, A. N. (1977). *J. Gen. Virol.* **34,** 475–483.

Covey, S. N., and Hull, R. (1981). *Virology* **111,** 463–474.

Daubert, S. D., Richins, R. D., and Shepherd, R. J. (1981). *Int. Congr. Virol. 5th* W18/05.

De Mejia, G. V. (1984). Ph.D. Dissertation, University of Florida.

De Mejia, G. V., Hiebert, E., and Purcifull, D. E. (1984). In preparation.

De Zoeten, G. A., Assink, A. M., and van Kammen, A. (1974). *Virology* **59,** 341–355.

Dougherty, W. G., and Hiebert, E. (1980a). *Virology* **101,** 466–474.

Dougherty, W. G., and Hiebert, E. (1980b). *Virology* **104,** 174–182.

Dougherty, W. G., and Hiebert E (1980c). *Virology* **104,** 183–194.

Dunn, D. B., and Hitchborn, J. H. (1965). *Virology* **25,** 171–192.

Edwardson, J. R. (1974). *Fla. Agr. Exp. Stn. Monogr. Ser.* No. 4.

Edwardson, J. R., and Christie, R. G. (1978). *Annu. Rev. Phytopathol.* **16,** 31–55.

Edwardson, J. R., and Christie, R. G. (1979). *Fitopatol. Bras.* **4,** 341–373.

Emmons, R. W., and Riggs, J. L. (1977). *Methods Virol.* **6,** 1–28.

Falk, B. W., and Tsai, J. H. (1982). *Phytopathology* **72,** 953.

Falk, B. W., and Tsai, J. H. (1983). *Phytopathology* **73,** 1259–1262.

Fenner, F. (1976). *Intervirology* **7,** 4–115.

Fujisawa, I., Rubio-Huertos, M., and Matsui, C. (1974). *Phytopathology* **64,** 287–290.

Garnsey, S. M., Gonsalves, D., and Purcifull, D. E. (1979). *Phytopathology* **69,** 88–95.

Gingery, R. E., Nault, L. R., and Bradfute, O. E. (1981). *Virology* **112,** 99–108.

Gooding, G. V., and Bing, W. W. (1970). *Phytopathology* **6,** 1293.

Hapner, S. J., and Hapner, K. D. (1978). *J. Histochem. Cytochem.* **26,** 478–482.

Harrison, B. D., Finch, J. T., Gibbs, A. J., Hollings, M., Shepherd, R. J. Valenta, V., and Wetter, C. (1971). *Virology* **45,** 356–363.

Hellmann, G. M., Thornbury, D. W., Hiebert, E., Shaw, J. G., Pirone, T. P., and Rhoads, R. E. (1983). *Virology* **124,** 434–444.

Herbert, D. C., Weaker, F. J., and Sheridan, P. J. (1982). *Histochem. J.* **14,** 161–164.

Hibi, T., Rezelman, G., and van Kammen, A. (1975). *Virology* **64,** 308–318.

Hiebert, E. (1981). *Int. Congr. Virol. 5th* P2/06.

Hiebert, E., and McDonald, J. G. (1973). *Virology* **56,** 349–361.

Hiebert, E., Purcifull, D. E., Christie, R. G., and Christie, S. R. (1971). *Virology* **43,** 638–646.

Hitchborn, J. H., and Hills, G. J. (1968). *Virology* **35,** 50–70.

Jeffree, C. E., and Yeoman, M. M. (1981). *New Phytol.* **87,** 463–471.

Kessler, S. (1975). *J. Immunol.* **115,** 1617–1624.

Knuhtsen, H., Hiebert, E., and Purcifull, D. E. (1974). *Virology* **61,** 200–209.

Koenig, R. (1981). *J. Gen. Virol.* **55** , 53–62.

Laemmli, U. K. (1970). *Nature (London)* **227,** 680–685.

Lawson, R. H., and Hearon, S. S. (1980). *Phytopathology* **70,** 327–332.

Lee, R. F., Garnsey, S. M., Brlansky, R. H., and Calvert, L. A. (1982). *Phytopathology* **72,** 953.

Lima, J. A. A., and Purcifull, D. E. (1980). *Phytopathology* **70**, 142–147.
Lima, J. A. A., Purcifull, D. E., and Hiebert, E. (1979). *Phytopathology* **69**, 1252–1258.
McDonald, J. G., and Hiebert, E. (1974). *J. Ultrastruct. Res.* **48**, 138–152.
McDonald, J. G., and Hiebert, E. (1975). *Virology* **63**, 295–303.
McWhorter, F. P. (1965). *Annu. Rev. Phytopathol.* **3**, 287–312.
Martelli, G. P., and Russo, M. (1977). *Adv. Virus Res.* **21**, 175–266.
Matthews, R. E. F. (1982). *Intervirology* **17**.
Miller, T. J., and Stone, H. O. (1978). *J. Immunol. Methods* **24**, 111–125.
Muller, S., Himmelspach, K., and van Regenmortel, M. H. V. (1982). *EMBO J.* **1**, 421–425.
Nagel, J., Zettler, F. W., and Hiebert, E. (1983). *Phytopathology,* **73**, 449–454.
Nishiguchi, M., Motoyoshi, F., and Oshima, N. (1980). *J. Gen. Virol.* **46**, 497–500.
Odell, J. T., and Howell, S. H. (1980). *Virology* **102**, 349–359.
Purcifull, D. E., and Batchelor, D. L. (1977). *Fla. Agr. Exp. Stn. Tech. Bull.* No. 788.
Purcifull, D. E., and Gooding, G. V. (1970). *Phytopathology* **60**, 1036–1039.
Purcifull, D. E., and Shepherd, R. J. (1964). *Phytopathology* **54**, 1102–1108.
Purcifull, D. E., Edwardson, J. R., and Christie, S. R. (1970). *Phytopathology* **60**, 779–782.
Purcifull, D. E., Hiebert, E., and McDonald, J. G. (1973). *Virology* **55**, 275–279.
Purcifull, D. E., Zitter, T. A., and Hiebert, E. (1975). *Phytopathology* **65**, 559–562.
Purcifull, D. E., Edwardson, J., Hiebert, E., and Gonsalves, D. (1984). CMI/AAB Descriptions of Plant Viruses, in press.
Rezelman, G., Franssen, H. J., Goldbach, R. W., Ie, T. S., and van Kammen, A. (1982). *J. Gen. Virol.* **60**, 335–342.
Rhoads, R. E., McKnight, G. S., and Schimke, R. T. (1973). *J. Biol. Chem.* **248**, 2031–2039.
Rottier, P. (1980). Ph.D. Dissertation, Landbouwhogeschool, Wageningen, Netherlands.
Rubio-Huertos, M. (1972). *In* "Principles and Techniques in Plant Virology" (C.I. Kado and H. O. Agrawal, eds.), pp. 66–75. Van Nostrand-Reinhold, New York.
Schenck, E. A., and Churukian, C. J. (1974). *J. Histochem. Cytochem.* **22**, 962–966.
Shalla, T. A., and Shepard, J. F. (1972). *Virology* **49**, 654–667.
Shepard, J. F., and Shalla, T. A. (1969). *Virology* **38**, 185–188.
Shepard, J. F., Secor, G. A., and Purcifull, D. E. (1974a). *Virology* **58**, 464–475.
Shepard, J. F., Gaard, G., and Purcifull, D. E. (1974b). *Phytopathology* **64**, 418–425.
Shepherd, R. J., Richins, R., and Shalla, T. A. (1980). *Virology* **102**, 389–400.
Shockey, M. W., Gardner, C. O., Jr., Melcher, U., and Essenberg, R. C. (1980). *Virology* **105**, 575–581.
Talbot, D. N., and Yphantis, D. A. (1971). *Anal. Biochem.* **44**, 246–253.
Thornbury, D. W., and Pirone, T. D. (1983). *Virology* **125**, 487–490.
Tolin, S. A., and Roane, C. W. (1975). *Proc. Am. Phytopathol. Soc.* **2**, 129.
Toriyama, S. (1982). *J. Gen. Virol.* **61**, 187–195.
Towbin, H., Staehelin, T., and Gordon, J. (1979). *Proc. Natl. Acad. Sci. U.S.A.* **76**, 4350–4354.
van Regenmortel, M. H. V. (1967). *Virology* **31**, 467–480.
van Regenmortel, M. H. V. (1982). "Serology and Immunochemistry of Plant Viruses." Academic Press, New York.
Weidemann, H. L. (1981). *Phytopathol. Z.* **102**, 93–99.
Xiong, C., Balàzs, E., Lebeurier, G., Hindelang, C., Stoeckel, M. E., and Porte, A. (1982a). *J. Gen. Virol.* **61**, 75–81.
Xiong, C., Muller, S., Lebeurier, G., and Hirth, L. (1982b). *EMBO J.* **1**, 971–976.
Yeh, S.-D. (1984). Ph.D. Dissertation, Cornell University.
Zettler, F. W., Abo El-Nil, M. M., and Hartman, R. D. (1978). CMI/AAB Descriptions of Plant Viruses, No. 191.
Zurawski, D. B. (1979). M.S. Thesis, University of Florida.

7 Use of Mosquitoes to Detect and Propagate Viruses

Leon Rosen

I. Rationale for Use of the Technique

Parenteral inoculation of mosquitoes was first employed as a method of virus isolation in the case of the dengue viruses because other methods then available were insensitive (Rosen and Gubler, 1974). The attractiveness of the procedure was enhanced by the finding that male mosquitoes were as susceptible as females, and because the males are incapable of piercing skin, it could be employed without the safety precautions required for females.

There are at least three reasons why mosquitoes may be useful for isolation and propagation of viruses that replicate in them. First, and most important, they may be more sensitive to infection than any other assay system. Second, the method can be employed under relatively primitive laboratory conditions with a minimum of equipment. The cost of labor and

281

supplies also is usually less than for mice or cell cultures, and these are important considerations in developing countries where many arthropod-borne viruses are prevalent. Finally, replication of certain viruses can be studied under a wider range of temperature in intact mosquitoes than in vertebrates or in insect or vertebrate cell cultures.

Briefly, the method consists of inoculating anesthetized mosquitoes parenterally and then holding them long enough at an appropriate temperature for maximum viral replication. Since most viruses do not produce signs of infection in mosquitoes, viral replication is detected by other methods, e.g., the examination of mosquito tissue by immunofluorescence or the assay of mosquito triturates for viral content in cell cultures. In the latter instance, the mosquito is used as an amplifying host to increase the amount of virus to a level that can be detected easily in a less sensitive system. In a few instances, viral replication can be detected by the death of infected mosquitoes (Tesh, 1980) or their development of lethal sensitivity to CO_2 (Rosen, 1980).

Obviously, mosquitoes can be used only for assay of viruses that will replicate in them. Fortunately, however, the method is not limited to the use of mosquito species in which the viruses are found naturally or to the assay of viruses found in mosquitoes. Thus far, inoculation of mosquitoes has been used to detect and propagate alphaviruses (Rosen *et al.,* 1981), flaviviruses (Rosen and Gubler, 1974; Rosen *et al.,* 1980; Aitken *et al.,* 1979; Hardy *et al.,* 1980), bunyaviruses (Rosen, unpublished), and rhabdoviruses (Rosen, 1980). It can also be used for "indigenous" viruses of mosquitoes or other insects (e.g., sigma virus of *Drosophila melanogaster*) (Rosen, 1980; Rosen and Shroyer, 1981; Tesh, 1980).

II. *Technical Considerations*

A. CHOICE OF MOSQUITO SPECIES

Viruses that replicate in mosquitoes have a much wider mosquito host range when inoculated parenterally than when the mosquito is exposed to infection by the oral route. Nevertheless, there are limitations as to the mosquito species that can be employed in the case of certain viruses. For example, the dengue viruses will not replicate to a significant extent when inoculated into several common species of *Culex* (e.g., mosquitoes of the *Culex pipiens* complex) (Rosen, unpublished). On the other hand, a virus such as that of Japanese encephalitis will replicate in every mosquito species that has been tested (Rosen, unpublished). Consequently, one must test the species to be employed beforehand to determine whether it will be suitable.

Naturally, it is easiest to utilize species that have been colonized and are reared readily in the laboratory. However, mosquitoes collected in the field as larvae and pupae and allowed to complete their development in the laboratory could also be employed. Insofar as they have been tested, male mosquitoes are as susceptible to infection as females of the same species. They offer the advantage of not requiring certain safety precautions, but have the disadvantage of being more fragile and less long-lived than females.

In the case of the dengue viruses and some other flaviviruses, it has been found that certain species of *Toxorhynchites* are favorable species for viral isolation and propagation (Rosen, 1981). Despite the fact that mosquitoes of this genus do not feed on blood, they are as susceptible to parenteral infection with the dengue viruses as are the known hematophagous vectors. *Toxorhynchites* mosquitoes offer the important advantages of size (see Fig. 1) and hardiness as compared with blood-feeding mosquitoes. Also, as with the males of hematophagous species, safety measures are not necessary for either male or female *Toxorhynchites*. Moreover, because their larval stages are predacious on other mosquito larvae, there usually is no problem in

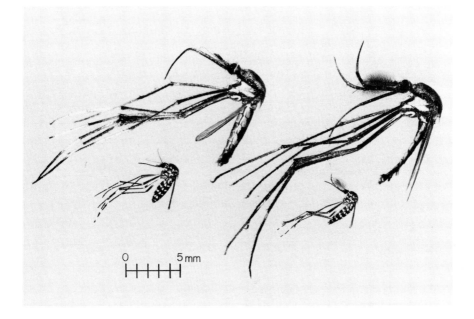

FIG. 1. *Toxorhynchites amboinensis* mosquitoes (above) photographed to the same scale as *Aedes albopictus* (below). Males are on the right and females are on the left. From Rosen (1981).

securing permission for the importation of *Toxorhynchites* into an area
where they do not occur naturally. *Toxorhynchites* are somewhat more difficult to raise in the laboratory than blood-feeding mosquitoes because the
larvae must be fed living food (usually larvae of other mosquito species) in
order to develop at a satisfactory rate. This drawback is more than compensated for by their size, hardiness, and safety.

B. MOSQUITO REARING

A description of the methods used to rear mosquitoes in the laboratory
is beyond the scope of this chapter and may be found in other publications.
In general, mosquitoes should be reared to produce large vigorous adults,
and the latter should be used as soon as possible after their exoskeleton has
hardened following eclosion. Although no difference in susceptibility to infection has been found in mosquitoes of different ages, it obviously is advantageous to have as high a proportion as possible survive the viral
replication period.

C. ANESTHESIA

Mosquitoes must be immobilized for inoculation, and the easiest way to
do this is to chill them in a test tube immersed in an ice bath. Certain mosquito species are insufficiently sensitive to cold and may require the use of
CO_2 gas either alone or in combination with cold to achieve anesthesia of
sufficient duration to allow inoculation with ease. This is especially important when inoculating females of blood-feeding species—to avoid the
possible escape of such mosquitoes from the stage of the microscope immediately after inoculation. Carbon dioxide can be used by allowing it to
flow into vertical tubes holding the mosquitoes (CO_2 is heavier than air) or
administered by continuous flow through a specially adapted stage. The use
of CO_2 and cold in combination for anesthesia led to the discovery of an
indigenous microbial agent of *Culex quinquefasciatus* mosquitoes that induces lethal sensitivity to CO_2 (Rosen and Shroyer, 1981). Such a possibility
should be kept in mind if inoculated insects do not recover as expected from
their anesthesia.

D. METHOD OF INOCULATION

Mosquitoes have been injected with viruses by one technique or another
for many years. However, most of the procedures either were not quantitative or were so cumbersome that it was not possible to inoculate large

numbers of insects within a short period of time. The technique described here is based on that employed by Rozeboom (Rosen and Gubler, 1974).

The method employs compressed air to force the inoculum through a "needle" made by drawing out glass tubing of known internal diameter of a fine point (Fig. 2). The "needles" used in our laboratory are prepared from borosilicate glass tubing having an approximate outside diameter of 0.7–1.0 mm, an approximate wall thickness of 0.2 mm, and an internal diameter averaging 0.47 mm. These measurements are not critical. The tubing is drawn to a point after heating over an alcohol lamp, and the tip is broken off at the appropriate diameter (see below) with a jeweler's forceps. The untapered portion of the tubing is then marked off at 1.0-mm intervals (either with a ballpoint pen or rubber stamp) to allow estimation of the amount inoculated by observing the movement of the fluid meniscus.

With the capillary tubing employed in our laboratory, a movement of the meniscus of 1.0 mm on the untapered portion of the tubing represents about 0.17 μl of fluid. This is the usual amount injected into a mosquito,

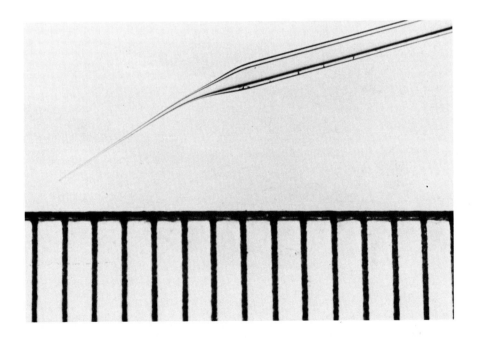

FIG. 2. Tip of glass "needle" used for inoculation; each division of the ruler is 1 mm. From Rosen and Gubler (1974).

but up to 10 times more can be injected into *Toxorhynchites* without undue mortality.

An airtight seal of the "needle" to a metal holder is made by using one or two O rings compressed by a nut (Fig. 3). The holders are not available commercially in the United States but can be made by most machine shops (see Fig. 4).

The entire injection apparatus and other details are pictured in Fig. 5. The "needle" holder is attached by plastic tubing to a 20-ml syringe held by a clamp on a ring stand. The tubing is attached to the syringe by a three-way stopcock that allows the plunger of the syringe to be manipulated, when necessary, without increasing or decreasing the pressure in the plastic tubing. The insertion of connectors in the plastic tubing allows the "needle" and holder to be removed for decontamination after use without changing the entire apparatus.

The glass tubing is filled with inoculum by immersing the tip in the fluid and withdrawing the plunger of the syringe. The inoculum is injected by moving the plunger forward. Because of the small size of the "needle" tip, considerable suction or pressure is required to move fluid in or out of the

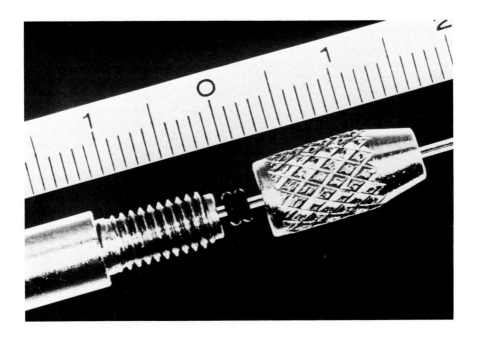

FIG. 3. "Needle" holder showing O rings used to obtain airtight seal.

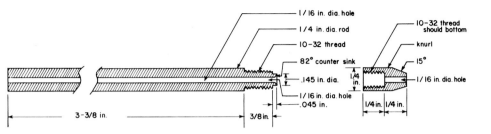

FIG. 4. Dimensions of "needle" holder.

glass tubing. Consequently, the amount of fluid injected can be controlled easily by varying the amount of pressure exerted with the syringe. Although the diameter of the "needle" tip is not critical, it should be made as small as possible and yet large enough to minimize plugging with debris contained in inocula. A small tip is advantageous for ease both in piercing the exoskeleton of the mosquito and in controlling the amount of inoculum. It also is advantageous to make the tapered portion of the "needle" as short as possible.

Anesthetized mosquitoes are placed on a large glass slide (1 × 3 in.) under a stereoscopic microscope and impaled with the tip of the "needle," with care taken not to transfix the insect. Because of the position usually assumed by the anesthetized specimens on the slide, it is convenient to inoculate mosquitoes through the membranous area anterior to the mesepisternum and below the spiracle or through the neck membrane. Once the mosquito is impaled, the marked portion of the glass tubing is brought into view under the microscope and the plunger of the syringe is depressed while the fluid meniscus is observed. When the desired amount has been inoculated, pressure on the plunger is released or slightly reversed and the mosquito is examined to ascertain that all the inoculum has entered the insect and that transfixation has not occurred. The mosquito is then brushed off the "needle" with a jeweler's forceps into a small cage. An experienced technician can inoculate at least 200 insects with the same inoculum in a 1-hr period.

E. Holding Time and Temperature

The appropriate holding time and temperature for mosquitoes after inoculation depend on the virus and the type of subsequent examination envisioned. Peak titers of infectivity in mosquitoes are reached well before peak antigen accumulation in certain tissues (see below). The maximum holding temperature and time are limited by the resistance of the mosqui-

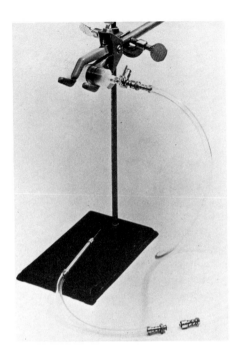

FIG. 5. Complete injection apparatus. From Rosen and Gubler (1974).

toes employed. Most species of mosquitoes do not thrive at temperatures in excess of 32°C, and they remain in better condition and survive longer at lower temperatures. It is generally believed that arthropod-borne viruses replicate faster the higher the mosquito temperature. However, recent data have shown that the peak viral titer attained may be lower at higher temperatures (Kramer *et al.*, 1983). Also, it is well established that at least some indigenous viruses of insects will not replicate at temperatures over 30°C. In the case of the dengue viruses, peak infectivity is attained in *Aedes albopictus* mosquitoes about a week after inoculation at a holding temperature of 26°C (Gubler and Rosen, 1977). Thus, this length of time would be an appropriate holding period for mosquitoes to be assayed for infectivity, or for mosquitoes to be used to prepare a virus pool. On the other hand, if the mosquitoes are to be examined by immunofluorescence, a longer holding time is desirable. Holding time and temperature are determined by taking into account the rate of viral replication and antigen accumulation at various temperatures, the ability of the mosquito to tolerate various possible time and temperature combinations, and the projected system of virus detection.

F. Methods of Examination

When mosquitoes were first employed to isolate dengue viruses they were used as an amplification system and the triturated mosquitoes were tested for the presence of viruses by plaque assay in cell cultures (Rosen and Gubler, 1974). Later, it was found that the presence of viral antigen could be detected easily in simple squashes of mosquito heads by a fluorescent antibody technique (Kuberski and Rosen, 1977a).

Assay for infectivity can be used for all mosquito–virus combinations, but obviously it is slower and more cumbersome than direct detection of viral antigen. In some cases, such assay requires that mosquito triturates be treated to remove bacteria and fungi that might interfere in cell culture systems. This can be done by filtration through small membrane filters, centrifugation, and/or treatment with antibiotics. The secondary assay system will depend on the virus under consideration.

Detection of viral antigen in the mosquito has been used successfully for all flaviviruses thus far tested, for bunyaviruses, and for rhabdoviruses. It has not been successful in our hands with alphaviruses. The usefulness of the technique depends on the fact that the nervous system in the head of the mosquito contains large amounts of antigen that can be visualized readily. While other parts of the mosquito might also be employed, it is far more time consuming to dissect such organs as the salivary glands or midgut than to decapitate the mosquito. And, while high infectious titers have been found in mosquitoes infected with alphaviruses, apparently antigen is not present in large enough quantity in the nervous tissue of the mosquito head to allow ready visualization. It should be noted that viral antigen is consistently present in the head of mosquitoes parenterally infected with flaviviruses, but not necessarily in those infected by the oral route (e.g., by feeding on a viremic vertebrate). It is not known if the presence or absence of viral antigen in large quantities in the mosquito head in the latter mosquitoes is correlated with virus in the salivary glands and the ability of the mosquito to transmit infection.

The fluorescent antibody system used to detect viral antigen in mosquito tissue is similar to that employed for cell cultures and will not be discussed here. However, there are two points that are of particular importance. First, more antibody is required to detect antigen in mosquito tissue than in cell culture. Consequently, one cannot determine the amount of antibody in a cell culture system and then apply it to the mosquito. The amount of antibody required must be determined on the mosquito tissue itself. Second, one cannot employ human or monkey antiviral antibody in the indirect fluorescent antibody technique with mosquito tissue. Normal human and monkey sera bind to the tissue, and consequently false-positive results are obtained. Human sera are satisfactory, however, in the direct fluorescent antibody technique. Monoclonal mouse antibody can be used in the indirect fluorescent antibody technique to type dengue viruses directly in mosquito tissues (Rosen, unpublished). However, as noted above, a higher concentration of antibody is required than for cell cultures.

In addition to infective assays and fluorescent antibody examination of mosquito tissue, other procedures can be employed to assay inoculated mosquitoes. These include the complement-fixation technique (Kuberski and Rosen, 1977b) and enzyme immunoassays (Hildreth *et al.*, 1982). In the complement-fixation technique it has been found necessary to use only male mosquitoes (except in the case of *Toxorhynchites*) because enzymes present in female mosquitoes lyse sheep red cells. An unusual assay method for certain viruses is to anesthetize the inoculated mosquitoes with CO_2 at low temperature to determine whether they become irreversibly paralyzed (Rosen, 1980; Turell and Hardy, 1980).

III. Possible Limitations and Pitfalls

A. LIMITATIONS ON AMOUNTS OF INOCULUM PER MOSQUITO

An obvious limitation of the use of mosquitoes for the detection of viruses is the relatively small amount of inoculum that can be employed in comparison with cell cultures or living vertebrates (100–1000 times less). This can be compensated for in part in several ways. First, the use of *Toxorhynchites* mosquitoes permits the inoculation of 10 times as much material as the use of smaller, blood-sucking species. Second, a large number of mosquitoes can be used for each specimen. Finally, the virsus content of certain types of specimens, such as human sera, can be concentrated by high-speed centrifugation. In the case of dengue viruses, the increased sensitivity of mosquitoes to infection with the viruses compared with other isolation methods results in a higher isolation rate from clinical specimens despite the testing of 1/500th of the amount of inoculum. On the other hand, in a situation such as the isolation of alphaviruses, in which the inoculated mosquitoes must be triturated and assayed in cell cultures, the degree to which sensitivity is increased is not great enough to compensate for the smaller inoculum (Rosen *et al.*, 1981). Basically, this is because cell cultures are relatively sensitive to alphaviruses, not because the mosquitoes are insensitive.

With certain types of specimens, namely, those that may contain relatively large amounts of homologous antibody in addition to virus, the intact mosquito may offer an advantage in addition to its inherent sensitivity. Apparently, when antibody–virus complexes are inoculated into the mosquito, the virus is more likely to dissociate and replicate than is the case when comparable specimens are inoculated into intact vertebrates or cell cultures. This may explain, in part, the greater efficacy of mosquito inoculation in isolating dengue virus from organs of patients dying of the disease as compared with other methods (Rosen, unpublished).

B. LACK OF SIGNS OF INFECTION

Another limitation on the use of mosquitoes to isolate viruses is that one must usually know what is being sought. A previously unknown virus would not normally be detected by the fluorescent antibody technique and would be picked up by secondary infectivity assay in cell cultures only if it happened to produce a cytopathic effect in that system. In some cases, this selectivity might be advantageous, for example, in trying to isolate one particular virus from specimens that frequently contained others.

C. POSSIBLE PICKUP OF INDIGENOUS MOSQUITO VIRUSES

When passing viruses in mosquitoes, as in other systems, there is always the possibility of picking up viruses already present in the insect. If this does happen, such agents can sometimes be eliminated by passing the original virus in a warm-blooded vertebrate or a cell culture from such an animal, since many indigenous mosquito viruses will not replicate at higher temperatures.

D. POSSIBLE PROBLEMS IN THE IMPORTATION OF EXOTIC MOSQUITO SPECIES

In some instances it may not be possible or desirable to establish colonies of mosquitoes in areas where they do not occur naturally. Thus far, it has been possible to solve this problem by employing *Toxorhynchites* mosquitoes, which pose no danger. It is possible, however, that it may be necessary in some cases to employ a hematophagous mosquito species.

E. SAFETY CONSIDERATIONS

Obviously, the use of female blood-feeding mosquitoes imposes the necessity for safety precautions to avoid their inadvertent escape, either during inoculation or afterward. In general, such mosquitoes should be inoculated in a "safety cage" so constructed that any mosquito that escapes from the microscope stage can be recaptured easily. Also, the mosquitoes should be held in double cages after inoculation in a room with limited access. Except for viral transmission experiments, there is little justification for the use of female blood-feeding mosquitoes for viral assay.

REFERENCES

Aitken, T. H. G., Tesh, R. B., Beaty, B. J., and Rosen, L. (1979). *Am. J. Trop. Med. Hyg.* **28**, 119-121.

Gubler, D. J., and Rosen, L. (1977). *J. Med. Entomol.* **13**, 496-472.

Hardy, J. L., Rosen, L., Kramer, L. D., Presser, S. B., Shroyer, D. A., and Turell, M. J. (1980). *Am. J. Trop. Med. Hyg.* **29**, 963-968.

Hildreth, S. W., Beaty, B. J., Maxfield, H. K., Gilfillan, R. F., and Rosenau, B. J. (1982). *J. Clin. Microbiol.* **15**, 879-884.

Kramer, L. D., Hardy, J. L., and Presser, S. B. (1983). *Am. J. Trop. Med. Hyg.* **32**, 1130-1139.

Kuberski, T. T., and Rosen, L. (1977a). *Am. J. Trop. Med. Hyg.* **26**, 533-537.

Kuberski, T. T., and Rosen, L. (1977b). *Am. J. Trop. Med. Hyg.* **26**, 538-543.

Rosen, L. (1980). *Science* **207**, 989-991.

Rosen, L. (1981). *Am. J. Trop. Med. Hyg.* **30**, 177-183.

Rosen, L., and Gubler, D. (1974). *Am. J. Trop. Med. Hyg.* **23**, 1153-1160.

Rosen, L., and Shroyer, D. A. (1981), *Ann. Virol. (Inst. Pasteur)* **132 E**, 543-548.

Rosen, L., Shroyer, D. A., and Lien, J. C. (1980). *Am. J. Trop. Med. Hyg.* **29**, 711-712.

Rosen, L., Gubler, D. J., and Bennett, P. H. (1981). *Am. J. Trop. Med. Hyg.* **30**, 1294-1302.

Tesh, R. B. (1980). *J. Gen. Virol.* **48**, 177-182.

Turell, M. J., and Hardy, J. L. (1980). *Science* **209**, 1029-1030.

8 Prions: Methods for Assay, Purification, and Characterization

Stanley B. Prusiner, Michael P. McKinley,
David C. Bolton, Karen A. Bowman, Darlene F. Groth,
S. Patricia Cochran, Elizabeth M. Hennessey,
Michael B. Braunfeld, J. Richard Baringer, and
Mark A. Chatigny

I. Introduction

The unusual molecular properties of the scrapie agent seem to distinguish it from both viruses and viroids and have led to the introduction of the term "prion" for the infectious particle (Prusiner, 1982). Procedures that hydrolyze or modify proteins produced a diminution of scrapie prion infectivity; in contrast, procedures that hydrolyze or modify nucleic acids do not alter the infectivity (Diener et al., 1982; McKinley et al., 1981; Prusiner, 1982; Prusiner et al., 1981b). Other experiments indicate that the smallest infectious form of the prion may be too small to contain even a single gene (Prusiner, 1982, 1984; Prusiner et al., 1983). These observations suggest that replication mechanisms different from those of viruses and viroids must be operative.

Besides scrapie, five other transmissible disorders of the central nervous system (CNS) may be caused by prions; however, only in the case of scrapie has firm evidence for a prion etiology been obtained (Prusiner, 1984). These transmissible diseases include transmissible mink encephalopathy (TME), chronic wasting disease (CWD), kuru, Creutzfeldt–Jakob disease (CJD), and Gerstmann–Sträussler syndrome (GSS) (Burger and Hartsough, 1965; Gajdusek, 1977; Masters et al., 1981; Williams and Young, 1982).

Development of an incubation time interval bioassay (Prusiner et al., 1980d, 1982b) has permitted us and others to devise methods for purifying the prion that causes scrapie (Diringer et al., 1983; Prusiner et al., 1981b, 1982a, 1983). Radioiodination of partially purified fractions led to the discovery of a protein, designated PrP, that is a structural component of the prion (Bolton et al., 1982). Studies have shown that the concentration of PrP is directly proportional to the titer of the prion, and that the kinetics of degradation of PrP by proteolytic enzymes are virtually identical to those for the infectious prion (McKinley et al., 1983b). Extensive purification of prions has resulted in preparations that contain one major protein, PrP (Prusiner *et al.*, 1983). Those preparations also contain numerous rod-shaped particles that can be observed by electron microscopy. The ultrastructural morphology of these rod-shaped particles suggested that they might be related to amyloid (Cohen *et al.*, 1982), for, like amyloid, they form large arrays of varying shapes and sizes. When extensively purified preparations of scrapie prions are stained with Congo red dye, they exhibit green color and birefringence by polarization microscopy that are characteristic of amyloid (Prusiner *et al.*, 1983).

In this chapter, we review in detail our methods for bioassay and purification, as well as molecular and morphological characterization, of the scrapie prion. In addition, we discuss practical aspects of biocontainment and decontamination.

II. Assays for Scrapie Prions

At present, the only methods for measuring scrapie prion infectivity remain the incubation time interval assay and the end-point titration. Both methods are extremely slow because they require waiting for the onset of clinical neurological dysfunction, following a prolonged incubation period. The length of the incubation period varies greatly with the animal species, route of inoculation, and dose of prions (Gajdusek, 1977; Prusiner *et al.*, 1980a, 1982b). The hamster, inoculated intracerebrally, has the shortest incubation period of any experimental host (Kimberlin and Walker, 1977; Marsh and Kimberlin, 1975).

The short incubation period in the hamster and the high titer of the infectious agent in its brain make the hamster the preferred animal for scrapie research. The disadvantages of the hamster compared to the mouse are (1) the lack of a large number of inbred hamster strains, (2) the greater cost of the hamster, and (3) the greater cost of animal care.

A. END-POINT TITRATIONS

An end-point titration for the scrapie prion is performed by serially diluting a sample at 10-fold increments. Each dilution is typically inoculated intracerebrally into four to six animals, and the waiting process ensues. Since the highest dilutions at which scrapie develops are the only observations of interest, 10–12 months must be allowed to pass before the titration may be scored if mice are used. From the score at the highest positive dilutions, a titer or concentration of the infectious agent in the original sample can be calculated. Not only are the resources great with respect to animals and time, but pipetting 10 serial 10-fold dilutions can create additional problems with accuracy of the measurements.

In a typical end-point titration, we have used female Swiss Webster mice. Serial 10-fold dilutions of samples are prepared with phosphate-buffered saline containing 5% fetal calf serum, 0.5 U/ml penicillin, 0.5 μg/ml streptomycin, and 2.5 μg/ml amphotericin. For each dilution, six mice are inoculated intracerebrally with 30 μl of diluted suspension. The inoculations are all made in the left parietal region of the cranium with a 26-gauge needle inserted to a depth of approximately 2 mm. Typically, 10 serial 10-fold dilutions are made for an end-point titration. After 3 months, the animals are examined weekly during the next 9 months for clinical signs of scrapie. These signs include bradykinesia, plasticity of the tail, waddling gait, and a coarse, ruffled appearance of the coat. Histological examination of the brain and spinal cord should be done occasionally to confirm the clinical diagnosis. Typically the animals die within 4–6 weeks after the onset of scrapie. The disease is rather stereotyped and progressive. At the time of onset, the back of the sick mouse is painted with a 1% (w/v) solution of picric acid (Mallinckrodt, St. Louis, MO). Titers are calculated according to the method of Spearman and Kärber (Dougherty, 1964). The standard errors of our titrations generally vary between 0.2 and 0.4 log unit.

Because of the increased cost, limited numbers of end-point titrations have been performed in hamsters. Five to six months must pass before the titration in hamsters may be scored. The principles of the methodology are the same with the following exceptions: (1) 50 μl is inoculated intracerebrally, (2) four hamsters are sometimes used for each dilution, (3) the clinical signs of scrapie are slightly different as discussed below, and (4) the abdomen of the sick hamster is painted with picric acid.

B. INCUBATION TIME INTERVAL ASSAY

The incubation time interval assay reduces the number of animals, the time required for bioassay, and potential pipetting errors compared to the endpoint titration (Prusiner et al., 1982b). With hamsters, studies on the scrapie agent have been dramatically accelerated by the development of a bioassay based on measurements of incubation times. It is now possible to assay samples with the use of four animals in 60–70 days, if the titers of the scrapie agent in the sample are $> 10^8$ ID_{50} U/ml.

1. Nomenclature

In discussions of slow infectious diseases, the incubation period is the time from inoculation to the onset of illness. The term "incubation time interval assay" has been adopted to describe our bioassay procedure. The assay utilizes two measurements: (1) the time from inoculation to the onset of illness (y), and (2) the time from inoculation to death (z). The difference between z and y is the duration of clinical illness.

The dilution (D) of the sample to be assayed refers to the fractional concentration of the agent or prion. Conventional serial 10-fold dilutions were made, and the dilution is expressed as $1/10^d$; thus, the log D equals $-d$. We recognize that the dilution process itself has a positive exponent. It is noteworthy that the median infective dose (ID_{50}) and median lethal dose (LD_{50}) values for a given sample are identical, since scrapie is uniformly fatal. The ID_{50} is expressed in units (U).

2. Preparation of Inoculum

Hamster-adapted scrapie agent in its sixth passage in an LHC/LAK inbred hamster was kindly provided by Dr. Richard Marsh and has been used in most of our experiments (Marsh and Kimberlin, 1975; Prusiner *et al.,* 1980d). The brain is homogenized in 320 mM sucrose, and the homogenate is clarified by centrifugation at 1000 g for 10 min. Random-bred LVG/LAK female weanling hamsters obtained from Charles River Laboratories are inoculated intracerebrally with 50 μl of 10^{-1} dilution. After approximately 60 days, the hamsters are killed upon developing signs of clinical scrapie. Homogenates of brains containing prions are prepared from these animals. The titer of the final inoculum used in many of our studies is a median infective dose (ID_{50}) of $10^{9.5}$ U/g of brain tissue as determined by end-point titration using the method of Spearman and Kärber (Dougherty, 1964).

3. Inoculation of Hamsters

Four weanling female hamsters are inoculated intracerebrally with 50 μl of a given sample at a specified dilution using a 26-gauge needle. The inoculations are all made in the left parietal region of the cranium with the needle inserted to a depth of approximately 3 mm. The diluent used is phosphate-buffered saline containing 0.5 U/ml of penicillin, 0.5 μg/ml of streptomycin, 2.5 μg/ml of amphotericin, and 50 mg/ml of recrystallized fraction V bovine serum albumin (Pentex-Miles Laboratories).

4. Care of Hamsters

The animals are housed in polyurethane cages 20 cm high, 20 cm wide, and 42 cm deep and covered with a tight-fitting wire top that holds the food. Eight animals are housed per cage, and groups of four are distinguished by notching both ears of one group. Water is supplied by an automatic watering system, and Purina rat chow is provided *ad libitum.* A generous supply of pine wood shavings is placed on the bottom of each cage and changed twice a week. All personnel entering the room are required to wear masks, gowns, rubber boots, gloves, head covers, and eye protection.

5. Clinical Signs of Scrapie

None of the animals show signs of neurological dysfunction before 7 weeks after inoculation. The onset of clinical scrapie is diagnosed by the presence of at least two of the following signs: generalized tremors, ataxia of gait, difficulty righting from supine position, or head bobbing. The bobbing movements of the head increase progressively and may result from visual difficulties due to degenerative changes in the retina (Hogan et al., 1981). Between 1 and 10% of the hamsters show generalized convulsions at this early stage of the illness. With further deterioration, the ataxia becomes so pronounced that balance is maintained with considerable difficulty. Kyphotic posture, bradykinesia, and weight loss appear 7–15 days after the onset of illness. Over the next week the hamsters become unable to maintain an erect posture; they lie quietly on their sides and exhibit frantic movements of the extremities when disturbed. Death follows in 3–5 days.

6. Computer Records

In order to organize and process the large amount of data obtained from these studies, computer programs in Fortran language were written for a Hewlett-Packard 1000 minicomputer. The programs permit recording the time of onset of illness for each animal as well as the time of death.

7. Calibration Curves

To use measurements of time intervals to predict scrapie titers, calibration curves must be constructed using samples with a wide range of titers. The titers are determined by end-point titrations, and the time intervals from inoculation to onset of illness and to death are measured for each dilution. Four hamsters are inoculated intracerebrally with a given sample at a specified dilution. After 55 days have elapsed, the animals are examined twice weekly for clinical signs of scrapie. From the number of animals positive at each dilution, the titer is calculated using the method of Spearman and Kärber (Dougherty, 1964). The injected dose is then calculated by multiplying the titer times the dilution.

Curves relating the injected dose to the time intervals from inoculation to onset of clinical illness as well as from inoculation to time of death are shown in Figs. 1 and 2. The titers of the samples used to construct these curves varied over a range from 10^3 to $10^{8.5}$ ID_{50} U/ml as determined by end-point titration. As illustrated in Fig. 1, the interval from inoculation

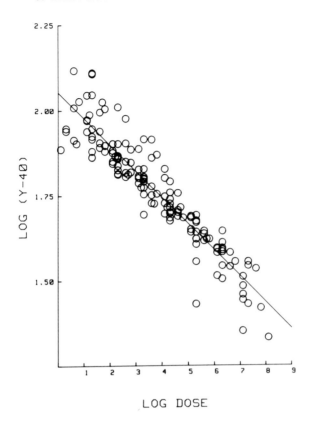

FIG. 1. Relationship between inoculated dose of scrapie prions and the time interval from inoculation to the onset of illness. Each symbol is the mean for four hamsters. The solid line was fitted by regression analysis with a coefficient of 0.87.

to onset of illness minus a time factor of 40 is a linear function of the inoculated dose. The time factor was determined by maximizing the linear relationship between time interval and dose. With a factor of 40, the regression coefficient for the line is 0.87.

While the onset of illness requires clinical judgment with respect to the diagnosis of scrapie, the time of death is a completely objective measurement. As shown in Fig. 2, the time interval from inoculation to death minus a time factor of 61 is a linear function of the injected dose. Like the analysis of data for onset of illness, the time factor was determined by maximizing the linear relationship between this time interval and the dose. With a factor of 61, the regression coefficient for the line is 0.86.

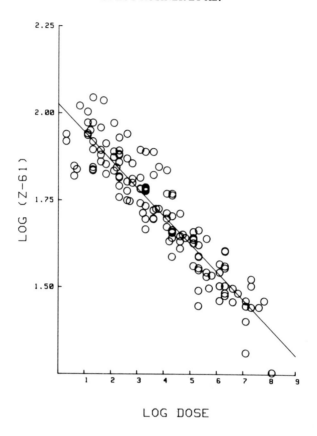

LOG DOSE

FIG. 2. Relationship between the inoculated dose of scrapie prions and the time interval from inoculation to death. Each symbol is the mean for four hamsters. The solid line was fitted by regression analysis with a coefficient of 0.86.

8. Dose–Response Relationships—Derivation of Equations

From the linear relationships just described, equations relating titer, dilution, and time intervals can be written as follows:

$$Log\ T_y = 26.66 - (12.99)\log(\bar{y} - 40) - \log D$$
$$Log\ T_z = 25.33 - (12.47)\log(\bar{z} - 61) - \log D$$

where T is the titer, expressed in ID_{50} U/ml, D is the dilution, defined as the fractional concentration of the diluted sample, $\bar{y}$ is the mean interval from inoculation to onset of clinical illness in days, and $\bar{z}$ is the mean interval from inoculation to death in days. The most precise estimate of titer is obtained by calculating a weighted average of T_y and T_z.

Similar linear relationships were obtained when the reciprocals of the time intervals were plotted as a function of the logarithm of the dose. Regression coefficients of 0.87 and 0.88 were obtained for lines relating $1/y$ and $1/z$ to the logarithm of the dose, respectively. Equations describing these functions gave similar results to those obtained with the two equations above.

C. A COMPARISON OF END-POINT TITRATION AND INCUBATION TIME INTERVAL ASSAYS

A comparison of the end-point titration and incubation time interval assays is shown in Table I. The economics of both time and resources afforded by the incubation time interval assay are highly significant. We estimate that our research has been accelerated by more than 100-fold by the use of the incubation time interval assay. It is doubtful that the purification and characterization methods described below could have been developed if the end-point titration method had been used to assay samples.

D. RADIOLABELING OF SCRAPIE PrP

Although the incubation time interval assay has been the cornerstone of our work to date, we are now using the radiolabeled PrP as an estimate of the concentration of the scrapie agent. Compelling evidence has been accumulated to show that PrP is, in fact, a structural component of the scrapie prion (McKinley et al., 1983b). A simple procedure for determining the presence and amount of PrP involves radiolabeling the purified native protein (PrP) with ^{125}I-labeled Bolton–Hunter reagent, digesting it with proteinase K for 30–60 min at 37°C, and then denaturing it prior to sodium dodecyl sulfate (SDS) electrophoresis on a 15% polyacrylamide gel. The details of these procedures are described in Section IV.

TABLE I

COMPARISON OF BIOASSAYS FOR SCRAPIE PRIONS

Condition	End-point titration	Incubation time interval
Rodent species	Mice	Hamsters
Time (days)	360	60–70
Number of animals	60	4
Sample dilution	10^{-1} to 10^{-10}	10^{-1}

III. Histopathology

Histopathological examination of the brains of scrapie-infected mice and hamsters is an important aspect of research on prions. The histopathology of scrapie rodents is quite stereotyped and has been well described (Baringer and Prusiner, 1978; Beck and Daniel, 1965; Fraser, 1979; Marsh and Kimberlin, 1975).

For bioassays, confirmation of the clinical disease by histopathological examination need not be done on every animal, but should be done on a routine basis.

A. Light Microscopy

Brains of hamsters or mice are fixed in formalin and then embedded in paraffin. Sections from paraffin blocks are then stained with hematoxylin and eosin for light microscopy. Because the details for these techniques can be found in numerous histology and histopathology texts, no detailed procedures are given here (Luna, 1968).

B. Electron Microscopy

For ultrastructural studies, nervous system tissue is perfusion fixed in order to minimize artifactual change that can potentially be confused with pathological changes in the tissue. Hamsters are anesthetized with ether, and anesthesia is maintained by application of a small ether-soaked sponge over the nose. The animal is held in a supine position to a small board. While the animal is anesthetized, the chest is opened by an incision on each side of the manubrium extending nearly to the axilla. The flap so created is reflected upward, exposing the heart. The pericardial sac is opened, and a tiny incision in the tip of the left ventricle is made using a small pair of scissors. Perfusion is accomplished utilizing a small (21-gauge) cannula connected via a perstaltic pump to a reservoir containing a quarter-strength, ice-cold Karnovsky solution. The system has previously been carefully flushed to assure that no bubbles are contained within the tubing or pump. The cannula is inserted into the incision in the left ventricle and maintained with its tip within the ventricular chamber by clamping it in the heart, utilizing a small pair of forceps. The perfusion is started immediately at a rate of approximately 5 ml/min. Immediately after the perfusion is initiated, a hole is cut in the right atrium to permit escape of the perfusion fluid. Although it might seem appropriate to cut the hole in the atrium before starting the perfusion, the procedure as outlined seems to provide a much higher frequency of good perfusions judged at the time of the perfusion, and subsequently, by viewing the tissue.

The entire operation of opening the chest, starting the perfusion, and opening the atrium can usually be accomplished within about 30 sec, thus minimizing the creation of pathological change by hypoxia or ischemia. The success of the perfusion may be judged by the fact that the neck and upper extremity muscles of the animal undergo a rapid stiffening, resulting in the neck becoming stiff and the muscular tissues firm to the touch within 30 sec to 1 min after initiation of the perfusion. The perfusion is continued for 5 min; longer perfusion times do not seem to result in any better preservation of tissue.

The skin over the scalp is reflected using sharp dissection, and the calvarium is removed using miniature rongeurs. By careful rongeuring of bone, it is possible to expose the entire forebrain, cerebellum, brain stem, and spinal cord. During this procedure, it is important to keep the nervous system tissues moist by repeated application of quarter-strength Karnovsky solution. Once the bone has been sufficiently removed, the cranial nerves and spinal roots at the base of the brain can be cut and the nervous system can be removed and further fixed in Karnovsky solution for a minimum of 2 hr, usually overnight.

The brain tissue is next cut into small blocks of representative specimens from cerebral cortex, hippocampus, basal ganglia, brain stem, cerebellum, and spinal cord. These tissue blocks are approximately 1.5–2 mm in each dimension. It is important in the handling of tissue to use the sharpest possible razor blades and to minimize crushing or squeezing of tissue by the careless use of forceps.

Once the tissue is cut into small pieces, it is transferred to vials containing 2% osmium tetroxide in a 0.1 M sodium phosphate buffer (pH 7.3). Osmium tetroxide solutions must be handled within a hood to avoid inhalation or exposure of the corneas or respiratory epithelium of the investigator. The tissues are fixed in osmium solution for a minimum of 3 hr, agitating the solution periodically. Next, the tissues are dehydrated through a series of graded alcohols starting with 35% alcohol in distilled water and progressing through 50, 70, 80, 90, 95, and 100% ethanol with 10-min immersion at each step. The immersion in the 100% alcohol is repeated in order to remove the last traces of water from the tissue. The tissue is next immersed in propylene oxide for a minimum of 0.5 hr, after which it is immersed for 0.5-hr minimum periods in a 1:1 mixture of propylene oxide and Epon 812 followed by a 1:3 mixture, and in turn followed by a 100% solution of Epon. The osmicated blocks are then transferred to 100% Epon in silicone rubber molds designed for electron microscopic applications; the embedded blocks are cured to appropriate hardness in a vacuum oven at 60°C.

Trimming and sectioning of the blocks for light and electron microscopy are done according to customary practice. Sections 1 or 2 μm thick are utilized for light microscopic examination of the tissue. The sections are air dried on glass slides and stained by covering them with a drop of 1% toluidine blue for 30 sec on a warm plate followed by washing of the section with distilled water. The sections are examined by bright-field microscopy. Alternatively, unstained sections can be examined by phase microscopy.

Areas of interest are identified on the 1-μm sections, and the corresponding blocks are trimmed for electron microscopic examination. Sections for electron microscopy are cut at 50- to 80-nm thickness and mounted on 200-mesh uncoated grids. The grids are stained in the conventional manner with uranyl acetate and lead citrate and examined in the electron microscope at accelerating voltages between 40 and 80 kV.

IV. Purification

Purification of the scrapie prion has been crucial in defining the molecular properties of the infectious particle. The need to purify the prion away from cellular molecules prior to meaningful characterization studies is similar to investigations on many other biological particles.

A. PURIFICATION STRATEGIES

The major problem confronting investigators attempting to purify the scrapie agent for nearly three decades was the bioassay. The assay system in sheep and goats, using end-point titrations (Pattison and Millson, 1960), was so long and required so many animals that no progress in purification could be made. The introduction of the mouse bioassay system in 1961 (Chandler, 1961), still using end-point titrations, made it possible to begin characterizing the scrapie agent, but progress in purification was still impractical—the reason being that development of each purification step depends so heavily on the results of earlier experiments. When so little was known about the molecular properties of the scrapie agent, development of a purification protocol was largely an empirical process.

In addition to the difficulties in bioassay, the extraordinary heterogeneity of the scrapie agent with respect to size, density, and charge was a complicating factor in the development of purification protocols (Prusiner *et al.*, 1978a,b, 1980b). Early studies clearly indicated that the apparent heterogeneity of the agent was so extreme that it was probably not due to intrinsic changes within the agent; rather, it seemed to be a consequence of its hydrophobicity (Prusiner *et al.*, 1978b, 1979).

With the development of the incubation time interval assay (Prusiner *et al.*, 1980d, 1982b), it became possible to assay a sufficient number of samples within a relatively short enough period of time in order to begin characterizing the infectious particles with respect to their range of sizes, densities, and charges as well as devising procedures for separating the infectious particles from cellular contaminants (Prusiner *et al.*, 1981b, 1982a, 1983).

Prior to the convincing identification of a protein component within the scrapie agent (Prusiner *et al.*, 1981b), investigators attempting to purify the agent did not even know how to express the level of purification; i.e., the denominator for specific infectivity measurements was unknown. With the identification of a protein, the extent of purification could then be measured and expressed in ID_{50} U/mg protein. This quantitation procedure is critical to evaluating the efficacy of a purification protocol.

B. GENERAL PRINCIPLES

Several general principles have proved to be useful in the purification of prions. First, prions are stable in a variety of detergent solutions that do not denature proteins; the infectivity is unstable in detergent solutions that denature proteins, such as SDS (Millson and Manning, 1979; Prusiner *et al.*, 1980d). Second, the infectivity can be separated from cellular components by detergent extractions followed by differential centrifugation or precipitation (Prusiner *et al.*, 1977, 1980e). While ultracentrifugation has been useful in pelleting the agent, it is cumbersome on a large scale; in the early stages of our large-scale purification we have substituted precipitation by polyethylene glycol (PEG) (Prusiner *et al.*, 1982a, 1983). Third, for many years the infectious particles in crude extracts have been found to be resistant to both nuclease and protease digestions (Hunter, 1979; Prusiner *et al.*, 1980d). We have found both of these hydrolytic enzymes to be valuable in purification. Omission of either micrococcal nuclease or proteinase K results in a diminished extent of purification. More specifically, omission of the proteinase K digestion diminished the purification by the sucrose gradient procedure by a factor of 10 (Prusiner *et al.*, 1982a). The enzyme digestions were carried out at 4°C because early studies showed that warming prions to 37°C caused increased aggregation (Prusiner *et al.*, 1978b).

After proteolytic digestion, some of the small peptides and oligonucleotides could be removed by ammonium sulfate precipitation (Prusiner *et al.*, 1980e). This was performed in the presence of cholate, since other investigators studying membrane-bound enzyme had shown that effective precipitation of hydrophobic proteins with ammonium sulfate required cholate (Tzagoloff and Penefsky, 1971). In one study on the scrapie agent, ammonium sulfate was used in the absence of cholate, but precipitation of the agent was nonselective (Malone et al., 1979).

Using sodium dodecyl sarcosinate (Sarkosyl) gel electrophoresis, the digested peptides and polynucleotides were separated from large forms of the prion (Prusiner et al., 1980c). Although the method was cumbersome, it did provide sufficient purification to allow us to show convincingly that infectivity depended upon a protein(s) using protease digestion and reversible chemical modification with diethylpyrocarbonate (DEP) (McKinley et al., 1981; Prusiner et al., 1981b). Without the purification provided by Sarkosyl gel electrophoresis, neither protease digestion nor chemical modification with DEP was found to diminish the titer of prions (Prusiner et al., 1981b).

The major drawback of gel electrophoresis for purifying the prion was the limited amount of material that could be processed. Only 3–4 ml of detergent-extracted, enzyme-digested material could be loaded on a preparative electrophoresis apparatus at any one time; electrophoresis for 6–8 hr was necessary (Prusiner et al., 1981b). This was followed by a cumbersome procedure in which electroelution of the infectious agent from the top of the gel was performed.

Because the size of our preparations was severely limited using Sarkosyl gel electrophoresis, we developed an alternative purification scheme using sucrose gradient sedimentation (Prusiner et al., 1982a). This procedure takes advantage of the hydrophobicity of the scrapie prions. An enzyme-digested ammonium sulfate-precipitated fraction was mixed with Triton X-100 and SDS prior to rate-zonal sedimentation through a sucrose step gradient containing no detergent. Presumably, the large forms of the prions are present either as aggregates that were not dissociated in the Triton X-100/SDS or as aggregates that are formed as the prions enter the gradient. Both of these now appear to be the case. The behavior of prions under these conditions seems to be analogous to that of the calcium ATPase, where this procedure was first used (Warren et al., 1974a,b). Omitting the Triton X-100–SDS mixture, or substituting octylglucoside or Triton X-100 alone, resulted in virtually no purification of the prion (Prusiner et al., 1982a).

The development of the discontinuous sucrose gradient method provided another major step in the level of purification as well as characterization of prions. Preparations were of sufficient purity to allow identification of a unique protein (PrP) within these preparations (Bolton et al., 1982). Even though the purification was substantially increased over that obtained with Sarkosyl gel electrophoresis, and the amount of material that could be processed in a single centrifugation was greater by factor of 10 than that achieved with Sarkosyl gel electrophoresis, the quantities of purified fractions were still insufficient to extend purification and provide a sufficient amount for characterization.

To overcome this problem, a large-scale purification procedure was developed (Prusiner *et al.*, 1983). Not only has the large-scale purification protocol increased amounts of purified fractions, but, equally important, it has yielded fractions that contain significantly fewer contaminants compared to those of the smaller-scale protocol. The zonal rotor sucrose gradient centrifugation used in the large-scale protocol gave increased resolution of the particles being separated due to its configuration and long path length, especially when compared to gradient centrifugation in a reorienting vertical rotor (Anderson, 1962). In addition, dynamic loading and edge-unloading of the rotor probably also helped to maximize the resolution of purification procedure.

In our early studies on scrapie, we reported that prions could exist in multiple molecular forms, with small forms having a sedimentation coefficient of 40 S or less, as well as a succession of larger forms; in fact, some larger aggregates had sedimentation coefficients of greater than 10,000 S (Prusiner *et al.*, 1978b, 1980b). All these data were derived from studies in which the infectivity of gradient fractions was determined by end-point titrations in mice. Studies now show that scrapie prions aggregate into rods, and that these rods form large arrays or clusters of varying size and shape (Prusiner *et al.*, 1983). Using the electron microscope, we now have ultrastructural evidence for the multiple molecular forms of prions. The aggregation of prions into polymeric structures of varying size and shape, as noted above, has made purification extraordinarily difficult.

We now believe that prions can exist in an extremely large number of polymeric states. The smallest, or monomeric, form of the prion appears to have a molecular weight (M_r) greater than 30,000 but less than 100,000 and is probably made up of three or fewer PrP molecules (Prusiner *et al.*, 1983). As many as 1000 PrP molecules can polymerize to form a single rod, and as many as 100 rods have been seen in a single array or cluster. These forms exist in a complex equilibrium, and it is possible that the zonal rotor profile shown in Fig. 3 does, in fact, result from the prolonged centrifugation. Once all the rods are sedimented from the top of the gradient, new rods may assemble from smaller forms of the prion because the concentration of the larger forms has been reduced by sedimentation into the gradient.

C. PURIFICATION PROTOCOLS

Details of three purification protocols are given below. The first scheme involves the use of anionic detergent extractions combined with differential centrifugation and followed by Sarkosyl gel electrophoresis (Prusiner *et al.*,

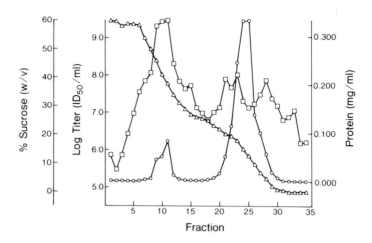

FIG. 3. Purification of scrapie prions by sucrose gradient centrifugation in a zonal rotor. The gradient was centrifuged at 32,000 rpm for 14.5 hr at 4°C. Fractionation of the gradient was from the outer edge of the Ti15 zonal rotor equipped with a B-29 liner. Sucrose concentration ($\triangle$), protein concentration ($\bigcirc$), and scrapie prion titer ($\square$) are shown. Fraction 1 was from the outer edge of the rotor. Each fraction was 40 ml.

1981b). The second and third procedures use combinations of anionic and nonionic detergents and employ discontinuous sucrose gradient centrifugation (Prusiner *et al.*, 1982a, 1983). The second and third procedures differ from each other in that the third procedure is performed on a large scale. The third procedure uses a zonal rotor for the sucrose gradient centrifugation step and yields extensively purified preparations of the scrapie agent.

Protein concentrations are determined by the method of Lowry *et al.* according to a modified procedure described by Peterson (1977). Bovine crystalline serum albumin is used as standard.

The sources of all the chemicals used in these purification procedures are listed at the end of this section.

1. Sarkosyl Gel Electrophoresis Method

1. In a typical preparation, 200 weanling female hamsters (LVG/LAK) are inoculated intracerebrally with $\sim 10^7$ ID_{50} U of hamster-adapted scrapie agent and are sacrificed 60–65 days later. Hamsters are anesthetized with CO_2 and sacrificed by cervical dislocation.
2. The brains are removed and immediately washed with ice-cold 320 mM sucrose. All solutions are degassed under reduced pressure and purged with argon to remove oxygen.
3. A 10% (w/v) homogenate in 320 mM sucrose is prepared using a

Polytron device equipped with PT35K generator. Homogenization is performed for 30 sec in a flask surrounded by ice water within a Baker Sterilgard biosafety hood. During homogenization the temperature of the suspension is not allowed to exceed 15°C. The titers of the homogenate vary between $10^{7.5}$ and $10^{8.5}$ ID_{50} U/ml, and the protein concentration is generally 8 mg/ml. All subsequent operations are performed at 4°C.

4. The homogenate is centrifuged at 1000 g for 10 min in a JA-10 rotor in a Beckman J-21C centrifuge housed within a Baker biosafety hood.

5. The supernatant fluid (S_1) is decanted and centrifuged again at 2100 g for 30 min in the same rotor.

6. To the supernatent fluid (S_2), ethylenediaminetetraacetic acid (EDTA) and dithiothreitol (DTT) are added to give final concentrations of 5 mM each.

7. The supernatant is centrifuged in a Ti15 zonal rotor at 32,000 rpm for 16 hr in a Beckman L5-65 ultracentrifuge housed within a Baker biosafety hood. The pellet (P_3) is recovered on the external wall of the rotor by using a specially designed core that permits removal of the supernatant fluid prior to opening the rotor.

8. The pellet is resuspended in 20 mM Tris-OAc (pH 8.3) containing 1 mM EDTA and 1 mM DTT and adjusted to a protein concentration of 10 mg/ml.

9. The suspension is then treated with sodium deoxycholate (DOC) at a final concentration of 0.5%.

10. The extract is centrifuged in a Beckman 50.2 Ti rotor at 50,000 rpm for 120 min.

11. The pellet (P_4) is resuspended in 20 mM Tris-OAc (pH 8.3), 1 mM EDTA, and 1 mM DTT to give a final protein concentration of 10 mg/ml. The P_4 pellet was stored under argon at -70°C.

12. Further purification is achieved by sequential digestions with microccocal nuclease at 12.5 U/ml and 4 mM $CaCl_2$ for 16 hr at 4°C followed by proteinase K at 100 μg/ml and 0.2% Sarkosyl for 8 hr at 4°C. Both digestions are performed with constant stirring.

13. Protease activity is terminated by addition of 0.1 mM phenylmethylsulfonyl fluoride (PMSF). A stock solution of 30 mM PMSF in n-propanol is stored at -20°C.

14. A 25% solution of sodium cholate is added slowly to give a final concentration of 2% (w/v) followed by addition of solid $(NH_4)_2SO_4$ (ultrapure grade) to 30% saturation.

15. Stirring of the sample is continued for 30 min prior to centrifugation at 20,000 rpm for 30 min at 4°C in a Beckman JA-20 rotor.

16. The pellet (P_5) is resuspended at 60 mM Tris-OAc (pH 8.3), 1 mM

EDTA, and 0.2% Sarkosyl. To ensure that it is uniformly resuspended, the pellet is sonicated twice for 10 sec in a Branson 350-W bath sonicator.

17. Residual $(NH_4)_2SO_4$ is removed by dialysis against the resusupension buffer. The titer of the P_5 fraction varied from $10^{9.0}$ to $10^{10.0}$ ID_{50} U/ml, and the protein concentration was 2–5 mg/ml.

18. Preparative agarose gel electrophoresis at 4°C is used for further purification. A 4-ml sample of P_5 is electrophoresed in a preparative horizontal apparatus toward the anode through an agarose gel containing Sarkosyl. The gel is 2 cm in length and has a height of 1.8 cm. The running buffer is 60 mM Tris-OAc (pH 8.3), 1 mM EDTA, 0.2% Sarkosyl. The 0.8% agarose gel is made by adding 0.8 g of Seakem agarose to 100 ml of electrophoresis buffer and boiling for 10 min. The agarose is then cooled to 55°C before it is poured into a flat-bed preparative gel electrophoresis apparatus (Polsky *et al.*, 1978). An electronic timing device provides automatic cycling of the electrophoresis system, in which an eluted fraction is collected at the end of the gel every 40 min. The electrophoresis is performed in the constant current mode at 100 mA using an ISCO power supply. After elution of a peak absorbing at 280 nm, the electrophoresis is terminated.

19. Initially, the scrapie agent was removed by pulverizing the gel with a Brinkmann Polytron and eluting overnight into a buffer containing 60 mM Tris-OAc (pH 8.3), 1 mM EDTA, and 0.2% Sarkosyl. The agarose was removed by pelleting in a JA-20 rotor at 20,000 rpm for 10 min.

20. In order to minimize the agarose contaminants in the eluted fraction, an E-C geluter apparatus is used to electroelute the scrapie agent from the preparative slab gel. The gel is sliced lengthwise into three parts and inserted into the geluter so that the scrapie agent will migrate toward the anode and elute into separate buffer chambers.

The titers of the gel eluates ranged from $10^{6.5}$ and $10^{8.5}$ ID_{50} U/ml, and the protein concentrations were 20–50 μg/ml. The degree of purification of the scrapie agent was generally 100-fold with respect to protein. In a few instances the degree of purification was as high as 1000-fold. The imprecision of the bioassay and a $10^{0.5}$–10^1 ID_{50} U/ml reduction in titer upon addition of DOC to fraction P_3 undoubtedly contribute to this variability in degree of purification.

2. Discontinuous Sucrose Gradient Method

1. In a typical preparation, 200 weanling female hamsters (LVG/LAK) are inoculated intracerebrally with $\sim 10^7$ ID_{50} U of hamster-adapted

scrapie agent and are sacrificed 60–65 days later. The hamsters are anesthetized with CO_2 and sacrificed by cervical dislocation.

2. The brains are removed, washed with sucrose, and a 10% (w/v) homogenate in 320 mM sucrose is prepared as described in step 3 of the Sarkosyl gel electrophoresis method. All operations are performed at 4°C unless otherwise noted, and only the final concentrations of all added chemicals are given.

3. The homogenate is centrifuged at 2700 rpm for 10 min in a JA-10 rotor in a Beckman J-21C centrifuge housed in a Baker biosafety cabinet.

4. The supernatant fluid (S_1) is decanted and centrifuged again at 4000 rpm for 30 min.

5. To the supernatant fluid (S_2) are added 1 mM EDTA and 1 mM DTT. Triton X-100 and DOC were added at ratios of 4:1 and 2:1 on a detergent:protein (w/w) basis, respectively, and stirred for 20 min in an ice bath.

6. To this detergent extract are added 0.03 M Tris-OAc (pH 8.3), 0.1 M KCl, 20% (v/v) glycerol, and 8% (w/v) polyethylene glycol (PEG) (approximate M_r 8000). The addition of these chemicals is dropwise, and stirring is continued for an additional 30 min.

7. The PEG precipitate is collected by centrifugation at 10,000 rpm in a JA-10 rotor for 30 min.

8. The pellet (P_3) is resuspended in 20 mM Tris-OAc (pH 8.3) containing 0.02% Triton X-100 and 1 mM DTT and adjusted to a protein concentration of 10 mg/ml. This buffer and all subsequent buffers are degassed under vacuum and saturated with argon prior to use.

9. Further purification is achieved by sequential digestions with 12.5 U/ml micrococcal nuclease in the presence of 2 mM $CaCl_2$ for 12 hr at 4°C, followed by 100 μg/ml proteinase K in the presence of 0.2% Sarkosyl and 2 mM EDTA for 8 hr at 4°C. Both digestions are performed with constant stirring.

10. Protease activity is terminated by addition of 0.1 mM PMSF from a 30 mM stock solution in n-propanol.

11. A 25% solution of sodium cholate is added slowly to give a concentration of 2% (w/v), followed by addition of solid $(NH_4)_2SO_4$ to 30% saturation. Stirring of the sample is continued for 30 min prior to centrifugation at 20,000 rpm for 60 min at 4°C in a Beckman JA-20 rotor.

12. The pellet (P_4) is resuspended in 20 mM Tris-OAc (ph 8.3), 1 mM EDTA, 1 mM DTT, and 0.2% Sarkosyl.

13. Twenty percent (v/v) Triton X-100 and 8% (w/v) SDS (specially pure) are mixed together and then added with stirring to fraction P_4 to give final concentrations of 2% Triton X-100 and 0.8% SDS. The

final protein concentration is adjusted to 0.75 mg/ml. A period of 15 min without agitation precedes layering 4 ml of suspension onto a sucrose gradient.

14. The gradient is formed in 40-ml self-sealing polyallomer tubes filled with 5 ml of 60% (w/v) and 31 ml of 25% sucrose; the sucrose solutions contain 20 mM Tris-OAc (pH 8.3) and 1 mM EDTA.

15. The tubes are centrifuged at 50,000 rpm in a vertical VTi50 rotor in a Beckman L5-65 ultracentrifuge housed in a Baker biosafety cabinet. After 120 min of centrifugation at 4°C, the speed was reduced to 3000 rpm with the brake, and the rotor was then allowed to coast to a stop.

16. The tubes are fractionated from the bottom after puncture with a Hoefer needle device. Fifteen fractions of 2.7 ml each are collected. To each fraction is added 0.05% (w/v) sulfobetaine 3-14 prior to freezing in a dry ice–ethanol bath and storage at −70°C.

The highest degree of purification is found in a fraction from the 25 to 60% interface near the bottom of the gradient. Generally, 15–25 ml of the purified gradient fraction containing ~30 µg/ml of protein and $10^{9.5}$–10^{10} ID_{50} U/ml can be obtained from 200 hamster brains. Purifications of 100- to 1000-fold were obtained by this method.

3. Large-Scale Discontinuous Sucrose Gradient Method

In order to prepare sufficient amounts of purified fractions containing scrapie prions to continue our studies, it was necessary to scale up the purification protocol described above. An outline of this large-scale protocol is depicted in Fig. 4.

1. Brains are taken from hamsters that are sacrificed 75 days after inoculation intracerebrally with 10^7 ID_{50} U of the scrapie agent. Hamsters are anesthetized with CO_2 and sacrificed by cervical dislocation. The brains are frozen in liquid nitrogen and stored at −70°C. For each preparation, 900–1000 brains are thawed.

2. Ten liters of a 10% (w/v) homogenate is made by suspending the brains in 0.32 M sucrose and disrupting the tissue with a Brinkmann Polytron equipped with a PT45 generator set at 8 for 2 min in a Baker Sterilgard biosafety hood. The homogenization vessel is a stainless steel flask with a capacity of 13.5 liters. The vessel measures 31.8 cm high and 26.7 cm in diameter. The lid is fastened to the cylinder with three screws with wing nuts. An O ring around the edge of the cylinder provides a gastight seal for the lid. The center of the lid has an opening 4.45 cm in diameter for the PT45 probe to be inserted. This opening is made nearly gastight with a gasket. The

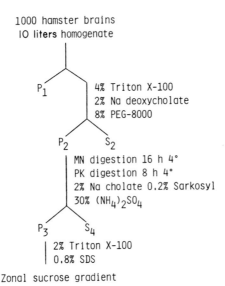

1000 hamster brains
10 liters homogenate

P_1 4% Triton X-100
     2% Na deoxycholate
     8% PEG-8000

P_2 S_2

MN digestion 16 h 4°
PK digestion 8 h 4°
2% Na cholate 0.2% Sarkosyl
30% $(NH_4)_2SO_4$

P_3 S_4

2% Triton X-100
0.8% SDS

Zonal sucrose gradient

FIG. 4. Scheme for large-scale purification of scrapie prions.

homogenization vessel is kept in an ice bath during use. The preparation is kept at 4°C throughout the purification. All solutions are purged with argon to remove any oxygen remaining after evacuation.

3. The homogenate is clarified by continuous-flow centrifugation at 10,000 rpm using a Beckman JCF-Z rotor equipped with a large pellet core. The centrifugation is performed in a Beckman J21-C centrifuge housed in a Baker biosafety cabinet. The homogenate is pumped through the rotor at 180 ml/min.

4. One millimolar EDTA and 1 mM DTT are added to the supernatant (S_1), to which Triton X-100 and DOC are then added at ratios of 4:1 and 2:1 on a detergent:protein (w/w) basis, respectively.

5. Once the volume of the S_1 with the detergent is determined, a PEG-8000 cocktail is made from stock solutions of 1 M Tris-OAc (pH 8.3) at 4°C, 3.5 M KCl, glycerol, and 50% PEG with an approximate M_r of 8000. The cocktail gives the extract a final concentration of 0.03 M Tris-OAc, 0.1 M KCl, 20% glycerol, and 8% PEG-8000. The detergent extract is precipitated by adding the PEG-8000 cocktail and stirred at 4°C for 15 min.

6. The pellet (P_2) is collected by continuous-flow centrifugation at 15,000 rpm using the JCF-Z rotor equipped with a standard pellet core. The fraction S_1 is pumped through the rotor at 160 ml/min.

7. Fraction P_2 is resuspended in 20 mM Tris-OAc (pH 8.3) containing 0.2% (v/v) Triton X-100 and 1 mM DTT to give a protein concentration of 10 mg/ml. It is digested with 12.5 U/ml of micrococcal nuclease in the presence of 2 mM CaCl$_2$ for 12 hr at 4°C followed by 100 µg/ml proteinase K for 8 hr at 4°C in the presence of 2 mM EDTA and 0.2% (w/v) Sarkosyl. The digestion was terminated with 0.1 mM PMSF.

8. The digest is precipitated with $(NH_4)_2SO_4$ (ultrapure) at 30% saturation in the presence of 2% (w/v) sodium cholate. This is centrifuged in a Beckman type 19 rotor in a Beckman L5-65 ultracentrifuge housed in a Baker biosafety cabinet at 19,000 rpm for 1 hr at 4°C.

9. The pellet (P_3) is adjusted to 0.825 mg/ml, and 20% Triton X-100 and 8% SDS (specially pure) is added to a final concentration of 2% Triton X-100 and 0.8% SDS. The protein concentration is reduced to 0.75 mg/ml. Typical preparations yield approximately 500 ml of the P_3 fraction.

10. A Ti15 zonal rotor with a B-29 liner is filled through the edge at a pump speed of 25 ml/min while rotating at 2000 rpm. The centrifugation is performed in a Beckman L5-65 ultracentrifuge housed in a Baker biosafety cabinet. The rotor is loaded with 600 ml of 25% sucrose in 20 mM Tris-OAc, 1 mM EDTA (pH 8.3) at 4°C, followed by 800 ml of 56% sucrose in 20 mM Tris-OAc, 1 mM EDTA (pH 8.3) at 4°C. Each zonal centrifugation uses 140 ml of the P_3 sample, which is 10% of the total rotor volume. Sucrose is added to 6% to fraction P_3 prior to loading it into the center of the rotor while displacing 56% sucrose through the outer edge. The pump speed is now 15 ml/min. An overlay of 320 ml of 20 mM Tris-OAc, 1 mM EDTA (pH 8.3) follows the sample through the center line. The rotor is accelerated to 5000 rpm and then capped. The speed is increased to 32,000 rpm for a period of 14.5 hr.

11. Fractions of 40 ml each are collected from the edge by replacing the rotor volume with 20 mM Tris-OAc, 1 mM EDTA (pH 8.3) at 4°C. All fractions are kept on ice. Sulfobetaine 3–14 is added at a final concentration of 0.05% to each fraction. All fractions are frozen in a dry ice and ethanol bath and stored at −70°C.

Each zonal gradient yields two or three fractions of 40 ml each that have specific infectivities of $10^{10.5}$–10^{11} ID$_{50}$/mg protein representing a purification of 3000- to 10,000-fold over the homogenate (Table II). As much as 95% of the infectivity loaded onto the gradient is recovered in the two or three fractions containing highest titers (Fig. 3). The overall recovery of scrapie infectivity from this protocol is generally greater than 20%. Dis-

TABLE II
LARGE-SCALE PURIFICATION OF SCRAPIE PRIONS

Fraction	Specific infectivity (log ID_{50} U/mg protein)	Purification (fold)	Recovery (%)
H	7.1	—	—
S_1	7.7	3.4	230
P_1	7.5	2.4	38
S_2	7.2	1.1	110
P_2	8.6	32.0	180
S_3	7.7	3.5	18
P_3	9.6	95.0	95
GF^a	10.7	3600.0	59

[a] GF denotes pooled gradient fractions with peak of scrapie prion infectivity.

aggregation of scrapie prions during the purification procedure may contribute to this high level of recovery.

4. *Sources of Chemicals*

$CaCl_2$, J. T. Baker, Philipsburg, NJ
Dithiothreitol (DTT), Calbiochem-Behring, La Jolla, CA
Ethylenediaminetetraacetic acid (EDTA), Sigma Chemical, St. Louis, MO
Glycerol, Mallinckrodt, St. Louis, MO
KCl, Mallinckrodt, St. Louis, MO
Micrococcal nuclease, Sigma Chemical, St. Louis, MO
$(NH_4)_2SO_4$ (ultrapure), Schwarz-Mann, Spring Valley, NY
n-Propanol, Mallinckrodt, St. Louis, MO
Phenylmethylsulfonyl fluoride (PMSF), Sigma Chemical, St. Louis, MO
Polyethylene glycol (PEG), Sigma Chemical, St. Louis, MO
Proteinase K, Merck, Rahway, NJ
Seakem agarose, Marine Colloids, Rockland, ME
Sodium cholate, Calbiochem-Behring, La Jolla, CA
Sodium deoxycholate (DOC), Calbiochem-Behring, La Jolla, CA
Sodium dodecyl sarcosinate (Sarkosyl), Sigma Chemical, St. Louis, MO
Sodium dodecyl sulfate (SDS) (specially pure), BDH Chemical, Poole, England
Sucrose (enzyme grade), Schwarz-Mann, Spring Valley, NY
Sulfobetaine 3-14, Calbiochem-Behring, La Jolla, CA
Tris(hydroxymethyl)aminomethane (Tris), Sigma Chemical, St. Louis, MO
Triton X-100, Sigma Chemical, St. Louis, MO

V. Identification and Detection of PrP

A. STRATEGY

Until recently, no macromolecule had been specifically associated with the scrapie agent or the disease process. Many attempts to identify a specific protein or nucleic acid in partially purified preparations of scrapie prions showed no differences between scrapie fractions and analogous fractions purified from normal brain. Likewise, attempts to find a protein(s) associated with the disease process by examination of scrapie-infected and normal brain homogenates were not successful.

Once convincing data were obtained showing that scrapie infectivity depended upon protein molecules (McKinley *et al.*, 1981; Prusiner *et al.*, 1981b, 1982a), a concerted effort to identify a protein specific to scrapie fractions was made. As the methods for purification improved, it became clear that any scrapie-specific protein would be present in low concentration. Thus, it seemed that sensitive methods would be required to detect a scrapie-specific protein.

The preferred method for detecting specific macromolecules associated with infectious pathogens is incorporation of radioactive precursors during the period of maximal biosynthetic activity. However, this strategy was not likely to be useful with the scrapie prion for two reasons: (1) no system for prion replication to high titers in cultured cells has been identified and (2) the kinetics of replication of prions in the hamsters are too slow to radiolabel efficiently a scrapie-specific protein. It was for these reasons that we turned our attention to the use of radiochemical labeling procedures for the detection of scrapie-specific proteins.

Although many radiolabeled compounds are commercially available for chemical modification of proteins, many of these are of too low a specific activity to offer much hope in the identification of a specific protein. Since the specific activity that would be required for detection of a scrapie-specific protein was not known, we presumed that reagents of the highest theoretical specific activity should be used. Thus, we chose techniques that couple ^{125}I to proteins in our search for a scrapie-specific protein.

B. IDENTIFICATION OF PrP

Radioiodination of partially purified fractions derived from the discontinous sucrose gradient procedure (purification method 2) led to the discovery of the protein PrP, which is unique to scrapie prion preparations (Bolton *et al.*, 1982). Since denaturation of purified samples prior to SDS-gel electrophoresis results in a significant loss of titer, we could not determine whether PrP is infectious when eluted from SDS gels (Bolton *et al.*, 1984). PrP, in its native or nondenatured conformation, is resistant to protease digestion, whereas proteins of similar M_r from normal preparations are readily digested. PrP exhibits microheterogeneity and has an apparent M_r of 27,000–30,000.

Studies clearly indicate that PrP is a structural component of the prion (McKinley *et al.*, 1983b). The evidence for PrP being a structural component of the prion is summarized in Table III. PrP has been found only in the brains of animals infected with a prion (Bolton *et al.*, 1982, 1984). It has been found both prior to and after neuropathological changes have occurred, as assessed by light microscopy. The prion and PrP have been found to copurify by two different procedures (purification methods 1 and 2), and PrP is the only major protein in extensively purified preparations (purification method 3). The concentration of PrP has been shown to be directly proportional to the titer of the prions (McKinley *et al.*, 1983b). Several properties of PrP have been found to be similar to those of the prions: poor imunogenicity, digestion by proteases, and chemical modification by DEP (Bolton *et al.*, 1982; McKinley *et al.*, 1981). Changing the properties of PrP results in a corresponding alteration in the prion. For example, denaturation of PrP by boiling in SDS renders PrP sensitive to protease digestion and results in a diminution in titer (Bolton *et al.*, 1982, 1984).

C. CONCENTRATION OF PRIONS AND PrP BY PRECIPITATION

Concentration of proteins by precipitation prior to radiochemical labeling is advantageous for two reasons: (1) it permits removal of proteins from a buffer containing components that would interfere with the labeling procedure, and (2) it increases the concentration of the proteins, which usually leads to more efficient labeling. In addition, precipitation of the radiola-

TABLE III

EVIDENCE THAT PrP IS A STRUCTURAL PROTEIN OF THE PRION

PrP is found only in animals infected with the prion, both prior to and after neuropathological changes

Prion and PrP copurify by two different procedures; PrP is the only major protein in extensively purified preparations

Concentration of PrP is directly proportional to the titer of prions

Several properties of PrP are similar to those of the prion (poor immunogenicity, resistance to proteases, chemical modification by DEP)

Changing the properties of PrP results in a corresponding alteration in the prion

beled proteins from suspensions containing unreacted or quenched compounds may give better results in subsequent analyses, i.e., polyacrylamide gel electrophoresis. We have used several methods for the precipitation of the scrapie prion and its protein (PrP).

1. SDS-Quinine Hemisulfate

This procedure was originally designed for the isolation of denatured proteins eluted from SDS–polyacrylamide gels (Durbin, 1983). The procedure makes use of the insolubility of SDS–quinine complexes at acidic pH. The procedure is very efficient at precipitating proteins, particularly proteins likely to be strongly associated with SDS micelles. Usually, the prion titer is reduced by a factor of 10 by this precipitation procedure.

1. Add 10% SDS to the sample to a final concentration of 0.2% (w/v) (i.e., 20 μl per milliliter of sample). Mix thoroughly on a Vortex mixer or other rapid-mixing device.
2. Add 0.1 M quinine hemisulfate, 0.1 N HCl to the sample (200 μl per milliliter of the original sample) and mix. A white precipitate of SDS–protein and quinine forms immediately. Pellet the precipitate by centrifugation (1000 g for 10 min is usually sufficient). If the original sample is very dense (i.e., if it contains a significant concentration of sucrose, glycerol, or other materials used to form density gradients), it will be necessary to dilute the precipitated suspension prior to centrifugation. Alternatively, the sample may be diluted prior to precipitation and the volumes of the reagent increased appropriately.
3. Wash the pellet with 80% acetone by vigorously mixing, and sediment the precipitated proteins to a pellet by centrifugation as described above. This step removes residual SDS and quinine, which may interfere with subsequent analyses of the resuspension of proteins in some buffer systems.
4. Resuspend the acetone pellet in buffer.

2. Sodium Deoxycholate–Trichloroacetic Acid

Precipitation of PrP by sodium deoxycholate and trichloroacetic acid (TCA) provides a useful alternative to the SDS–quinine hemisulfate precipitation technique (Peterson, 1977). In this case, the infectivity of the prion is reduced by a factor of 1000 or more.

1. Place the protein suspension in a suitable tube; make it 0.015% with respect to DOC by adding 100 μl of 0.15% DOC per milliliter of suspension. Mix the suspension thoroughly.
2. Add TCA to a final concentration of 10% (w/v). This can be accom-

plished by adding the solid or 250 µl of a 50% solution (w/v) per milliliter of protein suspension. Hold in an ice-water bath for 20 min.

3. If the protein to be precipitated is suspended in a dense medium (i.e., high concentrations of sucrose, CsCl, etc.), it may be necessary to dilute the suspension with cold 10% TCA before proceeding with the centrifugation step.

4. Sediment the precipitated material by centrifugation in a low-speed centrifuge.

5. Remove and discared the supernatant. Excess TCA may be removed from the pellet by vigorous washing with acetone followed by sedimentation to a pellet.

3. Alcohols and Acetone

PrP can also be precipitated from suspension by dilution with methanol, ethanol, or acetone. Virtually all of the scrapie infectivity can be recovered in the precipitate when ethanol is used as the solvent. Lower recoveries of infectivity have been found with methanol and acetone.

1. Make the protein suspension to be precipitated 80% (v/v) in methanol, ethanol, or acetone (i.e., add 4 volumes of the desired solvent).

2. Cool to 0°C in an ice water bath and let stand for 20 min or more.

3. Sediment the precipitated proteins by centrifugation in a microcentrifuge for 10 min.

4. Remove the supernatant and resuspend the pellet in buffer as required.

4. Sources of Chemicals

Acetone, Mallinckrodt, St. Louis, MO
Ethanol, Gold Shield Chemical, Hayward, CA
HCl, Mallinckrodt, St. Louis, MO
Methanol, Mallinckrodt, St. Louis, MO
Quinine Hemisulfate, Sigma Chemical, St. Louis, MO
Sodium dodecyl sulfate (SDS), BDH Chemical, Poole, England
Sodium deoxycholate (DOC), Calbiochem-Behring, La Jolla, CA
Trichloroacetic acid (TCA), Mallinckrodt, St. Louis, MO

D. RADIOLABELING METHODS

Three methods for introducing radioiodine into proteins have been used for detecting PrP. These are the Bolton-Hunter method (Bolton and Hunter, 1973), chloramine-T oxidation (Greenwood et al., 1963), and Iodobead oxidation (Markwell, 1982).

1. *Iodination with Bolton–Hunter Reagent*

Iodination of proteins with *N*-succinimidyl-3-(4-hydroxy[5-^{125}I] iodophenyl) propionate (Bolton–Hunter reagent).

a. Preparation of the Reagent. This procedure is designed to iodinate *N*-succinimidyl 3-(4-hydroxyphenyl) propionate, also known as Tagit or NSHPP, with 20 mCi of ^{125}I to produce a final yield of 4.0 ml of Bolton–Hunter reagent containing >6 × 10^7 cpm per 100-μl aliquot. The reagent is used in 50- or 100-μl amounts to iodinate proteins.

Materials

Unstable solutions (make fresh each time)

Chloramine-T (5 mg/ml) in 0.25 *M* Na phosphate, pH 7.5

Na metabisulfite (12 mg/ml) in 0.05 *M* Na phosphate, pH 7.5

NaI (200 mg/ml) in 0.05 *M* Na phosphate, pH 7.5

Stable solutions

NSHPP: 40 μg/ml in dry toluene:ethyl acetate (1:1). Make up and aliquot in airtight tubes (Eppendorf microfuge tubes work well for this). Store desiccated at −20°C. It is stable for at least 1 year if properly stored.

Dimethylformamide (DMF) (dry with molecular sieve)

Benzene (dry with molecular sieve)

Radionuclide: Na^{125}I (carrier free); should be Amersham IMS 300 or equivalent

Procedure

1. Dry down 100 μl of NSHPP in a glass reaction tube.
2. Prepare individual addition syringes (or micropipets) with 100 μl of chloramine-T (5 mg/ml); 100 μl of Na metabisulfite (12 mg/ml); 100 μl of NaI (200 mg/ml). Have ready 20 μl DMF (dry), 4–5 ml benzene (dry).

   Note: Additional syringes can be made using 1 ml tuberculin syringes, a blunt needle (18 gauge), and PE 60 tubing (24 cm = 100 μl).
3. Spin Na^{125}I to the bottom of the shipping vessel and add 20 mCi to the reaction vessel containing dried NSHPP. Mix well.
4. Add 100 μl of chloramine-T and mix for 10–20 sec.
5. Rapidly add: 100 μl Na metabisulfite (12 mg/ml) and 100 μl NaI (200 mg/ml) and mix for 5 sec. Add 20 μl DMF and mix; then add 1.0 ml benzene and mix for 20–30 sec.
6. Recover the benzene phase (top) being careful not to contaminate it with the aqueous phase. Collect in a second glass tube. Repeat the benzene extraction and pool the extract with the first. Remove the aqueous phase and repeat the benzene extraction of the *reaction vessel* twice more, each time using 1.0 ml of benzene. Pool all benzene frac-

tions and aliquot in 0.6- to 0.8-ml amounts into airtight containers suitable for storage. Pierce Reactivials work well. Store aliquots as a benzene solution in a desiccated, lead-shielded container.

7. The reagent should be tested for composition of mono- and diiodo product by TLC on silica gel with a toluene:ethyl acetate (1:1) solvent. The aqueous phase can be analyzed by this method for purposes of comparison. The aqueous phase and benzene-soluble product should be counted in the gamma counter to estimate recoveries. These samples are very highly radioactive, and appropriate dilutions must be made to eliminate the problem of coincident radiation detection (10-μl aliquot at a 1:100 dilution should be sufficient).

b. Labeling of Proteins
Materials

0.1 M Na borate (pH 8.5) or 0.1 M Na borate, 0.1% SDS (pH 8.5); other buffers may be used but should be free of reactive amines such as Tris

^{125}I Labeled Bolton-Hunter reagent in benzene *N*-succinimidyl 3-(4-hydroxy-5-[^{125}I]iodophenyl) propionate

0.2 M glycine, 0.1 M Na borate (pH 7.5)

Procedure

1. Using nitrogen gas, dry down an aliquot of ^{125}I-labeled Bolton–Hunter reagent in a suitable reaction vessel.

2. Add 10–100 μl of the protein suspension (1–2 mg/ml) in one of the buffers described above). Mix well and incubate the reaction mixture at 0°C for 30 min, with occasional mixing. Longer reaction times will produce a more heavily labled product but may introduce labeling artifacts.

3. Add 500 μl of 0.2 M glycine, 0.1 M Na borate (pH 7.5) and mix well. Let stand at 0°C for 5 min.

4. The iodinated proteins may be separated from the unreacted reagent and iodinated glycine by precipitation as described above.

2. Iodination with Chloramine-T

1. Place 0.1–1 mCi of Na^{125}I in a fresh 1.5-ml microfuge tube.

2. Add 10 μl of 0.25 M sodium phosphate buffer (pH 7.5).

3. In rapid succession add each of the following: (1) 20 μl of concentrated protein in 50 mM sodium phosphate, 0.1% SDS (pH 7.5); (2) 10 μl of chloramine-T (5 mg/ml in 50 mM sodium phosphate, pH 7.5). Mix the reactants for 10–20 sec; (3) 100 μl of sodium metabisulfite (1.2 mg/ml in 50 mM sodium phosphate, pH 7.5); (4) 100 μl of NaI (20 mg/ml in 50 mM sodium phosphate, pH 7.5).

4. Mix thoroughly to stop the reaction. The radioiodinated proteins may be separated from unreacted iodine by a variety of methods, including the precipitation method described above.

3. *Iodination with Iodobeads*

1. Iodobeads [N-chlorobenzenesulfonamide (sodium salt)-derivatized polystyrene beads] are purchased from Pierce Chemical Co.
2. Place 250 μl of the protein suspension to be labeled in a 1.5-ml microfuge tube or other suitable reaction vessel. The protein may be in any of a number of buffers. We have found that the most efficient labeling is achieved when buffers which promote protein denaturation are employed. One such buffer is 100 mM sodium phosphate, 2% SDS, 8.0 M urea (pH 6.5).
3. Add one Iodobead to the suspension. The bead must be washed in a suitable buffer (100 mM sodium phosphate, 0.1% SDS, pH 6.5, works well) and dried on a clean filter paper prior to use.
4. Add 0.1–1.0 mCi of ^{125}I to the reaction mixture.
5. Allow the mixture to react at room temperature for 30 min with occasional mixing.
6. Transfer the suspension to a fresh microfuge tube and wash the bead twice with 125-μl aliquots of the reaction buffer or any suitable buffer.
7. Pool the washes with the labeled product. The radioiodinated proteins can be separated from the unreacted iodine and concentrated as described above.

4. *Ethoxyformylation by Diethylpyrocarbonate*

Treatment of partially purified preparations of prions with diethylpyrocarbonate (DEP) caused a reduction in titer (McKinley *et al.*, 1981). The ethoxyformylated, inactive prion was reactivated upon removal of the ethoxyformyl moiety with hydroxylamine. Samples containing PrP are suspended in 60 mM sodium phosphate (pH 7.2) containing 0.2% Sarkosyl. Ten millimolar [^{14}C]DEP (53 mCi/mmol) is reacted with each sample for 1 hr at room temperature. Excess DEP decomposes rapidly in water. Methods for the synthesis of radiolabeled DEP have been published (Melchior and Fahrney, 1970).

5. *Sources of Chemicals*

Benzene, Mallinckrodt, St. Louis, MO
Chloramine-T, Eastman-Kodak, Rochester, NY
Dimethylformamide, Sigma Chemical, St. Louis, MO
Ethyl acetate, Mallinckrodt, St. Louis, MO
Iodobeads, Pierce Chemical, Rockford, IL

N-Succinimidyl 3-(4-hydroxyphenyl)propionate (Tagit, NSHPP), Cal-
biochem-Behring, La Jolla, CA
Silica gel TLC plates, Analabs, North Haven, CT
Sodium iodide, BDH Chemical, Poole, England
Sodium metabisulfite, J. T. Baker Chemical, Phillipsburg, NJ
Toluene, Mallinckrodt, St. Louis, MO

E. GEL ELECTROPHORESIS

1. Polyacrylamide Gel Electrophoresis

Radiolabeled proteins are heated to 100°C for 2 min in electrophoresis
buffer containing 2% SDS and 5% 2-mercaptoethanol. The proteins are
separated by electrophoresis in 15% polyacrylamide gels employing the discontinuous buffer system of Laemmli (1970). Gels were stained in 25% 2-
propanol and 10% acetic acid containing 0.03% Coomassie brilliant blue
R-250 and destained in 10% 2-propanol and 10% acetic acid (Fairbanks et
al., 1971). Alternatively, gels may be stained using either of the silver staining methods described below. Gels are dried with a Hoefer slab gel dryer.
Autoradiographs of gels containing ^{125}I-labeled proteins are made at room
temperature with Kodak XAR-5 film. Gels containing proteins labeled with
[14C]DEP were processed for fluorography, and the autoradiograph was
exposed at −70°C (Bonner and Laskey, 1974).

2. Densitometry

Autoradiograms or stained gels are scanned in the region where PrP migrates using an LKB 2202 Ultrascan laser densitometer; the output is displayed on a 3390A Hewlett-Packard integrator. Each electrophoresis lane
is scanned 10 times at a spacing interval of four. The autoradiographic
densities are recorded in arbitrary units.

F. SILVER STAINING OF SDS–POLYACRYLAMIDE GELS

Two silver-staining procedures have proved to be useful in our studies.
The first procedure (Merril et al., 1981) is rapid, but has the disadvantage
of underestimating the amount of PrP, which is not stained well by this
method. The second procedure (Morrissey, 1981) is considerably more time
consuming, but PrP is more readily stained.

1. Procedure of Merril et al.

This procedure will easily detect <1 μg of protein and as little as 10 ng
for some proteins. Not all proteins stain with silver at the same intensity.
This procedure should be performed immediately after electrophoresis of

the gel because some proteins have a tendency to elute from the gel before being completely stained and fixed. All chemicals used in this procedure were analytical grade and purchased from Mallinckrodt unless otherwise noted.

1. Fixation step: immerse the gel in 50% methanol, 12% acetic acid for 20 min with shaking (100–200 ml per gel).
2. Removal of excess SDS: replace fixative solution with 10% ethanol, 5% acetic acid. Perform three washes for 10 min each with shaking.
3. Oxidation step: place the gel in a solution containing 0.0034 M potassium dichromate, 0.0032 N nitric acid for 5 min; variations in the length of time affects the darkness of the background.
4. Wash the gel with distilled water two or three times before the staining with silver. Shake with water briefly and remove the gel.
5. Silver stain: immerse the gel in a solution containing 0.012 M silver nitrate (Spectrum Chemicals, reagent grade) prepared fresh before staining. Stain for 20–30 min with shaking.
6. Image developer: rinse the gel twice with a solution containing 0.28 M Na_2CO_3 and 0.5 ml of formaldehyde per liter (a 37% solution). Then immerse the gel in 100–200 ml of image developer and develop to desired background. Usually 5–20 min are required. The formaldehyde solution (37%) must be fresh, or stain will not develop. A dark yellow background is sometimes desirable in order to see minute amounts of proteins.
7. Stop solution: 2% acetic acid is used to terminate the stain development.
8. Gels may be preserved in distilled H_2O in the cold.

In the case the image turns out to be too dark, a photo reducer can be used to adjust the image intensity. The photo reducer is prepared by mixing together the following components: (1 part) 50 ml $Na_2S_2O_3$ (450 g/liter) sodium bisulfite; (1 part) 50 ml $CuSO_4$–NaCl (37 g/liter–37 g/liter); (3 parts) 150 ml H_2O.

2. *Morrissey Procedure*

This technique is very sensitive and can be used to give uniform staining of proteins in gels 1 mm thick or less. For standard 0.75- or 1-mm gels (~ 9 cm $\times$ 13 cm), the volume of each reagent used is 100 ml per gel except for the 10% glutaraldehyde, which is 50 ml per gel. All chemicals are analytical grade and are purchased from Mallinckrodt unless otherwise noted.

Several points described below have been found to be helpful. Gels should be handled only with plastic gloves rinsed with H_2O to avoid stained fingerprints. Gentle but thorough agitation is important during each step of the staining procedure. Solutions may require filtering to prevent contamination with dust. Degassing of solutions seems to improve results. Several gels can be stained simultaneously in the same container. Coomassie blue staining followed by destaining may be done before silver staining. If this is done, steps 1 and 2 described below should be omitted.

Materials for this procedure include 50% methanol containing 10% acetic acid, 5% methanol containing 7% acetic acid, 10% glutaraldehyde, 5 μg/ml DTT (Calbiochem-Behring), 0.1% silver nitrate, 2.3 M citric acid. The developer is made by mixing 50 ml of 3% sodium carbonate with 100 ml of 37% formaldehyde; it should be made just prior to use.

1. Prefix the gel in 50% methanol containing 10% acetic acid for 30 min (100 ml is used for each gel).
2. Prefix gel in 5% methanol containing 7% acetic acid for 30 min.
3. Fix the gel in 10% glutaraldehyde for 30 min.
4. Rinse the gel in distilled H_2O; soak the gel in distilled H_2O with agitation for a minimum of 2 hr. Gels can be left overnight in distilled H_2O if desired.
5. Soak the gel in 3 μg/ml DTT for 30 min.
6. Pour off DTT and, without rinsing, add 100 ml of 0.1% silver nitrate for 30 min.
7. Pour off the silver nitrate, rinse once rapidly with distilled H_2O, rinse twice with a small volume (20–50 ml) of developer, and soak in developer (100 ml) until bands appear at the desired staining level. Usually about 5 min is required.
8. Add 5 ml of 2.3 M citric acid to stop the staining reaction. Usually 10 min is required.
9. Wash the gel with several changes of distilled H_2O. Gel may be preserved by drying or soaking in 0.03% sodium carbonate to prevent bleaching and sealed in cellophane.

G. Kinetics of PrP Digestion by Proteinase K

The identification of PrP as a structural component of the scrapie prion was facilitated by its resistance to protease digestion (Bolton *et al.*, 1982; McKinley *et al.*, 1983b). This resistance to protease digestion is relative, not absolute. We were able to take advantage of this relative resistance to remove contaminating proteins and leave PrP intact. It is noteworthy that PrP becomes sensitive to protease digestion after denaturation by boiling in SDS (Bolton *et al.*, 1982, 1984).

Samples for digestion were made 0.2% (w/v) with Sarkosyl and then digested between 0 and 30 hr at 37°C with either 100 or 500 μg/ml of freshly prepared proteinase K. Those samples analyzed by gel electrophoresis were iodinated with ^{125}I-labeled Bolton–Hunter reagent as described above, whereas those inoculated into hamsters for bioassay were chemically labeled with ^{127}I-labeled Bolton–Hunter reagent by the same procedure. After digestion, proteins labeled with ^{125}I-labeled Bolton–Hunter reagent were heated to 100°C for 2 min in electrophoresis sample buffer containing 2% SDS and 5% 2-mercaptoethanol. Samples for inoculation were diluted into 1 ml with phosphate-buffered saline containing 50 mg/ml of BSA and 0.1 mM PMSF. These samples were frozen in a dry ice–ethanol bath and stored at $-70°C$ until bioassay.

An example of a digestion experiment with proteinase K is shown in Fig. 5 and 6. These experimental results not only illustrate the methodology, but also are important in the design of other experiments using amino acid nonspecific proteases such as proteinase K, because they show the kinetics of digestion for PrP and the infectious prion. Aliquots of a fraction (F_2), partially purified by the discontinuous sucrose gradient method, were subjected to prolonged digestion with proteinase K. Control aliquots, not exposed to proteinase K, are shown by autoradiography in Fig. 5A; the amount of PrP, did not change over a 30-hr period. Addition of 100 μg/ml of proteinase K rapidly hydrolyzed almost all the proteins except PrP. More than 2 hr of digestion at 37°C was required for proteinase K to degrade PrP (Fig. 5B). A 5-fold increase in the concentration of proteinase K accelerated the hydrolytic cleavage of PrP (Fig. 5C). The results of the prion bioassays and quantitation of PrP by laser densitometry are plotted in Fig. 6. Neither the concentration of PrP nor the prion titer changed in the control samples from which proteinase K was omitted (Fig 6A). Addition of 100 μg/ml of proteinase K resulted in a progressive decrease in PrP and prion infectivity after 2 hr of incubation (Fig. 6B). The kinetics of PrP and prion reduction were virtually identical. Increasing the concentration of proteinase K to 500 μg/ml caused a shift to the left in the PrP and prion curves as expected (Fig. 6C). Again the kinetics of PrP disappearance were unique under these experimental conditions; all other proteins were digested within the first 30 min (Figs. 5B and 5C).

VI. Electron Microscopy

Electron microscopic examination of partially purified samples containing scrapie agent has shown numerous rodlike structures (McKinley et al., 1983; Prusiner et al., 1982a, 1983). The rods measure predominantly 100–200 nm in length; their width is 25 nm in rotary-shadowed preparations and 10–20 nm in negatively stained preparations (Fig. 7). Examination of extensively purified preparations revealed large aggregates of varying size and

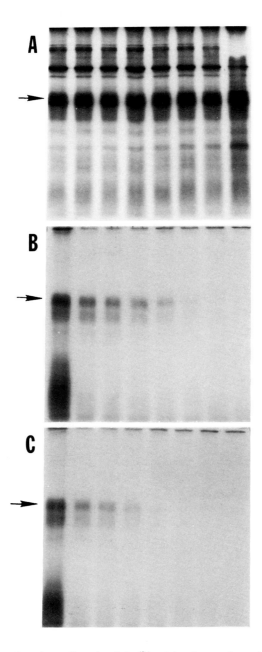

FIG. 5. Gel electrophoresis profiles of radioiodinated fraction F_2 digested with proteinase K for increasing periods of time. (A) Control; (B) 100 μg/ml proteinase K; (C) 500 μg/ml proteinase K. All digestions were performed at 37°C. Shown from left to right are samples incubated for 0, 0.5, 1, 2, 4, 8, 16, and 30 hr. The autoradiograms were exposed for 66 hr at 25°C.

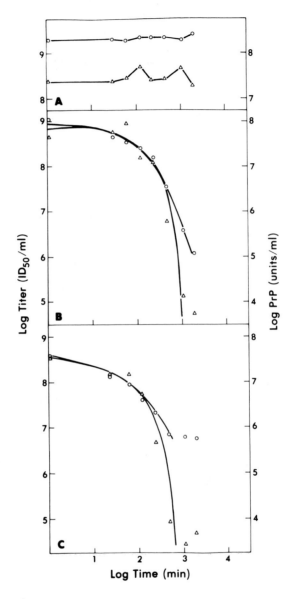

FIG. 6. Kinetics of PrP degradation and prion inactivation by digestion with proteinase K. The concentration of PrP (△) was determined by laser densitometry, and the prion titer (○) was measured by bioassay. (A) Control; (B) 100 μg/ml proteinase K; (C) 500 μg/ml proteinase K.

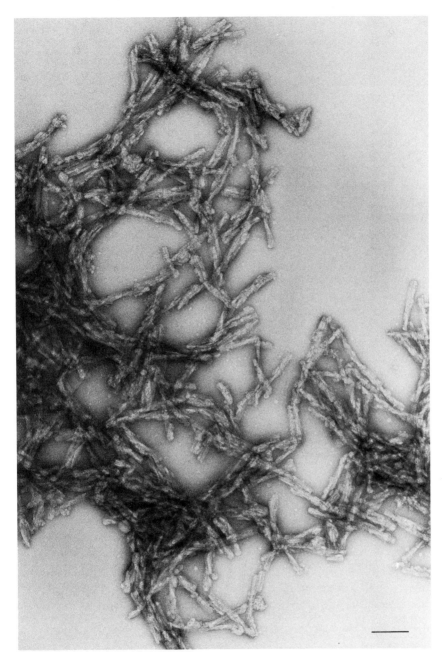

FIG. 7. Electron micrograph of an extensively purified fraction of prions, negatively stained with uranyl formate. Bar is 100 nm.

shape composed of numerous rods. In some cases, the rods seem to associate to form fibrillar structures. Quantitation of the rods has not been possible because conditions for dispersing the rods into single structures have not been found, thus far.

A. Preparation of Support Films

Electron microscope specimen grids, 400-mesh Cohen-Pelco handle type (Ted Pella, Inc.), are filmed with 0.25% Formvar and coated with a thin layer of carbon in an Edwards E306 evaporator. The films are then made hydrophilic by subjecting them for about 5 sec to a glow discharge at 70–80 ml (9–11 Pa) air pressure (Williams, 1977). The handles of the grids are prevented from becoming hydrophilic during the glow-discharge procedure by covering them with small pieces of metal. This precaution prevents any liquid drop subsequently applied to the grid surface from spreading to the tips of the forceps which hold the grids. Five microliters of polylysine (1 μg/ml) is pipetted onto the grid and drained off 30 sec later by touching the edge of the grid with a vacuum aspirator made from a flame-drawn Pasteur pipet. These pipets have a final internal diameter of <0.5 mm. Poly-L-lysine (Sigma, St. Louis, MO) of M_r 2000 has been used. The residual liquid on the grid can be inspected under a 10× stereomicroscope as it dries. The grid surface should be sufficiently hydrophilic that drying takes several seconds with the trailing edge of the liquid film showing one or two complete orders of interference colors. Thorough cleanliness, as assessed by hydrophilicity of the specimen films, is essential for the consistent staining of specimens (Williams, 1983). Thus, filter paper should not be used for removal of liquid drops from the grid surface. After the residual polylysine is dried, the grids are ready for specimen application.

B. Adsorption of Samples to Treated Films; Detergent Removal

A sample (6 μl) is applied to the treated films and allowed to adsorb for about 1 min. The unadsorbed portion of the 6-μl drop is washed away from the grid surface by slowly passing the grid (sample surface down) 10 times across the surface of a solution of 0.1 M ammonium acetate (Mallinckrodt, St. Louis, MO). This step is immediately followed by washing with 0.01 M ammonium acetate. Any remaining ammonium acetate is removed by the vacuum pipet described above. These washing steps can be used to remove detergents after adsorption of the sample. For samples to be rotary shadowed, 5 μl of 1% uranyl acetate is applied to the grid between the 0.1 and 0.01 M ammonium acetate washing steps (Williams, 1983).

C. NEGATIVE STAINING

Uranyl formate is the negative stain that we have primarily used in our studies. The stain is applied to samples adsorbed to grids and allowed to stand for 10 sec; excess stain is removed by the vacuum pipet. The procedure described below for the preparation of the stain should be followed carefully for best results. A 3–5% solution of uranyl formate is prepared by adding 0.6–1 g of powdered uranyl formate (Eastman-Kodak, Rochester, NY) to 20 ml of double-distilled water without mixing. After 5 min, an aliquot of the solution is placed in a 1-ml cuvette and its A_{450} is determined. The absorbance should be < 0.10. An aliquot of the solution is monitored every 30 sec. When an A_{450} of 0.10 is reached, the remainder of the solution is filtered through a 0.22-μm filter and placed into tubes of 0.5-ml fractions. Tubes containing the stain are then wrapped in foil and kept at 4°C until use. Stain should be used within 60 days of preparation. This procedure appears to minimize the impurities that contaminate the uranyl formate.

The other negative stain used in our studies is potassium phosphotungstate (Mallinckrodt, St. Louis, MO). It is prepared by adjusting phosphotungstic acid (1–3%) to pH 7.2 with potassium hydroxide. It is important that, during all above operations, the film be prevented from drying until after the final step.

D. ROTARY SHADOWING

The dried grids are rotary shadowed with tungsten as described by Williams (1954). A 20-mil tungsten wire, of 3 cm free length, is initially heated with a 31.5-A current, after which the applied voltage is left constant until burnout (~ 90 sec). This schedule evaporates about 10 mg of tungsten. An oblique angle of 8–10°, with a source-to-sample distance of ~ 7 cm, is satisfactory for contrast enhancement. The plate to which the grids are secured in the Edwards E306 evaporator rotates at 100 rpm.

E. SAMPLE CONCENTRATION

To facilitate the electron microscopic studies of scrapie prions, it has been frequently necessary to concentrate the samples. Four different procedures have been used.

1. Samples to be negatively stained or rotary shadowed are centrifuged in a Beckman airfuge using an EM-90 electron microscopy particle-counting rotor. The sector-shaped cells with walls that conform to the radii of the rotor minimize the formation of convection currents and the subsequent uneven distribution of sedimented particles. Additionally, this rotor offers an ideal small-cell volume (0.1 ml) and rapid sedimentation at a maximum force of 120,000 g. Particles are sedimented directly onto film-covered grids supported at the periphery of each sector on a piece of Parafilm (Wolinsky, 1983).

2. Samples are also centrifuged in the Beckman airfuge using the A-100 fixed-angle rotor. A sample volume of 0.15 ml is concentrated 15- to 30-fold for negative staining or rotary shadowing by resuspending the pellet in 5–10 μl of buffer.

3. Samples are concentrated by the SDS-quinine hemisulfate precipitation protocol described in Section V,C,1.

4. Samples are concentrated by precipitation with methanol (80%) for 20 min at 0°C followed by centrifugation in a microfuge for 10 min as described in Section V,C,3.

F. ELECTRON MICROSCOPY—MINIMAL BEAM EXPOSURE

Specimen damage by the electron beam is minimized by restricting irradiation of the region to be micrographed (Williams and Fisher, 1970). The region selected for photography is irradiated only during photographic exposure. Focusing and stigmatic correction are done on an adjacent specimen region. A minimum-size, condensed-beam spot ($\sim 2 \mu$m) is first placed near the edge of the viewing screen and visible in the binocular. After focusing, the beam is suitably expanded and the photographic exposure is promptly made. After each photomicrograph, the beam is recondensed and repositioned during specimen scanning; no area traversed by the condensed beam is later photomicrographed.

VII. Polarization Microscopy

Congo red staining of scrapie prions represents a new approach to the study of purified preparations (Prusiner et al., 1983). Congo red has been used extensively in research on amyloid; after staining with alkaline Congo red, amyloid fibers exhibit green birefringence when viewed under polarized light (Puchtler et al., 1962). Additional data by X-ray crystallography indicate that the conformation of amyloid responsible for the properties described here is an antiparallel β-pleated sheet (Glenner et al., 1974).

Polarization microscopy of Congo red-stained preparations provides a convenient and rapid assay system for structural studies of polymerized scrapie prions.

A. REAGENTS

1. An 80% ethanol–NaCl-saturated solution is made by adding 60 g of NaCl to 2 liters of 80% alcohol. After stirring overnight, the solution is filtered through Whatman No. 1 paper.
2. Congo red dye [sodium diphenyldiazobis(α-naphthylamine) sulfonate] is purchased from Eastman-Kodak, Rochester, NY. Stock Congo red is prepared by adding 0.2 g of Congo red dye to 100 ml of the ethanol–NaCl solution. This is also stirred overnight and filtered using a Nalgene 0.22-μm filter. The solution is stable for approximately 2 months.
3. The Congo red stain was made immediately prior to use by adding 1% NaOH to the stock Congo red solution to give a final NaOH concentration of 0.01%. This solution must be used within 15 min of preparation since Congo red dye is slowly hydrolyzed in alkali.
4. A fixative solution is prepared using the ethanol–NaCl solution and 1% NaOH to give a final NaOH concentration of 0.01%.
5. Mayer's egg albumin fixative is made by Harleco and purchased from American Scientific Products, Sunnyvale, CA.

B. PREPARATION AND STAINING

The protocol used for Congo red staining of the scrapie agent is a modification of a published procedure (Glenner, 1983; Puchtler et al., 1962). Ethanol-cleaned standard 7.6 × 2.5-cm glass microscope slides are covered with a thin layer of Mayer's egg albumin fixative. Six or seven drops of the fixative are applied to the slide and spread using the edge of a pipet. Excess fixative is drained off by touching the side of the slide to a paper towel. The fixative is then allowed to dry for approximately 1 hr. Samples applied to slides are air dried in a Baker Sterilgard biosafety hood overnight. Fixation is accomplished by pipetting the NaOH–ethanol–NaCl solution onto the slides and allowing it to equilibrate for 10 min. Slides are drained to remove the excess solution, and freshly prepared Congo red stain is added. Staining proceeds for 20 min, and the Congo red is washed off with successive changes of distilled water until no color is seen in the washes. The preparations are next dehydrated for 2–3 min in three successive changes of 100% ethanol. The slides are air dried followed by mounting with Permount (Fisher Scientific, San Francisco, CA) using No. 1 22 × 22-mm coverglasses. Slides are viewed by bright field and polarization microscopy using a Leitz Orthoplan microscope equipped with strain-free lenses, optimally aligned crossed polarizers, and a 100-W quartz–halide light source. The binding of Congo red dye to arrays of prion rods is shown in Fig. 8A by bright-field microscopy. The green birefringence seen by polarization microscopy is shown in Fig. 8B.

VIII. Biocontainment and Decontamination Procedures

The development of appropriate biocontainment and decontamination procedures prior to starting work on prions is mandatory. Considerable experience or appropriate training with standard microbiological practices and techniques is necessary for all laboratory personnel.

The practices, techniques, facilities, and decontamination procedures described below probably represent an extreme with respect to biosafety, but they have been adopted because there are so many unknowns surrounding prions. We have found these procedures to be both rational and convenient. They are all based upon known principles of biocontainment and have evolved from our cognizance of laboratory-acquired infections with pathogens other than prions (Pike, 1979). There is no documented case of a laboratory-acquired prion infection in humans.

A. Principles and Philosophy

All aspects of research on prions are complicated by the prolonged incubation periods associated with prion diseases. Not only do these long incubation periods impede laboratory research efforts, but they also cause difficulty in evaluating biosafety procedures (Chatigny and Prusiner, 1980). The short incubation times associated with typical infectious pathogens permit detection and assessment of biohazards within a reasonable time frame. The analysis of exposure to infectious pathogens such as prions with long incubation periods is far more difficult. Thus, it seems prudent to reduce the possibility of exposure to a minimum level. Case reports of accidental exposure to the CJD agent indicate that direct inoculation is the major biohazard that must be considered (Beronoulli et al., 1977; Duffy et al., 1974; Gajdusek et al., 1977). The possibility of airborne infection continues to seem remote (Chatigny and Prusiner, 1980).

In the design of a biocontainment facility in which to work, as well as in the design of our experimental protocols, we have considered the following observations. The scrapie prion is clearly not contagious or communicable for human beings. There is no evidence that human beings in contact with scrapie prions have an increased incidence of any neurological

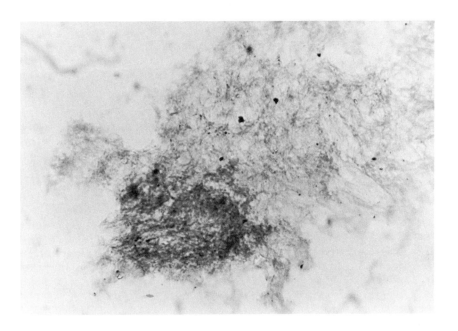

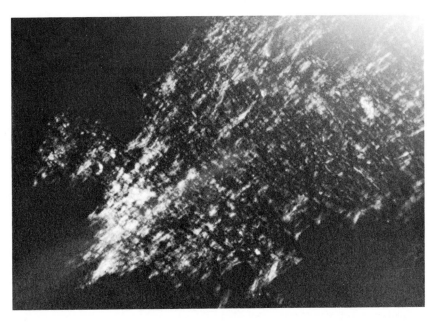

FIG. 8. Light micrographs of an extensively purified fraction of prions stained with Congo red dye. (A) Bright field, ×500; (B) polarized light, ×500.

disorder (Masters *et al.,* 1978). However, the great similarities between the agents causing scrapie and CJD (Gajdusek *et al.,* 1977; Hadlow *et al.,* 1980) suggest to us that prudent biosafety measures are warranted, since the CJD agent invariably causes a fatal neurological illness.

For many years the extreme resistance of the scrapie agent to inactivation by procedures that readily inactivate conventional viruses as well as microorganisms such as bacteria, parasites, and fungi has been well known (Gajdusek *et al.,* 1977; Prusiner, 1982). These unusual properties of the scrapie agent are a consequence of its molecular structure.

The safety and decontamination procedures outlined in this section reflect a desire (1) to contain prions for analytical procedures and animal operations at a P2 level (National Institutes of Health, 1982), (2) to contain prions to preparative procedures and aerosol producing operations at a P3 level (National Institutes of Health, 1982), (3) to contain any aerosol generated by experimental procedures, and (4) to minimize the risk of accidents caused by unnecessary, cumbersome biocontainment measures.

B. Government Requirements

In the United States, the Department of Agriculture (USDA) has a series of regulations governing scrapie research. The shipping of scrapie-containing materials within the United States is governed by these regulations. A veterinary permit must be obtained from the USDA, Animal and Plant Health Inspection Service, Beltsville, Maryland, prior to transfer of scrapie materials. The USDA requires that the work be limited to laboratory studies and that inoculated animals be held in isolated insect- and rodent-proof facilities. Disposal of experimental animals requires incineration, and waste and bedding must be autoclaved or incinerated. Work on scrapie is to be performed in USDA-inspected facilities, and laboratory personnel are not to have contact with sheep or goats.

Work on the CJD and kuru agents is governed by guidelines issued by the Centers for Disease Control (CDC) and the National Institutes of Health (NIH) (Richardson and Barkley, 1983). Biosafety level 2 practices, equipment, and facilities are recommended for all activities with the CJD agent. Extreme care is dictated for avoidance of accidental autoinoculation or other traumatic parenteral inoculations. Avoidance of aerosols or droplets as well as the use of gloves is recommended.

Biosafety level 2 demands standard microbiological practices, laboratory coats, decontamination of all infectious wastes, and limited access to the laboratory and animal facilities, as well as gloves (Richardson and Barkley,

1983). Biological safety cabinets are required for mechanical or manipulative procedures that may generate aerosols.

Biosafety level 3 requires those practices cited above for level 2 as well as protective laboratory garments and biosafety cabinets for all manipulations of infectious materials (Richardson and Barkley, 1983).

Our practices with scrapie prions conform to biosafety level 3 for all preparative and aerosol-producing procedures. Our analytical practices with prions represent a level of biosafety between levels 2 and 3.

Those precautions used in caring for patients with hepatitis are recommended for use in caring for CJD patients (Centers for Disease Control, 1982).

C. LABORATORY PRACTICES AND TECHNIQUES

A knowledge of and adherence to excellent microbiological practices is our first line of containment. All personnel are aware of potential hazards and are trained in the safe handling of materials containing prions. Biological safety cabinets (Class II) are readily available for performing any procedure that could intentionally or accidentally generate aerosols. Standard microbiological practices and biological safety cabinets are our primary barriers, and our biocontainment facility is the secondary barrier.

Preparative procedures are performed within a P3 level biocontainment facility described below.

Disposable laboratory clothing with closures in the back and solid fronts is used for preparative procedures. The clothing either has elastic cuffs or the cuffs are closed with tape. Hats, safety glasses, masks, and two pairs of rubber gloves are worn if large preparations are being handled. With homogenization procedures, during which the arms of the gown are within a hood and there is a possible exposure to considerable aerosols, the gowns and gloves are changed immediately. In all preparative procedures double gloves are always worn.

For analytical laboratory work, where small amounts of the infectious particles are being analyzed, normal laboratory coats or virological smocks are used as well as gloves. No hats, masks, or safety glasses are routinely employed for these analytical procedures.

D. BIOSAFETY CABINETS

Two procedures commonly used in our work are capable of producing aerosols and significantly contaminating the environment. These are homogenization of infected tissues and prolonged centrifugation used for concentration of prions (Dimmick *et al.,* 1973). In order to contain these aerosols, biosafety cabinets manufactured by the Baker Company (Sanford, ME), were installed. Biosafety cabinets for general laboratory procedures as well as specially designed cabinets for housing high-speed and ultracentrifuges are included within the facility. The centrifuge cabinets are positioned in a line on each side of an open-front Baker Sterilgard cabinet; the entire assembly is manufactured as a unit.

Actual operational tests on the centrifuge cabinet were performed to determine whether the centrifuge and the cabinet could satisfactorily withstand a major catastrophe. The system was tested by exploding a loaded rotor turning at 53,000 rpm. Under these conditions, some leakage occurred from the centrifuge, which was displaced within the cabinet, but the cabinet integrity was maintained and no leak to the environment was detected (Chatigny et al., 1979).

Air exhausted from these cabinets is filtered directly within the cabinet by HEPA (99.97 + % efficient) filters and discharged into a separate exhaust system. The remainder of the ventilating air exhausted from the laboratory is not filtered. However, the supply air into the laboratory is filtered to 95% ASHRAE (52–68) standards. As a result, the laboratory is remarkably dust free and the filters in the safety cabinets have been usable for extended periods of time, far exceeding that projected by the manufacturer (approximately 2 years).

It should be noted that the successful design, construction, and use of this system required close collaboration among the laboratory director, centrifuge manufacturer, safety cabinet manufacturer, and bioengineer.

E. BIOCONTAINMENT FACILITIES

Since our research has been focused on concentrating and purifying infectious prions, we designed a facility in which purification procedures could be performed safely. We examined the major sources of massive aerosol generation that could contaminate the laboratory and possibly the surrounding area (Chatigny and Prusiner, 1980). Our primary concerns are (1) accidentally generated aerosols as direct routes of infection, (2) potential accumulation of contaminants with time, and (3) lack of an effective gas or vapor phase decontaminant.

As noted above, our analytical laboratories generally comply with requirements for P2 facilities (National Institutes of Health, 1982) and biosafety level 2 operation (Richardson and Barkley, 1983). Our preparative laboratory complies with requirements for a P3 facility and biosafety level 3 operation.

Having developed a safe primary barrier consisting of biosafety cabinets (see Section VII, D) capable of containing major aerosol-producing operations, it was necessary to design a laboratory facility that would facilitate proper function of these devices and provide a secondary barrier. The secondary barrier consists of a laboratory with a tight perimeter and an effective ventilation system. Ventilation is accomplished by a once-through system with high-efficiency filters on the incoming air, providing a clean makeup for the biological safety cabinets. The volume of air needed for the safety cabinets determined the overall ventilation requirements. Our decision to have the exhausts from these cabinets go directly outside meant that air exchange rates as high as 40–50 air changes per hour are encountered in the laboratories. The installation of the safety cabinet equipment, and ensuring its proper functioning, can be extremely difficult. Mechanical design engineers and contractors frequently encounter severe problems in integrating all the equipment and mechanical systems.

We did tests of the systems with fluorescent tracers and aerosol detection equipment to assure ourselves that the secondary barrier laboratory was functioning as designed to contain the air from within the space and to draw a slight amount of air from the public corridors into an anteroom in which clean air was supplied to prevent contamination of the laboratory work. Every effort must be made to seal all joints to the floors, ceilings, and walls, as well as to assure that no penetrations permit back-leakage through piping, electrical conduits, or other entry points.

The use of many filters in the air-supply system, and in the exhaust from the safety cabinets, meant that there was the possibility for change in air flow quantities as the facility aged. We do not recommend installation of complicated control dampers or flow-volume controls. Our system was designed to age gracefully within the limits we had established and to require a minimum of maintenance to remain effective.

The overall level of containment between the public hallways and the containment laboratory with personnel traversing was such that less than 1 airborne particle in 10^7 would have a reasonable probability of escaping the containment facility. These tests were repeated after 5 years of operation, with minimal maintenance, and the facility—while showing lower air exchange rates because of filter loading—functioned at about the same level of effectiveness as it had when originally constructed.

The design, development, and testing of this relatively simple small-facility ventilating system required a strong sustained effort by senior scientists in the program, a consulting mechanical engineer, building engineers, and speciality laboratory safety consultants. The capability for assessment of the degree of containment needed and the ability to develop measurement techniques to evaluate the success of the design are critical. We cannot overemphasize the need for cooperation among users, designers, and evaluators throughout the entire design and construction phases.

The availability of this facility has helped immeasurably in generating within the work force a spirit of cooperation, a willingness to work safely, and a clear realization that all participants need to think constantly about containment procedures. We believe that the operation is safe and that the precision of the laboratory work has not been compromised.

F. Containment of Aerosols

All procedures that generate aerosols are performed within biosafety hoods. These include homogenization, centrifugation, especially with continuous flow and zonal rotors, and sonication (Chatigny and Prusiner, 1980; Dimmick *et al.*, 1973). Homogenization of brains from scrapie-infected animals is performed in closed vessels or vessels covered with Saran wrap. The initial homogenization procedure in our purification protocols usually involves a Polytron apparatus. Two Polytrons have been used in our studies, the routine laboratory size as well as an industrial-scale Polytron. Specially designed stainless steel vessels, which allow the hub of the Polytron generator to seat onto an O ring, are used to minimize aerosols. After the homogenization, the homogenate is allowed to sit for 15–30 min prior to opening the vessel. Transfer of liquids is done either within a hood or by pumping from one vessel to another.

Vacuum apparatus is connected to an NaOH trap and HEPA filter in order to prevent contamination of the house vacuum.

Sonication procedures are always performed within a closed vessel. Either a cup horn is used with closed tubes containing the infectious prion or a gastight vessel surrounding a sonicator probe is employed. Sonication procedures are all performed within a hood.

All centrifugation procedures are done within centrifuges that are housed in Baker biocontainment hoods. These hoods are especially important when continuous-flow and zonal rotors are used in the production of large amounts of purified prions.

G. Decontamination Procedures

All dry waste is autoclaved for 4.5 hr before being considered sterile. Liquid waste is autoclaved for 90 min in the presence of 1 N NaOH (Prusiner *et al.*, 1981a). It is clear that the autoclaving of liquid waste for 90 min at 121°C is insufficient to attain sterility (Table IV). More than 10^3 ID_{50} U/ml remained after autoclaving; thus, some additional procedure is required. Although NaOH alone is not sufficient, in combination with autoclaving it seems to be adequate; however, further studies are urgently needed.

In cases in which equipment cannot be autoclaved, the use of 1 N NaOH

TABLE IV

INACTIVATION OF SCRAPIE PRIONS BY NaOH AND HEAT[a]

Temperature (°C)	Time (hr)	Control	1.0 N NaOH	1.0 N NaOH[b] (scrapie hamsters)
		(log ID_{50} U/ml ± SE)		
25	0.5	9.4 ± 0.21	<0	1/4
25	1.5	8.3 ± 0.21	<0	0/4
25	4.0	8.6 ± 0.25	<0	2/4
25	24.0	7.9 ± 0.25	<0	1/4
121	1.5	3.4 ± 0.21	<0	0/4

[a] Homogenates of scrapie-infected hamster brains were diluted with an equal volume of 0.32 M sucrose or 2.0 N NaOH. The samples were incubated at the designated temperatures for varying periods of time. The incubation was terminated by adding a 10-fold excess of assay diluent, and the diluted samples were stored at −70°C until assay.

[b] Number of animals developing scrapie between 200 and 400 days after inoculation divided by the number of animals inoculated.

three times is probably quite satisfactory. Apparatus such as Pipetman shafts, Hamilton repeating syringes, and Dewar flasks cannot be autoclaved, but can be sterilized with 1 N NaOH. Three changes of 1 N NaOH are used, and each change is allowed to sit for 30 min. Afterward, the equipment is thoroughly rinsed with water.

When equipment cannot be autoclaved or treated with NaOH, 10% SDS is used at a temperature between 60° and 100° C, depending on the apparatus. This decontamination procedure does not render the equipment sterile, and it must be considered contaminated from that time on and stored in appropriate bags to indicate that this is the case. The primary reason for the SDS decontamination procedure is to remove almost all of the infectious particles so that they do not interfere with subsequent experiments. Plastic gel electrophoresis devices are treated with 10% SDS at 60°C and plastic centrifugation rotor cores are treated with SDS at 80°C. Polytron generators and centrifuge bottles that are rated by their manufacturers to be capable of tolerating 30 min of autoclaving are treated with 10% SDS at 100°C instead. To remove the SDS, the apparatus is washed three times with distilled water. All the liquid is then autoclaved and the equipment is stored in bags, noting that it is still contaminated, i.e., not sterile. The seal assemblies of our zonal rotors are cleaned with Beckman rotor cleaning concentrate solution No. 555 diluted 1:6. After a zonal rotor centrifugation is completed, 300 ml of this solution heated to 40°C is pumped through each line of the seal assembly. To remove the cleaning solution, 1.5 liters of water is pumped through each line of the seal assembly twice. After a large preparation is carried out in a biocontainment hood, the walls of the hood are washed in NaOH and rinsed with water. HPLC columns are decontaminated with six column volumes of 4 N guanidine-HCl followed by water (Prusiner *et al.,* 1981a).

H. ANIMAL FACILITY OPERATION

In the animal facility, three principles guide biocontainment procedures. First, we want to protect the animal from any infectious pathogens that human caretakers might bring into the facility. Second, we want to protect the humans from any pathogens, whether they be scrapie or others, that the animals might bring to them, although this seems to be rather unlikely. And third, we wish to comply with USDA guidelines. All animal caretakers wear disposable jumpsuits, masks, hats, glasses, and double gloves. Rubber boots are worn. As personnel enter each animal room, they wear surgical gowns. The masks, gloves, and surgical gowns are all changed before entering a single room.

In the animal facility, the animals are handled with long forceps that are kept in a solution of Prepodyne that has been diluted 1:5. At the entrance to each animal room there is a footbath containing FSD-4 disinfectant. All animal records are written on double forms: the original remains in the animal room and the carbon copy is sent to the laboratory. For work with scrapie, the cages are not autoclaved, but are washed and then reused. Bedding, however, is either autoclaved or incinerated. All the cage cards are kept, and these are autoclaved before storage. All animals are incinerated for disposal. For CJD studies in rodents, the procedures are the same as those described above, except that the cages are autoclaved with the bedding in them prior to being dumped and washed.

I. ACCIDENTS

The most serious accident is a puncture wound. Should such a wound occur, we would attempt to let it bleed profusely, and either wash the wound thoroughly with straight bleach (5% sodium hypochlorite) or 2 N NaOH. The washing is repeated several times. Because the scrapie agent is so resistant to inactivation by chemical procedures, it is important to protect the eyes in any situation where an aerosol or spill might contaminate the cornea or conjunctiva. Should this happen, wash the eye thoroughly with water, but extreme efforts are taken to prevent this from happening by the use of safety glasses.

In case of a spill, if it is associated with an aerosol, the area is evacuated for at least 30 min prior to reentry. If no aerosol was associated with a spill, it is wiped up three times with a solution of 1 N NaOH and then the area is rinsed thoroughly with water.

J. LABORATORY SUPPLIES FOR BIOCONTAINMENT

Prepodyne Scrub for decontamination of animal forceps (West Chemical Products, Lynbrook, NY)

FSD-4 disinfectant for foot baths (quaternary germicidal cleaner) (Dubois Chemicals, Cincinnati, OH)

All-purpose goggles with wide curved lenses for greater impact safety and side protection (Lexan plastic manufactured by General Electric and available from VWR Scientific Company, San Francisco, CA)

Triflex sterile latex surgeons' gloves (50 pairs per box) (Travenol Laboratories, Deerfield, IL)

Nonsterile vinyl examination gloves (100 per box) (Travenol Laboratories, Deerfield, IL)

Isolation protector gowns (disposable surgical gowns) (American Hospital Supply Corporation, Evanston, IL)

Disposable jumpsuits (made of Dupont TYVER, Durafab, Cleburne, TX).

ACKNOWLEDGMENTS

The authors thank Mss. F. Elvin, R. Mead, and L. Gallagher for editorial and administrative help, as well as Mss. P. Glenn, E. Espanol, and J. Fairley for expert technical assistance. M. P. M. was the recipient of an Alzheimer's Disease and Related Disorders Association Award. D. C. B. was supported by a postdoctoral fellowship from the National Institutes of Health. This research was supported by research grants from the National Insititutes of Health (NS14069 and AG02132), as well as by gifts from R. J. Reynolds Industries, Inc., Sherman Fairchild Foundation, and W. M. Keck Foundation.

REFERENCES

Anderson, N. G. (1962). *J. Phys. Chem.* **66**, 1984.

Baringer, J. R., and Prusiner, S. B. (1978). *Ann. Neurol.* **4**, 205.

Beck, E., and Daniel, P. M. (1965). *In* "Slow, Latent, and Temperate Virus Infections" (D. C. Gajdusek, C. J. Gibbs, Jr., and M. Alpers, eds.), NINDB Monograph 2, pp. 203–206. U.S. Gov. Printing Office, Washington, D.C.

Bernoulli, C., Siegfried, J., Baumgartner, G., Regli, F., Rabinowicz, T., Gajdusek, D. C., and Gibbs, C. J., Jr. (1977). *Lancet* **1**, 478.

Bolton, A. E., and Hunter, W. M. (1973). *Biochem. J.* **133**, 529.

Bolton, D. C., McKinley, M. P., and Prusiner, S. B. (1982). *Science* **218**, 1309.

Bolton, D. C., McKinley, M. P., and Prusiner, S. B. (1984). *Biochemistry,* in press.

Bonner, W. M., and Laskey, R. A. (1974). *Eur. J. Biochem.* **46**, 83.

Burger, D., and Hartsough, G. R. (1965). *J. Infect. Dis.* **115**, 393.

Centers for Disease Control (1982). Annual Summary 1979, Issued March 1982. National Nosocomial Infections Study Report, pp. 1–38. U.S. Department of Health and Human Services, Atlanta, Georgia.

Chandler, R. L. (1961). *Lancet* **1**, 1378.

Chatigny, M. A., and Prusiner, S. B. (1980). *Rev. Infect. Dis.* **2**, 713.

Chatigny, M. A., Dunn, S., Ishimaru, K., Eagleson, J. A., and Prusiner, S. B. (1979). *J. Appl. Microbiol.* **38**, 934.

Cohen, A. S., Shirahama, T., and Skinner, M. (1982). *In* "Electron Microscopy of Proteins" (J. R. Harris, ed.), Vol. 3, pp. 165–206. Academic Press, New York.

Diener, T. O., McKinley, M. P., and Prusiner, S. B. (1982). *Proc. Natl. Acad. Sci. U.S.A.* **79**, 5220.

Dimmick, R. L., Fogl, W. F., and Chatigny, M. A. (1973). *In* "Biohazards in Biological Research" (A. Hellman, M. N. Oxman, and R. Pollack, eds.), pp. 246–266. Cold Spring Harbor Laboratory, Cold Spring Harbor, New York.

Diringer, H., Hilmert, H., Simon, D., Werner, E., and Ehlers, B. (1983). *Eur. J. Biochem.* **134**, 555.

Dougherty, R. M. (1964). *In* "Techniques in Experimental Virology" (R. J. C. Harris, ed.), pp. 169–224. Academic Press, New York.

Duffy, P., Wolf, J., Collins, G., Devoe, A., Streeten, B., and Cowen, D. (1974). *N. Engl. J. Med.* **290**, 692.

Fairbanks, G., Steck, T. L., and Wallach, D. F. H. (1971). *Biochemistry* **10**, 2606.

Fraser, H. (1979). *In* "Slow Transmissible Diseases of the Nervous System" (S. B. Prusiner and W. J. Hadlow, eds.). Vol. 1, pp. 387–406. Academic Press, New York.

Gajdusek, D. C. (1977). *Science* **197**, 943.

Gajdusek, D. C., Gibbs, C. J., Jr., Asher, D. M., Brown, P., Diwan, A., Hoffman, P., Nemo, G., Rohwer, R., and White, L. (1977). *N. Engl. J. Med.* **297**, 1253.

Glenner, G. G., Eanes, E. D., Bladen, H. A., Linke, R. P., and Termine, J. D. (1974). *J. Histochem. Cytochem.* **22**, 1141.

Greenwood, F. C., Hunter, W. M., and Glover, J. S. (1963). *Biochem. J.* **89**, 114.

Hadlow, W. J., Prusiner, S. B., Kennedy, R. C., and Race, R. E. (1980). *Ann. Neurol.* **8**, 628.

Hogan, R. N., Baringer, J. R., and Prusiner, S. B. (1981). *Lab. Invest.* **44**, 34.

Hunter, G. D. (1979). *In* "Slow Transmissible Diseases of the Nervous System" (S. B. Prusiner and W. J. Hadlow, eds.), Vol. 2, pp. 365–385. Academic Press, New York.

Kimberlin, R., and Walker, C. (1977). *J. Gen. Virol.* **34**, 295.

Laemmli, U. K. (1970). *Nature (London)* **227**, 680.

Luna, L. G. (1968). "Manual of Histologic Staining Methods of the Armed Forces Institute of Pathology," pp. 1–258. McGraw-Hill, New York.

McKinley, M. P., Masiarz, F. R., and Prusiner, S. B. (1981). *Science* **214**, 1259.

McKinley, M. P., Bolton, D. C., and Prusiner, S. B. (1983a). *Proc. Electron Microsc. Soc. Am.* **41**, 802.

McKinley, M. P., Bolton, D. C., and Prusiner, S. B. (1983b). *Cell* **35**, 57.

Malone, T. G., Marsh, R. F., Hanson, R. P., and Semancik, J. S. (1979). *Nature (London)* **278**, 575.

Markwell, M. A. K. (1982). *Anal. Biochem.* **125**, 427.

Marsh, R. F., and Kimberlin, R. H. (1975). *J. Infect. Dis.* **131**, 104.

Masters, C. L., Harris, J. O., Gajdusek, D. C., Gibbs, C. J., Jr., Bernoulli, C., and Asher, D. M. (1978). *Ann. Neurol.* **5**, 177.

Masters, C. L., Gajdusek, D. C., and Gibbs, C. J., Jr. (1981). *Brain* **104**, 559.

Melchior, W. B., and Fahrney, D. (1970). *Biochemistry* **9**, 251.

Merril, C. R., Goldman, D., Sedman, S. A., and Ebert, M. H. (1981). *Science* **211**, 1437.

Millson, G. C., and Manning, E. J. (1979). *In* "Slow Transmissible Diseases of the Nervous System" (S. B. Prusiner and W. J. Hadlow, eds.), Vol. 2, pp. 409–424. Academic Press, New York.

Morrissey, J. H. (1981). *Anal. Biochem.* **117**, 307.

National Institutes of Health (1982). NIH Guidelines for Research Involving Recombinant DNA Molecules. *Fed. Regist.* **47**, 38048.

Pattison, I. H., and Millson, G. C. (1960). *J. Comp. Pathol.* **70**, 182.

Peterson, G. L. (1977). *Anal. Biochem.* **83**, 346.

Pike, R. M. (1979). *Annu. Rev. Microbiol.* **33**, 41.

Polsky, F., Edgell, M. H., Seidman, J. G., and Leder, P. (1978). *Anal. Biochem.* **87**, 397.

Prusiner, S. B. (1982). *Science* **216**, 136.

Prusiner, S. B. (1984). *Adv. Virus Res.,* in press.

Prusiner, S. B., Hadlow, W. J., Eklund, C. M., and Race, R. E. (1977). *Proc. Natl. Acad. Sci. U.S.A.* **74**, 4656.

Prusiner, S. B., Hadlow, W. J., Eklund, C. M., Race, R. E., and Cochran, S. P. (1978a). *Biochemistry* **17**, 4987.

Prusiner, S. B., Hadlow, W. J., Garfin, D. E., Cochran, S. P., Baringer, J. R., Race, R. E., and Eklund, C. M. (1978b). *Biochemistry* **17**, 4993.

Prusiner, S. B., Garfin, D. E., Baringer, J. R., Cochran, S. P., Hadlow, W. J., Race, R. E., and Eklund, C. M. (1979). *In* "Slow Transmissible Diseases of the Nervous System" (S. B. Prusiner and W. J. Hadlow, eds.), Vol. 2, pp. 425–464. Academic Press, New York.

Prusiner, S. B., Cochran, S. P., Groth, D. F., Hadley, D., Martinez, H. M., and Hadlow, W. J. (1980a). *In* "Aging of the Brain and Dementia" (L. Amaducci, A. N. Davison, and P. Antuono, eds.), pp. 205–216. Raven, New York.

Prusiner, S. B., Garfin, D. E., Cochran, S. P., McKinley, M. P., and Groth, D. F. (1980b). *J. Neurochem.* **35**, 574.

Prusiner, S. B., Groth, D. F., Bildstein, C., Masiarz, F. R., McKinley, M. P., and Cochran, S. P. (1980c). *Proc. Natl. Acad. Sci. U.S.A.* **77**, 2984.

Prusiner, S. B., Groth, D. F., Cochran, S. P., Masiarz, F. R., McKinley, M. P., and Martinez, H. M. (1980d). *Biochemistry* **19**, 4883.

Prusiner, S. B., Groth, D. F., Cochran, S. P., McKinley, M. P., and Masiarz, F. R. (1980e). *Biochemistry* **19**, 4892.

Prusiner, S. B., Groth, D. F., McKinley, M. P., Cochran, S. P., Bowman, K. A., and Kasper, K. C. (1981a). *Proc. Natl. Acad. Sci. U.S.A.* **78**, 4606.

Prusiner, S. B., McKinley, M. P., Groth, D. F., Bowman, K. A., Mock, N. I., Cochran, S. P., and Masiarz, F. R. (1981b). *Proc. Natl. Acad. Sci. U.S.A.* **78**, 6675.

Prusiner, S. B., Bolton, D. C., Groth, D. F., Bowman, K. A., Cochran, S. P., and McKinley, M. P. (1982a). *Biochemistry* **21**, 6942.

Prusiner, S. B., Cochran, S. P., Groth, D. F., Downey, D. E., Bowman, K. A., and Martinez, H. M. (1982b). *Ann. Neurol.* **11**, 353.

Prusiner, S. B., McKinley, M. P., Bowman, K. A., Bolton, D. C., Bendheim, P. E., Groth, D. F., and Glenner, G. G. (1983). *Cell* **35**, 349.

Puchtler, H., Sweaf, F., and Levine, M. (1962). *J. Histochem. Cytochem.* **10**, 355.

Richardson, J. H., and Barkley, W. E. (1983). Biosafety in Microbiological and Biomedical Laboratories (Draft), pp. 1–90. U.S. Department of Health and Human Services, Atlanta, Georgia.

Tzagoloff, A., and Penefsky, H. S. (1971). *Methods Enzymol.* **22**, 219–231.

Warren, G. B., Toon, P. A., Birdsall, N. J. M., Lee, A. G., and Metcalfe, J. C. (1974a). *Biochemistry* **13**, 5501.

Warren, G. B., Toon, P. A., Birdsall, N. J. M., Lee, A. G., and Metcalfe, J. C. (1974b). *Proc. Natl. Acad. Sci. U.S.A.* **71**, 622.

Williams, E. S., and Young, S. (1982). *J. Wildl. Dis.* **18**, 465.

Williams, R. C. (1954). *Adv. Virus Res.* **2**, 183.

Williams, R. C. (1977). *Proc. Natl. Acad. Sci. U.S.A.* **74**, 2311.

Williams, R. C., and Fisher, H. W. (1970). *J. Mol. Biol.* **52**, 121.

9 Detection of Genome-Linked Proteins of Plant and Animal Viruses

Stephen D. Daubert and George Bruening

I. Introduction

In the mid-1970s, several types of 5' end groups of virion nucleic acids had been discovered. Circular forms of virus nucleic acids also were known. However, the virion RNA of poliovirus had none of 5' end groups known at that time, yet the RNA was not circular. In 1977, Flanegan et al. and Nomoto et al. demonstrated that the 5' ends of the virion RNA and the replicative intermediate RNAs of poliovirus are masked by linkage to a low

347

molecular weight protein. This protein was designated VPg (Lee *et al.,* 1977), for "virion protein, genome-linked." It was found to be attached to the 5'-terminal pUp residue previously identified in studies of the polyribosomal poliovirus RNA, which is not protein linked.

Subsequently VPg-linked genomic RNAs were discovered for another group of single-stranded RNA animal viruses, the caliciviruses. Two groups of DNA animal viruses, one double-stranded RNA virus of animals, and six groups of single-stranded RNA viruses of plants also are considered to have genomic nucleic acids covalently coupled to proteins (e.g., Salas and Viñuela, 1980; Wimmer, 1982). The known 5'-linked proteins of virion DNAs and double-stranded RNAs are larger than those attached to viral single-stranded RNAs, and these larger proteins have been referred to as terminal proteins (TPs). For simplicity, and because the terminal proteins and the VPgs appear to share at least some functions, we refer to all the 5' covalently linked proteins of virion nucleic acids as VPgs.

Several morphologically similar bacteriophages have protein-bearing genomic double-stranded DNAs. These are φ29, φ15, GA-1, Nf, and M2Y of *Bucillus subtilis* (Yoshikawa and Ito, 1981) and Cp-1 of *Streptococcus pneumoniae* (García *et al.,* 1983a). Since the VPg of bacteriophage φ29 is one of the best characterized of all VPgs in terms of both structure and function, it too is discussed in this review of plant and animal virus VPgs. A bacteriophage of a different group, one that has lipid containing virions, also has been characterized as having a terminally linked protein (Bamford *et al.,* 1983).

As will be indicated below, VPgs are not limited to viral genomic nucleic acids, but have also been found linked to a subgenomic messenger RNA and a satellite RNA, as well as to nonviral linear nucleic acids.

II. Detection of a Genome-Linked Protein (VPg)

The search for a VPg becomes an effort to demonstrate that a given protein is associated with the virion nucleic acid, that the association is covalent, and that the association is sequence specific and does not have an explanation that is unrelated to genome expression. Strong, but not sequence-specific, binding of protein to virion nucleic acids is common. The nonspecific association may become covalent. For example, protein and nucleic acid in a plant virus may become cross-linking, due to the reactive phenolic derivatives in plant extracts (Loomis *et al.,* 1979). Even noncovalently bound coat protein can be difficult to remove. If 99.9% of the coat protein of a typical small, single-stranded RNA virus has been removed, the residue corresponds in mass to what would be expected for a VPg. Demonstration of a protein that can be released only after nuclease-catalyzed digestion of the purified virion nucleic acid indicates the possible presence of a covalent VPg–nucleic acid complex but does not constitute a proof for such a complex.

Sensitivity of the infectivity of isolated virus nucleic acid to proteinase treatment has been an indicator of the presence of a VPg (e.g., Mayo *et al.*, 1982a). However, such sensitivity alone is insufficient evidence of a VPg. For example, several RNA and DNA bacteriophages have proteins that enter the cell along with the virion nucleic acid during the normal infection initiation process. These proteins are not covalently attached to the nucleic acid but can stimulate the infectivity of the isolated nucleic acid by several orders of magnitude. Such noncovalently bound proteins can render the infectivity of the viral nucleic acid operationally proteinase sensitive (e.g., Chase and Benzinger, 1982). The opposite generalization also can mislead: lack of sensitivity of infectivity to proteinase does not necessarily indicate the absence of a VPg (see examples in Table I).

A. DETECTION OF NON-VPg 5' TERMINI

The investigations that led to the earliest discoveries of a VPg were stimulated when specific tests for certain well known 5' end groups proved to be negative. Neither a cap structure (m⁷GpppNp; Shatkin, 1976) nor a phosphorylated 5'-terminal group was detected on the virion RNA of poliovirus and other picornaviruses (Hewlett *et al.*, 1976; Nomoto *et al.*, 1976; Harris, 1979) or of cowpea mosaic virus (CPMV; Klootwijk *et al.*, 1977). In these experiments the virion RNA was labeled with ^{32}P *in vivo* and was digested with one ribonuclease or a mixture of ribonucleases (ribonucleases T1, T2, etc.) to degrade most of each molecule to nucleoside 3'-monophosphates (Nps). RNAs with known cap structures were similarly digested as positive controls. A cap structure or mono- (Hewlett *et al.*, 1976; Nomoto *et al.*, 1977), di- (Wimmer and Reichmann, 1968), or triphosphate (Roblin, 1968) terminus or any structure that is not susceptible to ribonuclease digestion should produce an entity or entities that are chromatographically distinct from Nps. Ion-exchange chromatography and electrophoresis have been the methods of choice for resolving the multiply phosphorylate compounds. None of these entities were detected in analyses of the end groups of mature RNA or nascent chains of poliovirus replicative intermediate RNA (reviewed by Kitamura *et al.*, 1981; Baron and Baltimore, 1982a). Analyses for 2'-*O*-methyl groups provide a secondary assay for a possible cap structure (Klein and Klootwijk, 1976; Klootwijk *et al.*, 1977).

Translation experiments also may indicate the absence of a cap structure.

The cap structure analog $p^{7m}G$ did not inhibit the translation of the VPg-bearing subgenomic coat protein mRNA of Southern bean mosaic virus (SBMV), whereas translation of the capped coat protein mRNA of brome mosaic virus was strongly inhibited by similar concentrations of the compound (Ghosh *et al.*, 1981). A system from sea urchin eggs failed to translate VPg-bearing or proteinase-treated CPMV RNA, but readily accepted capped mRNAs (Winkler *et al.*, 1983).

Resistance to exonuclease also may reveal an unusual 5'-linked group. Bjornsti *et al.* (1982) found that the DNA of the bacteriophage $\phi29$ could be modified by incubation with exonuclease III or with terminal nucleotidyltransferase and dNTPs, indicating accessibility of the 3' end. However, the DNA was not modified by the 5'-specific nucleases of bacteriophages λ or T7 or by polynucleotide kinase and ATP, regardless of whether it had been exposed to phosphatase or proteinase. For studies on the exonuclease susceptibilities of adenovirus DNA, see Sharp *et al.* (1976) and Dunsworth-Browne *et al.* (1980).

Detection of a 5' hydroxyl presents a different problem. Upon digestion by ribonucleases an RNA with a free 5' hydroxyl group will give rise to Nps only, giving no direct information about the structure at the 5' end. Most such RNAs become phosphorylated when they are incubated with bacteriophage T4 polynucleotide kinase (Donis-Keller *et al.*, 1977) and [γ-^{32}P]ATP. However, readily detectable amounts of ^{32}P can be incorporated at other than the 5' end. An example is the incorporation at the sites of "hidden breaks" in CPMV RNA. In this case the radioactivity migrated more rapidly than the two genomic RNA zones during gel electrophoresis (Daubert *et al.*, 1978), and, after complete digestion by the guanylate-specific ribonuclease T1, the radioactivity appeared in several oligonucleotides (Klootwijk *et al.*, 1977). Similar results were obtained with the VPg-linked subgenomic coat protein mRNA of SBMV (Ghosh *et al.*, 1981). If the radioactivity comigrates only with intact virion nucleic acid, the 5'-terminal nucleoside residue of the intact RNA or DNA can be identified by digesting the nucleic acid with nuclease P1 or snake venom diesterase and chromatography of the nucleoside 5'-phosphates (Silberklang *et al.*, 1977; Ambros and Baltimore, 1978; Nishimura, 1972; Hewlett *et al.*, 1976; Rekosh *et al.*, 1977).

B. Sensitive Detection of Nucleic Acid-Bound Protein

Positive evidence of a VPg requires highly purified virion nucleic acid and a sensitive test for protein. The VPgs of RNA viruses tend to be small proteins, present in only trace amounts compared to the mass of genomic

RNA. The amino acid composition may be unfavorable for convenient *in vivo* or *in vitro* labeling. Poliovirus VPg, for example, lacks sulfur amino acids and histidine, and its only tyrosine residue is esterified to the viral RNA and therefore is not directly susceptible to labeling with ^{125}I. Perhaps the most diagnostic radiolabel for initial detection of a VPg, or rather a VPg–nucleotide or VPg–oligonucleotide, is either ^{32}P (e.g., Rekosh *et al.,* 1977; Daubert *et al.,* 1978) or radioactive nucleoside, incorporated *in vivo* or possibly *in vitro* into the virion nucleic acid. The presence of this label in purified VPg is an indicator of the covalent bond that is essential to the definition of a VPg.

Of greater convenience and greater sensitivity in detection, but usually of less diagnostic value, are *in vitro* labeling agents. The Bolton and Hunter (1973) acylating reagent has been widely employed for sensitive detection of nucleic acid-bound protein (e.g., Rekosh *et al.,* 1977; Stanley *et al.,* 1978; King *et al.,* 1980; Persson and MacDonald, 1982). This reagent (*N*-hydroxysuccinimide ester of 3-[^{125}I]iodo-4-hydroxyphenylpropionic acid) and other *N*-hydroxysuccinimidyl esters of radioactively labeled carboxylic acids (e.g., Kummer *et al.,* 1981) are readily attacked by aliphatic amino groups and hence specifically label protein in the presence of excess nucleic acid.

Direct iodination has been useful for VPg-RNAs with one or more underivatized tyrosine and/or histidine residues, such as those of several plant comoviruses and nepoviruses (e.g., Daubert and Bruening, 1979; Mayo *et al.,* 1982a; Koenig and Fritsch, 1982). Both soluble (Hunter and Greenwood, 1962) and virtually insoluble (Iodogen; Fraker and Speck, 1978; Salacinski, 1981) chloramides promote the specific (or nearly specific, Reisman and De Zoeten, 1982) labeling of proteins by iodine in the presence of nucleic acid. Direct iodination of nucleic acid-bound VPg was successful with bacteriophage ϕ29 DNA (Yoshikawa and Ito, 1981) but not with adenovirus DNA (Rekosh *et al.,* 1977).

Other useful procedures for labeling small amounts of protein *in vitro* in the presence of nucleic acids are reductive methylation of amino groups (Ghosh *et al.,* 1981; Rice and Means, 1971) and carboxyethylation by treatment with diethylpyrocarbonate (Prusiner *et al.,* 1982). The radioactive protein derivatizing agents mentioned above generally are not influenced by common detergents, such as sodium dodecyl sulfate (SDS) and sodium *N*-lauryl sarcosinate (SLS), so a detergent frequently has been a component of the reaction mixture.

Many of the most widely used and successful procedures for recovering virion nucleic acid free of protein contaminants employ proteinase(s). Such protocols preclude the direct detection of a VPg. A VPg-bearing RNA or DNA that has been purified by a procedure that involves proteinase treatment nevertheless will lack any of the usual 5′-terminal groups and can be expected to react (e.g., Stanley *et al.*, 1978) with Bolton–Hunter reagent or similar reagents that are specific for aliphatic amino groups.

Variations on the detection methods described above are available. For a comprehensive review of potential protein derivatizing agents see Feeney *et al.* (1982). The great sensitivity of silver stains (Merril *et al.*, 1981) and other protein-specific stains may make radiolabeling unnecessary in some systems. Antibodies raised against a synthetic peptide that corresponds to a portion of the VPg also can be used to detect a VPg or a polypeptide that contains VPg sequences without prior isolation or detection of the VPg. That is, nucleotide sequences of a suspected VPg gene may provide sufficient information to synthesize the immunogenic peptide (Section IV).

C. Distinguishing Covalently and Noncovalently Bound Proteins

The benefits of sensitive radiolabeling methods may be lost if contaminating proteins remain in the nucleic acid preparation. As was indicated above, the usual procedures of detergent disruption of virions and extraction with phenol, or phenol and chloroform, may not be sufficient to remove the last traces of noncovalently bound protein from virion nucleic acid. Zabel *et al.* (1982), in the process of raising antibodies against the VPg of CPMV, made multiple injections of purified and base-hydrolyzed virion RNA. The resulting immunoglobulins reacted not only with VPg, but also with the two CPMV coat proteins. Similarly, sensitive radiolabeling methods also revealed coat protein contamination of extensively purified poliovirus RNA (Rothberg *et al.*, 1978). Operations such as heating the nucleic acid solution in the presence of detergent and mercaptan, and incubation with strong denaturing agents such as guanidinium thiocyanate (Chirgwin *et al.*, 1979), may remove more of the contaminating proteins. More aggressive approaches use multiple or continuous washing steps without, however, any guarantee of removing the last tenacious traces of tightly but noncovalently bound proteins. Examples of such procedures are rate-zonal sucrose density gradient centrifugation through detergent solution, isopycnic (equilibrium or quasi-equilibrium) density gradient centrifugation under denaturing conditions, and gel electrophoresis or exclusion chromatography in the presence of detergent, urea, or a similar protein denaturing agent.

Ambros and Baltimore (1978) treated highly purified poliovirus virions with a phenol–chloroform–isoamyl alcohol two-phase system and sedimented the resulting RNA through a sucrose gradient that contained 5 mg/ml SDS. Several analyses revealed no evidence for RNA-bound proteins other than the poliovirus VPg. Persson and MacDonald (1982) isolated pancreatic necrosis virus double-stranded RNA by heating virions in SDS solution and subjecting the released RNA to two cycles of centrifugation through sucrose gradients containing SLS or SDS. The RNA was treated with Bolton–Hunter reagent and recovered. The label was present entirely in low-molecular-weight materials after incubation with proteinase K. Digestion of the labeled complex with double-stranded and single-stranded specific ribonucleases released the 110,000-molecular weight VPg as the major labeled product. However, they detected lesser amounts of coat and other protein contaminants even in these highly purified RNA preparations.

Rekosh *et al.* (1977) disrupted adenovirus particles with 4 *M* guanidinium chloride and centrifuged the solution for 47 hr in a gradient of CsCl, 4 *M* guanidinium chloride, and 1 mg/ml SLS. The DNA, derivatized with Bolton–Hunter reagent, released only one radiolabeled, proteinase-sensitive zone (analyzed by polyacrylamide gel electrophoresis) upon digestion with nucleases. Cesium chloride–guanidinium chloride gradient centrifugation also has been used to purify the double-stranded form of poliovirus RNA (Wu *et al.,* 1978).

Single-stranded RNAs generally will not form zones in CsCl gradients. Poliovirus RNA (Lee *et al.,* 1977) and CPMV RNA (Daubert *et al.,* 1978) have been centrifuged on gradients of freshly prepared cesium trichloroacetate. This procedure takes advantage of the denaturing powers of the trichloroacetate anion. The only protein zone detected in the gradient-purified RNAs corresponded to the VPg. Cesium trichloroacetate slowly hydrolyzes in solution to form chloroform and cesium bicarbonate. Possibly, cesium trifluoroacetate (commercially available) will be useful as a more stable yet denaturing salt for isopycnic density gradient centrifugation of nucleic acid–VPg complexes. Cs_2SO_4 density gradient centrifugation in SDS solution (Colpan *et al.,* 1983) also can result in the formation of an RNA zone under protein denaturing conditions. An alternative approach is centrifugation of RNA through a 50% (w/w) solution of CsCl in detergent, which can give a protein-free or nearly protein-free RNA precipitate below a protein–detergent zone in the gradient (Stanley *et al.,* 1978).

Figure 1 illustrates the tight binding of what is apparently coat protein to the virion RNA of the nepovirus tobacco ringspot virus (TobRV). Chromatography of the purified RNA in the presence of SDS on a column that can readily separate bulk quantities of SDS–protein complexes from virion RNA might be expected to remove all detectable amounts of noncovalently bound coat protein because the equivalent of many individual wash steps should have occurred during chromatography. Instead, coat protein readily was detected after the RNA was digested with nuclease (Fig. 1). Such a contaminant was not consistently observed, however. Lanes 1 and 2 of Fig.

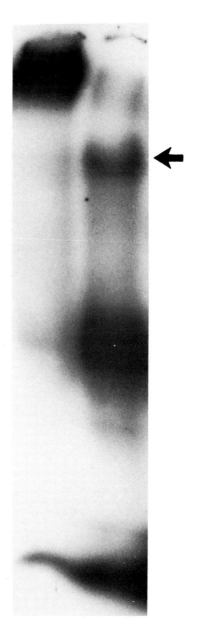

FIG. 1. Demonstration of a VPg linked to the RNA of tobacco ringspot virus (TobRV). Virions were purified by the method of Schneider *et al.* (1972). RNA was isolated by heating in SDS, phenol extraction, and chloroform extraction. It was concentrated by ethanol precipitation and then chromatographed on 4% agarose beads in a buffered solution of 2 mg/ml SDS, 2.5 m*M* mercaptoethanol (Kiefer *et al.*, 1982). The RNA was incubated with Iodogen and Na[125]I (Fraker and Speck, 1978) and recovered by multiple precipitations with ethanol.

8 of Kiefer *et al.* (1982) show RNA of the same TobRV strain that has been purified, labeled, and analyzed by the same procedure. In the latter experiment virtually all the radioactivity was associated with what must be presumed to be the VPg zone.

In general, a conclusion that a VPg is present is suspect if it is based only upon the appearance of a new proteinase-sensitive zone after digestion with a nuclease. Nevertheless, a protein that survives all attempts to remove it from the intact nucleic acid must be considered to be a candidate VPg. Currently recognized groups of viruses that have a proven or candidate VPg are listed in Table I. The notation "VPg not essential for infectivity" indicates that proteinase-treated VPg–nucleic acid complex retains its infectivity. It is possible that other viral nucleic acids also do not require a VPg for infectivity and would be infectious if the VPg were removed completely, not leaving the residual oligopeptide that remains after proteinase treatment. Adenovirus DNA sequences cut from recombinant plasmids lack a VPg and have extraneous terminal nucleotide sequences derived from the plasmid, yet they are infectious (Berkner and Sharp, 1983).

Relationships of viruses within the groups in Table I and relationships between groups are indicated by Dobos *et al.* (1979), Matthews (1982), Marion and Robinson (1983), and Strauss and Strauss (1983). The occurrence of a VPg attached to a plant virus satellite RNA and to the subgenomic messenger RNA of a plant virus is discussed in Section V. Specific plant viruses that have VPg-genomic RNAs but are not extensively discussed elsewhere in this chapter are the comoviruses cowpea severe mosaic virus, squash mosaic virus, and Echtes Ackerbohnenmosaik-Virus (Daubert and Bruening, 1979; Mayo *et al.*, 1982a) and the nepoviruses arabis mosaic virus, raspberry ringspot virus, strawberry latent ringspot virus, tobacco ringspot virus, tomato blackring virus, and tomato ringspot virus (Mayo *et al.*, 1979, 1982a,c; Kiefer *et al.*, 1982).

D. QUANTITATIVE ANALYSIS OF VPg

Virus nucleic acids that are not known to be VPg linked may nevertheless become at least slightly radioactive in an *in vitro* labeling reaction (e.g., Reisman and de Zoeten, 1982). One test of the authenticity of an *in vitro*

The RNA was applied to an SDS-permeated polyacrylamide gel [20% acrylamide, 1% methylenebisacrylamide; buffer system of Laemmli (1970)] before (left lane) or after (right lane) treatment with nuclease P1 (Daubert *et al.*, 1978). This procedure apparently released not only a VPg, but also coat protein (arrow). Migration of a VPg as a broad zone during electrophoresis is common and has been observed even for chemically synthesized VPg of poliovirus (Baron and Baltimore, 1982a).

TABLE I

GROUPS OF VIRUSES WITH GENOME-ASSOCIATED PROTEINS

Virus group	Example	Genome	Comment (References)[a]
Plant viruses			
Comovirus	Cowpea mosaic virus	Two ssRNAs	VPg not essential for infectivity (1)
Luteovirus	Potato leafroll virus	ssRNA	(2)
Nepovirus	Tobacco ring-spot virus	Two ssRNAs	(3)
Potyvirus	Tobacco etch virus	ssRNA	VPg not essential for infectivity (4)
Sobemovirus	Southern bean mosaic virus	ssRNA	(5)
	Pea enation virus	ssRNA	(6)
Animal viruses			
Adenovirus	Adenovirus	dsDNA	(7)
Calicivirus	Vesicular exanthema virus	ssRNA	(8)
Hepadnavirus	Hepatitis virus B	Partially dsDNA	VPg attached to negative strand (9)
Picornavirus	Poliovirus	ssRNA	VPg not essential for infectivity (10)
	Infectious pancreatic necrosis virus	dsRNA	(11)
Bacterial viruses			
	Bacteriophage ϕ29	dsDNA	(12)

[a] Key to references: (1) Daubert et al., 1978; Stanley et al., 1978; Daubert and Bruening, 1979; Mayo et al., 1982a; (2) Mayo et al., 1982b; (3) Mayo et al., 1979, 1982a; Kiefer et al., 1982; Jones et al., 1983; (4) Hari, 1981; (5) Ghosh et al., 1979; Veerisetty and Sehgal, 1980; (6) Reisman and de Zoeten, 1982; (7) Rekosh et al., 1977; (8) Burroughs and Brown, 1978; Schaffer et al., 1980; (9) Gerlich and Robinson, 1980; (10) Flanegan et al., 1977; Lee et al., 1977; Sangar et al., 1977; Hruby and Roberts, 1978; (11) Persson and MacDonald, 1982; (12) Salas et al., 1978.

labeled, candidate VPg uses semiquantitative analysis of the extent of labeling. The problem is to determine how the amount of radioactivity incorporated into the VPg moiety compared to the minimum expected incorporation, that is, the amount that would be found if a single group were derivatized per genomic nucleic acid molecule. A level of incorporation that is considerably less than the molar amount is suspect. If the amount of nucleic acid can be readily determined, the problem is reduced to one of determining the amount of incorporated radioactivity and the specific radioactivity of the derivatizing reagent. However, the amount of highly purified, radioactive virion nucleic acid may be too small to quantitate easily. In this circumstance, an internal control will be useful.

One approach to obtaining the required internal control is to derivatize the nucleic acid portion of the putative VPg–nucleic acid in such a way that a fixed number of radioactive groups are incorporated. A virion RNA with free 2′,3′ hydroxyl groups, or that can be treated to produce them, will be well suited to such a derivatization. The procedure of Fahnstock and Nomura (1972) and Shih *et al.* (1974) has been used to modify quantitatively the 3′ end of ribosomal and viral RNAs. As a test of the potential usefulness of a derivative that is a tyrosine analog, as an internal control for iodination, an RNA that lacks a VPg was oxidized with periodate (J. Poen and G. Bruening, unpublished observations). The oxidized RNA was incubated for 40 min at room temperature in a solution of 25 mM sodium cyanoborohydride and 1 mM 4-(2-aminoethyl)phenol (tyramine) buffered with 100 mM N-hydroxyethylpiperazine-N′-ethylsulfonic acid (HEPES) 60, mM NaOH. This reaction forms a six-membered (morpholino) ring through reductive alkylation of the tyramine nitrogen by the biscarbonyl groups of the oxidized 3′ ribose. An Iodogen (Fraker and Speck, 1978)-promoted reaction introduced ^{125}I as a terminal label, as indicated by partial sequencing according to the procedure of Donis-Keller *et al.* (1977).

Presumably an excess of a diamine could be substituted for tyramine to introduce a single aliphatic primary amine per RNA molecule, which would be subject to derivatization by Bolton–Hunter reagent or reductive methylation. A VPg may have one or more underivatized tyrosine/histidine or lysine/α-amino residues. Therefore, derivatization of a VPg–nucleic acid as described above may result in only a fractional increase in the extent of radiolabeling, compared to the radiolabeling of VPg–nucleic acid that has not been derivatized. The amount of radioactivity associated with a presumed VPg zone should be unchanged by prior derivatization of the nucleic acid portion of the VPg–nucleic acid, assuming that the VPg does not possess groups subject to the derivatization reaction, such as periodate-sensitive carbohydrate sidechains.

Other possible approaches to quantitating the amount of candidate VPg present per nucleic acid molecule include (1) comparison of radioactivity, from incorporated amino acids, that migrates during electrophoresis with virion nucleic acid and with coat protein(s) or other virion proteins (Persson and MacDonald, 1982), correcting as necessary for differences in amino acid composition and abundances of virion proteins; and (2) comparison of radioactivity in Nps and in pNp (Hewlett *et al.*, 1976), in those circumstances in which hydrolysis procedures are available that release pNp from the 5′ end of the VPg–nucleic acid and Nps from the remainder of the molecule (e.g., Stanley *et al.*, 1978).

Quantitative analyses also may be useful for comparative purposes. King *et al.* (1980) discovered that the VPg population of the RNA of multiply plaque-purified foot-and-mouth disease virus (FMDV) consists of at least two species. They compared the relative amounts of protein in the two fractions by the use of dual radioisotopic labels.

The amount of material in a VPg zone of an electrophoresis gel may give a low estimate of the amount of VPg in the original sample. Although relatively large amounts of synthetic poliovirus VPg can be precipitated by the use of traditional protein-precipitating agents (Baron and Baltimore, 1982a), radiochemical amounts of both VPg, VPg–pUp, and VPg–pUpU are not readily precipitated and were lost from gels washed with commonly used "fixing solutions" such as 10% acetic acid, 30% isopropanol (Lee *et al.*, 1977; Crawford and Baltimore, 1983).

E. Terminal Attachment of VPg

Since every known VPg attachment is to the 5' end of the genomic nucleic acid, sequence-specific cleavage of the nucleic acid should produce one (or two in the case of double-stranded DNA with both 5' termini blocked) VPg-oligonucleotide complex(es). In contrast, a noncovalently attached protein, or a protein that has become covalently attached by a nonspecific reaction, should not yield a specific oligonucleotide–protein complex. For a small VPg at least, the mobility of the VPg–oligonucleotide complex during electrophoresis in detergent-permeated gels can be expected to be markedly influenced by the length of the attached oligonucleotide chain. Demonstrable changes in protein–oligonucleotide mobility after digestion of the nucleic acid with different nucleases, or application of some other procedure that alters the oligonucleotide chain length, is the basis of a reliable test for the presence of an authentic VPg.

An example of such an analysis is given in Fig. 2 (see also Daubert *et al.*, 1978). The VPg–oligonucleotide produced by incubation of the two CPMV genomic RNAs with ribonuclease T1 is expected to have 17 nucleotide residues (Stanley and van Kammen, 1979), whereas the principal iodinated products from the nuclease P1 digestion should be VPg–pU and VPg–pUpA. Figure 2 shows the expected nuclease-specific alteration in mobility. Poliovirus VPg, VPg–pU, and VPg–pUpU (Takegami *et al.*, 1983b; Crawford and Baltimore, 1983) and FMDV VPg–pUp and VPg–pUpUpGp (King *et al.*, 1980) show distinct mobilities during electrophoresis in gels. The VPg-oligonucleotides of CPMV and poliovirus are proteinase sensitive, as expected (e.g., Daubert *et al.*, 1978; Stanley and van Kammen, 1979); the resultant oligonucleotide–oligopeptide complex showed a mobility that was drastically altered from that of the VPg–oligonucleotide.

Presumably the terminal location of a DNA-bound VPg would similarly

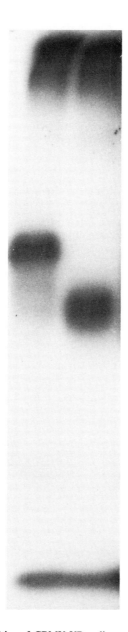

Fig. 2. Electrophoretic mobilities of CPMV VPg–oligonucleotides. Virions were purified by CsCl density gradient centrifugation (Bruening, 1969). A mixture of the two genomic RNAs was isolated from virions by heating in SDS solution and extraction of the solution with phenol and then chloroform (Daubert et al., 1978). The RNA was incubated with Iodogen and Na^{125}I and reisolated (Daubert and Bruening, 1979). The RNA was digested with ribonuclease T1 (left lane) or with nuclease P1 (right lane) before electrophoresis in an SDS-permeated polyacrylamide gel (19% acrylamide, 0.62% methylenebisacrylamide).

be indicated by alterations in the mobilities of the specific VPg–oligodeoxynucleotides generated by cleavage of the genomic DNA with restriction endonuclease (e.g., Rekosh *et al.,* 1977; Kemble and Thompson, 1982). An elegant example of the variation in electrophoretic mobilities of VPg–oligodeoxynucleotides with oligonucleotide length was seen in the studies on the synthesis of bacteriophage ϕ29 DNA (Peñalva and Salas, 1982; Fig. 3 of Watabe *et al.,* 1983). DNA replication *in vitro* is dependent upon the incorporation of the VPg into the nascent chain, presumably as the primer. The growth of individual chains in the reaction mixture was arrested at various sites by the incorporation of dideoxynucleotide residues from dideoxynucleoside triphosphates. Each additional deoxynucleotide residue reduced the mobility of the VPg–oligodeoxynucleotide complex (VPg apparent molecular weight 27,000) to form a series of discrete zones. Similarly, adenovirus VPg with an attached dideoxyguanylate-terminated oligonucleotide of 26 residues migrated more slowly than free VPg in SDS-permeated polyacrylamide gels (Lichy *et al.,* 1982).

The terminal position of VPg also has been demonstrated by electron microscopy. The proteins (presumably VPg) of poliovirus and Mengovirus RNAs were allowed to react to with the *N*-hydroxysuccinimidyl ester of biotin (Richards *et al.,* 1979; Thornton *et al.,* 1981). The RNA was allowed to react with avidin-coupled spheres that readily were detected by electron microscopy of shadowed preparations. A similar series of reactions was performed with dinitrophenylated VPg–RNA and anti-2,4-dinitrophenylgroup antibodies (Wu *et al.,* 1978). The images located the VPg at one end of single-stranded RNA and at both ends of RF molecules. The larger VPg of adenovirus (Rekosh *et al.,* 1977) and of infectious pancreatic necrosis virus (Persson and MacDonald, 1982) cause the double-stranded genomic DNA and RNA of those respective viruses to end-aggregate in very specific patterns without the introduction of any exogenous binding protein.

III. Characterization of Genome-Linked Proteins and Adjacent Nucleotide Sequences

A. IDENTIFICATION OF NUCLEOTIDYL–AMINO ACID BONDS

The most direct proof of the covalent nature of a VPg–nucleic acid structure is the identification of the covalent bonds (e.g., -C-O-P-O-C-) that connect protein to nucleic acid. These bonds can be considered as resulting from the condensation of a functional group of an amino acid residue side chain from the VPg with the 5′-terminal phosphoryl group of the nucleic acid. The stability of this bond to enzymatically or chemically catalyzed hydrolysis provides clues as to the particular type of bond involved. For example, snake venom phosphodiesterase will hydrolyze a phosphorus–oxygen but not phosphorus–nitrogen bond.

Proteins can be linked to nucleotide residues through the hydroxyl groups of serine, threonine, or tyrosine residues, as phosphodiesters, or through aliphatic amino functions, as phosphoramides. More reactive linkages, such as that of a mixed carboxylic acid–phosphoric acid anhydride or that of a phosphoryl group to an imidazole nitrogen or guanidino nitrogen, have not been detected as stable protein–nucleic acid structures.

It is difficult to make generalizations concerning the chemical reactivities of bonds between a phosphodiester and an amino acid residue embedded in protein (e.g., see Shabarova, 1970; Tabrosky, 1974) because the linkage is significantly influenced by neighboring groups of the polypeptide chain. However, tests of the stability of the protein–nucleic acid linkage, in the VPg–nucleic acid or some degradation product of it, can be useful. A comparison of the sensitivity of the amino acid–nucleotide linkage of the bacteriophage ϕ29 VPg–DNA to different reagents was useful as a clue to its identity (Peñalva and Salas, 1982). The phosphoric acid ester of an aliphatic hydroxyl group can be expected to be broken by treatment under mildly alkaline conditions. The bond should resist attack by hydroxylamine, unlike the bond in carboxylic acid esters. The lability of the bacteriophage ϕ29 VPg–DNA linkage to cleavage catalyzed by piperidine, but not by hydroxylamine, thus supported the eventual identification of the participating amino acid residue as serine.

The stability of the VPg–RNA linkages of picornaviruses to both hydroxylamine and mild alkali implicated an aryl phosphoester, and O^4-uridylyltyrosine was eventually identified as the core of the VPg–RNA linkage of poliovirus, FMDV, and EMC virus (e.g., Rothberg et al., 1978).

Phosphoramidates and phosphodiesters differ in stability as a function of pH (Tabrosky, 1974; McCollum and Tobrosky, 1983; Steiner et al., 1980). Diagnostic incubation conditions are given in Table II. These are derived from studies of model compounds and serve as a guide for the analysis of a VPg–oligonucleotide or peptide–oligonucleotide derived from a properly radioactively labeled (e.g., ^{32}P) RNA- or DNA-bound VPg. The lability in base of the serine and threonine phosphate esters is due to a beta-elimination reaction. This occurs more readily in a peptide or an intact protein, where the serine or threonine residues are involved in two peptide bonds, than in free serine phosphate or threonine phosphate (McCollum and Taborsky, 1983).

Hydrolysis and beta-elimination reaction products readily can be analyzed by paper or thin-layer chromatography or electrophoresis or by high-performance liquid chromatography (e.g., Ambros and Baltimore, 1978;

TABLE II
SUSCEPTIBILITY OF PHOSPHORYL BONDS TO HYDROLYSIS

Amino acid residue	Bond	Extent of hydrolysis in	
		Acid	Base
Tyr	Arylphosphodiester	[a]	+
Ser, Thr	Alkylphosphodiester	[a]	+ + +
Lys	Phosphoramidate	+ + +	[a]
His	Phosphoramidate	+ + +	[a]

[a]More than 95% of the model compound containing the indicated bond survives. The "acid" condition is 0.1 M HCl, 1 hr, 37°C, and the "base" condition is 1 M NaOH, 6 min, 95°C.

Rothberg *et al.,* 1978; Steiner *et al.,* 1980). Unfortunately, the sensitivity to acid and base hydrolysis will not distinguish the two types of aliphatic phosphodiesters and may even mislead in distinguishing aryl and alkyl phosphoric acid esters. The actual identification of the phosphodiester or other bond is in practice made by determining the structure of the amino acid–nucleotide, e.g., by chromatographic comparisons with synthetic standards.

Poliovirus VPg–RNA was digested with ribonuclease(s) to give VPg–pUp, with nuclease P1 to give VPg–pU, or with micrococcal nuclease to give VPg–p (Ambros and Baltimore, 1978; Rothberg *et al.,* 1978; Baron and Baltimore, 1982a). Free VPg was produced by digestion of VPg–pUp or VPg–pU with venom phosphodiesterase or of VPg–p with phosphatase. HCl-catalyzed hydrolysis at 110°C of VPg–pUp (5.6 M, 2 hr) and of VPg–pU and VPg–p (2 M, 24 hr) gave Tyr-O^4-pUp, Tyr-O^4-pU and O^4-phosphotyrosine, respectively. These results establish the VPg-to-RNA bonding. Phosphoserine and phosphothreonine also would be expected to survive the HCl-catalyzed hydrolysis procedures given above and to be recovered from the appropriate VPg–p.

The genomic DNA of bacteriophage ϕ29 was digested with micrococcal nuclease and the proteinase Pronase (Hermoco and Salas, 1980). The nucleotidyl-peptide remnant was purified by chromatography on DEAE-cellulose at pH 5.0. O-Phosphoserine was identified by paper electrophoresis of the products obtained after hydrolysis in 5.8 M HCl at 110°C for 90 min. Incubation of phosphatase-treated nucleotidyl peptide in 0.1 M NaOH for 3 hr at 37°C or with snake venom phosphodiesterase liberated 5′-dAMP. The results indicate a Ser–O^β–P–O-adenylate structure that is the core of the ϕ29 VPg-DNA linkage. Another example of a nucleotidyl amino acid derived from a VPg–nucleic acid complex is the Ser–O^β–P–O-cytidylate of adenovirus (Desiderio and Kelly, 1981). The VPg of adenovirus DNA can be removed by a relatively mild treatment: 0.5 M piperidine, 2 hr, 37°C (Tamanoi and Stillman, 1982). Enzymatic release of adenovirus VPg nucleotidyl peptide was accomplished by Berkner and Sharp (1983). They incubated proteinase-treated DNA with DNA polymerase and dTTP to remove a single deoxynucleotide residue from the 3′ end of each strand. This exposed the 5′ ends sufficiently to allow attack by nuclease S1 to give peptidyl pdC and adenovirus DNA lacking the terminal base pair of each end.

B. PARTIAL PURIFICATION OF VPgs

Methods have been developed for resolving VPg- from non-VPg-bearing nucleic acids and nucleic acid fragments. Molnar-Kimber et al. (1983) chromatographed the DNA of duck hepatitis B virus on benzoylated, naphthoylated DEAE (BND)-cellulose (Sedat et al., 1967). Double-stranded and single-stranded DNA fractions were eluted with NaCl and caffeine-containing buffers, whereas VPg–DNA remained bound until a buffered solution of 1 M urea, 10 mg/ml SDS was applied at 37°C. This 3000 base-pair DNA migrated at the expected position during agarose gel electrophoresis only if the DNA had been treated with proteinase K. The VPg–DNA migrated about 10% more slowly.

Adenovirus DNA did not enter a 1.4% agarose electrophoresis gel without prior treatment with proteinase (Sharp et al., 1976). The VPg–RNAs of infectious pancreatic necrosis virus did not enter a polyacrylamide gel unless SDS was present, and even then they migrated much more slowly than the proteinase-treated virion double-stranded RNAs. Similarly strongly altered electrophoretic mobilities were obtained with protein-linked DNA and protein-linked fragments of DNA from several sources (Sharp et al., 1976; Kemble and Thompson, 1982; Yoshikawa and Ito, 1981; García et al., 1983a).

Some VPg–nucleic acids exhibit slightly lower buoyant densities than the corresponding proteinase-treated nucleic acids (Persson and MacDonald, 1982; Rekosh et al., 1977), which at least in theory would allow a preparative separation given a gradient centrifugation method of sufficient resolving power. Their unusual solubility properties and/or affinities for surfaces (Baron and Baltimore, 1982a) also suggest separation methods. Separations of free and VPg-bearing nucleic acids and oligonucleotides have been based not only on specific affinity (e.g., for anti-VPg antibodies), but also on the tendency of VPg–nucleic acids to aggregate or bind to surfaces. Glass fiber filters have been useful in these systems for the specific adsorption of VPg–DNA and VPg–restriction fragment complexes (Crawford and Baltimore, 1983).

In the isolation of a VPg, advantage can be taken (1) of procedures that will yield pure or nearly pure VPg–nucleic acid complex, and (2) of the solubility of a VPg in the phenol-rich phase of a phenol–water two-phase system. Ambros and Baltimore (1978) digested radiolabeled poliovirus RNA with ribonucleases and partitioned the products. The back-extracted phenol-rich phase was diluted with acetone, and the resulting precipitate was dissolved in SDS solution and subjected to exclusion chromatography. This procedure gave apparently radiochemically pure VPg that was free of other poliovirus RNA-derived materials from the digest. However, such a VPg preparation can be expected to be contaminated with ribonucleases, since the ratio of ribonuclease to RNA and the ratio of VPg to RNA are not greatly different. To obtain chemically pure VPg, more steps (e.g., immunoabsorption) obviously would be needed, with risks of loss on vessel surfaces, etc. Procedures for more extensive purification of poliovirus VPg from immunoprecipitates are given by Crawford and Baltimore (1983).

The most abundant source of a VPg to date probably has been the chemical laboratory (Baron and Baltimore, 1982a), and for the small VPgs this is likely also to become a common source for chemical and biochemical studies. For larger VPgs, *in vivo* synthesis in cells bearing a recombinant expression plasmid seems to be the method of choice. García *et al.* (1983a) cloned and expressed the bacteriophage ϕ29 VPg gene under temperature control. They purified biologically active VPg in larger quantities than could have been obtained conveniently from the bacteriophage VPg–DNA.

C. CHARACTERIZATION OF VPgs

VPgs were first detected as zones during electrophoresis in paper or polyacrylamide gel. However, electrophoresis in SDS-permeated polyacrylamide gels and exclusion chromatography in the presence of protein denaturing agents have not proved to be reliable means of estimating VPg size. The electrophoretic mobility of CPMV VPg corresponded to a molecular weight of approximately 5000 (Daubert *et al.*, 1978), whereas the values from amino acid composition of the proposed possible VPg structures (see below) are 3538 and 4113, respectively. The molecular weight of poliovirus VPg similarly was estimated to be 7000–12,000 (e.g., Ambros and Baltimore, 1978) rather than the actual value of 2354. Even minor changes in amino acid sequence, especially in a small protein, can cause significant changes in apparent molecular weight estimated from migration distance (e.g., Nole *et al.*, 1979), and the cationic character of a VPg may lead to excessive SDS binding and hence abnormally slow migration. Excessively broad zones and doublet zones frequently have been observed after electrophoretic analyses of small VPgs (e.g., Crawford and Baltimore, 1983).

Analyses of proteolytically generated peptides have been useful in investigating systems that may have more than one VPg. Daubert *et al.* (1978) showed that the iodinated VPgs obtained after nuclease digestion of either RNA 1 or RNA 2 of CPMV gave very similar patterns of radioactive peptides after two-dimensional thin-layer chromatography, indicating that the same VPg is linked to each genomic RNA of the virus. Similarly, Mayo *et al.* (1982a) found that the trypsin-digested VPg of tomato blackring virus RNAs 1 and 2 and of the satellite RNA of tomato blackring virus all gave similar patterns of peptides. King *et al.* (1980) analyzed the VPg from two strains of FMDV. Both electrofocusing and electrophoresis in polyacrylamide gels gave two VPg–oligonucleotide zones for each FMDV strain. The proteins from these two zones differed not only in composition of tryptic peptides, analyzed by high-performance liquid chromatography, but also in amino acid composition.

Potentially the VPgs (terminal proteins) attached to the two ends of a virus DNA could be different. Yoshikawa and Ito (1981) digested iodinated VPg–DNA of bacteriophage φ29 with restriction endonuclease *Eco*RI and separated the large (left-hand end) VPg-DNA fragment from the small VPg-DNA fragment by sucrose density gradient centrifugation. The autoradiographic maps of tryptic peptides, separated by two-dimensional electrophoresis and chromatography, indicated that each end is coupled to an identical VPg.

A purified VPg can be expected to be available from virus particles in only limited amounts. For this reason, and because of the relative ease with which nucleotide sequences may be determined, the analysis of a VPg gene is the method of choice for determining the VPg amino acid sequences. However, partial amino acid sequences or other information usually is required to identify and/or verify the reading frame and amino-terminal and carboxyl-terminal amino acid residues. Kitamura *et al.* (1981) prepared poliovirus VPg that had been labeled *in vivo* with [^{3}H]lysine. Automated Edman degradation released radioactivity in significant amounts at cycles 9, 10, and 20, indicating lysine residues at positions 9, 10, and 20 from the amino terminus. A similar approach located single arginine, leucine, and valine residues. Among the oligonucleotides generated by digestion of poliovirus RNA by ribonuclease T1, only one had one leucine and two lysine codons in the proper relationship, and this one and other oligonucleotide were used to prime DNA synthesis with poliovirus cDNA as template. The amino acid sequence derived from the nucleotide sequence (Kitamura *et al.*, 1980, 1981) agreed with the Edman degradation results, establishing the VPg primary structure with only some uncertainty with regard to the carboxyl terminus. The identity of the carboxyl terminal residue was deduced from the specificity of a poliovirus-encoded proteinase (Hanecak *et al.,* 1982) and was confirmed by carboxypeptidase digestion and analysis of released amino acids (Adler *et al.,* 1983). The entire structure also was confirmed by chemical synthesis and comparison of the properties of synthetic and authentic poliovirus VPg (Baron and Baltimore, 1982a). The amino acid sequence for poliovirus VPg is given here:

Gly–Ala–Tyr–Thr–Gly–Leu–Pro–Asn–Lys–Lys–
Pro–Asn–Val–Pro–Thr–Ile–Arg–Thr–Ala–Lys–
Val–Gln

FMDV strains appear to have at least two VPgs (King *et al.,* 1980). VPg1 is the first, in gene position from the 5′ end of the virion RNA, of three apparently tandemly encoded VPgs of FMDV strain O_1K. These genes were identified on the basis of anticipated biochemical properties of the encoded amino acid sequences and analogies with poliovirus VPg. The predicted amino acid sequence (Forss and Schaller, 1982) of VPg1 is

Gly–Pro–Tyr–Ala–Gly–Pro–Leu–Glu–Arg–Gln–
Lys–Pro–Leu–Lys–Val–Arg–Ala–Lys–Leu–Pro–
Gln–Gln–Glu

The single tyrosine residue is the presumed attachment site, since in cannot be iodinated in VPg–pUp (King *et al.,* 1980).

Zabel *et al.* (1984) have correlated a partial amino acid sequence of CPMV VPg, determined by radiochemical microsequencing procedures, to a site on the nucleotide sequence of RNA 1 (Lomonossoff and Shanks, 1983). The carboxyl-terminal residue is uncertain, so that the VPg appears to be encoded by nucleotide residues 2964–3049 (28 amino acid residues) or 2964–3063 (33 amino acid residues) of CPMV RNA 1. The VPg would be generated by the cleavage of a Gln–Ser bond at its amino terminus and either a Gln–Met or a Gln–Ser bond at its carboxyl terminus. The corresponding amino acid sequences is

Ser–Arg–Lys–Pro–Asn–Arg–Phe–Asp–Met–Gln–
Gln–Tyr–Arg–Tyr–Asn–Asn–Val–Pro–Leu–Lys–
Arg–Arg–Val–Trp–Ala–Asp–Ala–Gln–(Met–Ser–
Leu–Asp–Gln)

The site of attachment of the RNA to this basic protein is unknown at present. However, since both tyrosine residues can be iodinated, it is unlikely that these are involved (unless the two residues alternate as donors of the hydroxyl group that becomes bonded to the first nucleotide residue of the VPg-RNA).

D. VPg-Linked Nucleotide Sequences

In several instances the sequence of nucleotide residues that is linked to the VPg has been inferred from the sequence at the 5' end of a form of the nucleic acid that lacks a VPg. For example, the extensive 5' sequence from the VPg-free poliovirus polyribosomal RNA was compared with a much more limited sequence from a VPg–nucleic acid complex (Rothberg et al., 1978). However, Stanley and van Kammen (1979) performed a more direct and extensive analysis of VPg–RNA complexes. They digested each of the generally ^{32}P-labeled genomic RNAs of CPMV to completion with ribonuclease T1 and identified the one radioactive product from each RNA that showed an altered electrophoretic mobility when the VPg–RNA had been incubated with proteinase. Each of these VPg–oligonucleotides was treated with phosphatase and coupled to [5'-^{32}P]pCp in a reaction catalyzed by bacteriophage T4 RNA ligase. Digestion with ribonuclease T1 and subsequently with proteinase resulted in the removal of the Cp residue and the production of a 3'-^{32}P-labeled oligonucleotide bearing variable remnants of the 5'-linked VPg. These [3'-^{32}P]oligonucleotide peptides were partially hydrolyzed in base, and the products were displayed by two-dimensional gel electrophoresis. From this "wandering-spot" analysis and the catalogs of oligonucleotides released by digestion with pancreatic ribonuclease, the 5'-terminal 17 nucleotide residues of each RNA were determined.

Najarian and Bruening (1980) took a different approach. They used synthetic oligodeoxyribonucleotides to prime DNA synthesis within the 5' region of CPMV RNA templates. Transcription was catalyzed by avian myelobastosis virus reverse transcriptase and proceeded to within one residue of the 5' terminus of the template regardless of whether most of the VPg had been removed by treatment with proteinase.

The stability of DNA to treatment with base, and the lability of serine phosphate residues in protein to base-catalyzed beta-elimination, facilitated the direct determination of the VPg-linked nucleotide sequences of adenovirus DNA (Desiderio and Kelly, 1981) and bacteriophage ϕ29 DNA (Hermoso and Salas, 1980; Yoshikawa and Ito, 1981). Adenovirus DNA restriction fragments were incubated in 50 mM NaOH at 70°C to release DNA strands bearing 5' phosphoryl groups [half-life of approximately 10 min; alternative condition 100 mM NaOH, 37°C, 3 hr (Hermoso and Salas, 1980)]. After phosphatase treatment two different strand ends were labeled with ^{32}P in a polynucleotide kinase-catalyzed reaction, whereas only one strand was labeled if the DNA restriction fragment had not been treated with base. A short portion of the sequence of the strand that was labeled only after base treatment was determined by partial, nucleotide residue-specific chemical cleavage. It agreed with the sequence previously determined using adenovirus DNA molecules that lack VPg.

IV. Biosynthesis of Genome-Linked Proteins

A. PRECURSORS OF VPgs

In only a few instances are the resemblances of a plant virus group and an animal virus group unusually close. The plant comoviruses and the animal picornaviruses form one such pair. The viruses of each group have a capsid composed of 60 copies of each coat protein, have genomic RNA that is VPg-linked and polyadenylated, and produce functional proteins by specific proteolysis of polyprotein precursors. Even gene order on the virion RNA(s) is similar (Goldbach and Rezelman, 1983; Takegami *et al.,* 1983a; references cited therein). The plant nepoviruses (Mayo *et al.,* 1982a) also show similarities to the comoviruses and picornaviruses.

The biosynthesis of the VPgs of picornaviruses and comoviruses, and presumably of nepoviruses also, may be very similar. The direct precursors of the VPgs of comoviruses and picornaviruses are themselves derived from an internal region of larger precursors translated from the continuous reading frame that constitutes most of the genomic RNA molecule(s). The polyprotein is cleaved in steps and probably by more than one pathway, largely by the action of virus-specified proteinase(s) (e.g., Kitimura *et al.,* 1980; Pelham, 1979). Both the VPg gene and the replicase gene of poliovirus RNA lie in the 3' region of poliovirus RNA, separated only by a proteinase gene (e.g., Hanecak *et al.,* 1982). Close linkage of the VPg and the replicase genes also seems to be characteristic of the comoviruses (Goldbach and Rezelman, 1983) and presumably of the nepoviruses. The larger of the two genomic RNAs of a comovirus or a nepovirus apparently encodes both these protein genes and whatever other genes are necessary for RNA replication (Goldbach *et al.,* 1980; Stanley *et al.,* 1980; Rezelman *et al.,* 1980; Robinson *et al.,* 1980).

Immunoglobulins raised against authentic VPg or chemically synthesized VPg or VPg peptide have been important tools for identifying VPg precursors. Zabel *et al.* (1982) purified CPMV virion RNA, hydrolyzed it with alkali, and injected the hydrolyzate into a rabbit. The antiserum was cross-absorbed with resin-bound CPMV virions to remove antibodies produced because of coat protein contaminants in the preparation. The antiserum was VPg specific, yet it detected a polypeptide of apparent molecular weight 60,000 (p60) in a membrane fraction from disrupted, CPMV-infected cells. Similarly, antibody against a synthetic poliovirus VPg or VPg peptide (Baron and Baltimore, 1982a; Semler *et al.,* 1982) reacted with larger than VPg-sized polypeptides in the membrane fraction of poliovirus-infected cells.

The most abundant VPg sequence-containing polypeptide of poliovirus-infected cells is the 109 amino acid P3-9. It is present in great excess over the amount that would be required for poliovirus RNA synthesis (Semler et al., 1982; Takegami et al., 1983b). The less abundant but much larger peptides X/9 and 3b/9, apparent molecular weights 45,000 and 70,000, respectively, also reacted with anti-VPg serum in a specific manner (reaction prevented by addition of VPg peptide). In early experiments poliovirus VPg itself was not detected in extracts of infected cells, but later was found when procedures for fixing proteins in polyacrylamide gels were modified (Crawford and Baltimore, 1983). Whether this VPg is a precursor to VPg–RNA or the product of the cleavage of VPg–RNA (Ambros et al., 1978), or both, is not known. The amounts of X/9 and 3b/9 increased when the infected cells were treated with an inhibitor of poliovirus-specified proteinase (Takegami et al., 1983a).

CPMV p60 and poliovirus 3b/9, X/9, and P3-9 all share the properties of being membrane associated in crude cell extracts and having the VPg sequence as the carboxyl-terminal portion of the polypeptide chain (Goldbach et al., 1982; Goldbach and Rezelman, 1983; Takegami et al., 1983a). Presumably each of the potential poliovirus VPg precursors is anchored to a membrane site by a hydrophobic sequence of 22 amino acids that is a part of P3-9. Poliovirus RNA replication activities are associated with membrane fractions from infected cell extracts (reviewed by Takegami et al., 1983a).

During the synthesis of adenovirus DNA, the first deoxynucleotide residue becomes attached to a protein of molecular weight approximately 80,000, which is designated here as pre-VPg80K (Lichy et al., 1982). Although the DNA-bound VPg of virus particles is of apparent molecular weight 55,000, the processing step that generated it is not a part of the DNA synthesis reactions. The protein processing probably occurs after replication is completed (Challberg and Kelly, 1981, 1982), and thus pre-VPg80K itself, rather than the VPg, must be considered the primer for virus DNA synthesis. The VPg of bacteriophage ϕ29 apparently is synthesized directly in its mature form (García et al., 1983b).

B. ACTIVE FORMS OF VPgs

For those viruses that have VPgs the first phosphodiester bond to be formed during virus nucleic acid synthesis is that between the VPg and the 5'-terminal nucleotide residue. The formation of this bond can be considered as the initial reaction of genomic nucleic acid synthesis. It is possible that in some systems this initiation reaction is not template dependent and that the form of the VPg that first participates in a template-dependent reaction already has one or more nucleotide residues attached to it. Crawford and Baltimore (1983) used immunoprecipitation to detect not only VPg but also VPg-pUpU in detergent-treated extracts of poliovirus-infected HeLa cells. VPg–pU derived from the VPg–pUpU of extracts and VPg–pU obtained by hydrolysis of virion RNA could not be distinguished, indicating that the VPg–pUpU structure is consistent with that at the 5′ end of completed virion RNA.

A membrane-rich preparation from infected cells formed VPg–pUpU *in vitro* in low yield when it was incubated with UTP (Takegami *et al.*, 1983b). The reaction occurred under conditions (low magnesium ion concentration) that do not promote the de novo poliovirus RNA synthesis *in vitro*. Thus the uridylylation of VPg may be unrelated to chain initiation *or,* if it is required for chain initiation, the reaction is not dependent upon template. An attractive but as yet unsupported model for poliovirus RNA synthesis begins with template-independent di-uridylylation of VPg (or of P3-9 with concerted proteolytic production of VPg–pUpU). Subsequent chain elongation would then be primed by VPgpUpU (Takegami *et al.*, 1983b; Crawford and Baltimore, 1983). In this model VPg–pUpU (or possibly P3-9–pUpU) is the first form of VPg to come in contact with the template.

As indicated in Sections IV,A and V, the form of the adenovirus VPg that first participates in a template requiring, deoxycytidylation reaction is pre-VPg80K. Similarly, the 31,000-molecular weight VPg of bacteriophage φ29 becomes coupled to a deoxyadenylate residue in a template-dependent reaction. The template-independent addition of nucleotide residues, which is an attractive hypothesis for the formation of primer in picornavirus RNA synthesis, thus does not appear to have a role in the synthesis of DNA in the adenovirus and bacteriophage φ29 systems.

V. Possible Functions of Genome-Linked Proteins

A. INITIATION OF VIRUS NUCLEIC ACID SYNTHESIS

All known nucleic acid polymerases catalyze polynucleotide chain formation in the 5′ → 3′ direction. Several RNA-dependent RNA polymerases (e.g., Van Dyke and Flanagan, 1980) and all DNA polymerases (Kornberg, 1980; Challberg and Kelly, 1982; Nossal, 1983) that have been studied *in vitro* require a primer with a free 3′ hydroxyl group. Generally, replicases might be expected to bind the template strand at a site that is least partially "upstream" of the template nucleotide residue that dictates the first residue of the transcript. Circular genomes are well suited to such replicase requirements. Linear genomes, however, have ends beyond which template-dependent replicases can find no purchase.

Thus the mechanisms by which an internally initiated copy strand might be completed, to be complementary to the template's 3' extreme, have been the subject of considerable discussion (e.g., Watson, 1972; Cavalier-Smith, 1974; Challberg and Kelly, 1982). Investigations of terminal structures of linear templates and transcripts have been rewarded with discovery of a great diversity of structures with actual and potential significance for the replication of virus nucleic acids. These structures include not only VPgs, but also the terminal hairpin loops that covalently connect the two strands of vaccinia virus DNA (Baroudy et al., 1982); similarly, the RNA that is complementary to (the encapsidated but non-VPg bearing) animal alphavirus RNA has an apparently noncoding 3'-terminal residue (Wengler et al., 1982).

The possibility of DNA nicking-closing, DNA ligase, and/or DNA topoisomerase activities has been postulated for the adenovirus VPg and its precursor, pre-VPg80K (reviewed by Ruben et al., 1983). However, no catalytic activity actually has been shown to be associated with any VPg. Rather, these proteins appear to have a role in one solution to the "replication of the ends" problem. The VPgs of both RNA and DNA viruses are almost certainly required for viral nucleic acid replication, acting as a primer either (1) directly or (2) after template-independent addition of one or two nucleotide residues or (3) in a precursor form.

Poliovirus VPg is located at the 5' end of both the virion RNA [messenger of (+)RNA] and its complement [(−)RNA]. These observations gave rise to notions about a primer function for poliovirus VPg when it first was discovered (Nomoto et al., 1977; reviewed by Wimmer, 1982). Free (e.g., pppNp. . .) nascent RNA 5' ends have not been detected on any poliovirus RNA. The presence of VPg on nascent strands of both (+) and (−) polarity and the primer dependence of purified, poliovirus-specified RNA polymerase also are consistent with primer function. That the VPg is required for poliovirus RNA synthesis is strongly supported by the specific action of anti-VPg serum in blocking poliovirus RNA synthesis in vitro (Baron and Baltimore, 1982b; Semler et al., 1983).

As was indicated in Section IV,B, the active form of poliovirus VPg in virion and (−)RNA synthesis may be VPg–pUpU–OH. Neither P3-9, the most abundant apparent precursor of VPg, nor potential precursors X/9 or 3b/9 have been detected as covalent complexes with nucleotides (Takegami et al., 1983b; Crawford and Baltimore, 1983).

That the polyprotein precursors of picornavirus VPg are not directly involved in priming RNA synthesis is also indicated by the demonstration of three different FMDV VPgs. The nucleotide sequence of FMDV RNA indicates that its polyprotein translation product contains three apparent VPg amino acid sequences linked in succession. This is the only known instance of a repeated gene in a viral genome (Forss and Schaller, 1982), although the possibility of two VPgs also has been raised for another picornavirus (Vartapetian *et al.*, 1980). The utility of a set of similar VPg genes for FMDV is not apparent because poliovirus, with one VPg gene, produces VPg in great excess over the amounts needed for virion RNA synthesis. The properties of the population of VPgs attached to FMDV genomic RNA molecules (King *et al.*, 1980) are consistent with the presence of the products of all three genes (VPgs 1, 2, and 3) in roughly equal amounts. Thus the exact position of the VPg gene in the precursor chain apparently is not crucial to its participation in the initiation of RNA synthesis. Since a polyprotein precursor of any or all of the three VPg sequences would have each VPg sequence in an environment at least slightly different, the mechanism for initiating RNA synthesis with a primer that also is a VPg precursor would have to be at least slightly different for each precursor/primer. Conceptually, the mature VPgs themselves would more easily be adapted to a uniform mechanism of priming RNA synthesis.

As was indicated in Section II, poliovirus RNA retains its infectivity when treated with proteinase to remove most of the VPg polypeptide chain. No enzyme activity capable of connecting a VPg to a mature RNA molecule has been discovered or is expected to be discovered on the basis of current information. Thus an attached VPg apparently is not required on the infecting virion RNA in order for the RNA to function as a template. An even more convincing indication that there is no requirement for template-bound VPg during the initiation of infection has been given by the production of poliovirus in cells into which a protein-free DNA copy of the viral genome has been introduced (Racarmillo and Baltimore, 1981). As expected, the progeny particles contained VPg bearing RNA molecules.

The most conclusive studies of the role of VPg as a primer in virus nucleic acid synthesis have been done *in vitro* with purified components. Such studies have been possible both with adenovirus and bacteriophage ϕ29 (e.g., Watebe *et al.*, 1982). Two copies of the VPg are necessary for ϕ29 DNA replication; one must be attached to the template DNA, and the other primes nascent-strand synthesis. DNA replication begins at either end of the molecule and proceeds by a strand-displacement mechanism (Sogo *et al.*, 1982). The formation of the first phosphodiester bond results in VPg–pdA. Even this first reaction requires a template with an attached, intact VPg; in addition, specific nucleotide sequences must be present in the 5'-terminal region adjacent to the VPg (Peñalva and Salas, 1982; Shih *et al.*, 1982; Salas *et al.*, 1983). *In vitro* complementation tests with components derived from infections by conditionally lethal $\phi29$ mutants show that functional forms of both the VPg (product of gene 3) and the gene 2 product are required for this reaction (Mellado *et al.*, 1980; Blanco *et al.*, 1983; Salas *et al.*, 1983). Furthermore, the initiation of DNA replication *in vitro* is inhibited by antiserum prepared against $\phi29$ VPg (Watabe *et al.*, 1983).

In vitro synthesis of adenovirus DNA, catalyzed by enzymes from infected cells, is dependent upon pre-VPg80K (Nagata *et al.*, 1983; Ostrove *et al.*, 1983). The DNA replication reactions are in several other ways similar to those of bacteriophage $\phi29$ DNA synthesis, including strand-displacement DNA synthesis and dependence upon template for formation of pre-VPg80K–pdC in a reaction that requires virus gene product(s). Host cell proteins also are required (Nagata *et al.*, 1982). Although active DNA templates need not have a attached VPg for activity *in vitro*, specific nucleotide sequences that are common to the two ends of the double-stranded adenovirus DNA are necessary (van Bergen *et al.*, 1983; Tamanoi and Stillman, 1983; Stillman, 1983; references cited therein). Although circular forms of adenovirus DNA have been found in infected cells (Ruben *et al.*, 1983), they are likely to be the product of recombinational events rather than intermediates in a DNA synthesis mechanism that does not require pre-VPg80K.

"Cores" of adenovirus particles, formed by treatment with mild detergent, also support both the deoxycytidylylation of pre-VPg80K and chain elongation of up to 300 residues *in vitro* (Goding and Russell, 1983). As in the case of poliovirus, adenovirus DNA sequences from plasmids are infectious (Berkner and Sharp, 1983). The plasmid-derived sequences lack VPg, and they have additional terminal sequences derived from the plasmid. However, the progeny virus have the correct terminal nucleotide sequence restored, and each 5' end linked to the VPg.

B. OTHER FUNCTIONS OF VPgs

VPgs appear to act at points in the virus replication cycle other than initiation of nucleic acid synthesis. The VPg-free poliovirus RNA of the polyribosomes is abundant in infected cells and yet does not become encapsidated. Similarly, all the genomic RNAs and DNAs found in the virions of the viruses indicated in Table I (with the exception of the hepadnaviruses) appear to bear a VPg. Bjornsti *et al.* (1982) showed that VPg–DNA complexes of a bacteriophage $\phi29$ mutant that is temperature sensitive in gene 3 (pre-VPg80K) were not efficiently packaged *in vivo* or *in vitro* at the nonpermissive temperature, whereas wild-type VPg–DNA complexes were encapsidated under the same conditions. Partial proteolytic digestion of the VPg of the VPg–DNA complex also prevented packaging of the $\phi 29$ DNA.

The adenovirus pre-VPg80K that primes DNA synthesis is specified by an early region of the virus DNA. However, its conversion to the 55,000-molecular weight VPg after completion of DNA synthesis is due to the action of a late viral function (reviewed by Stillman, 1983; Challberg and Kelly, 1981, 1982). Since proteolytic events are common in virion assembly pathways, it is possible that the pre-VPg80K-to-VPg conversion is a part of the adenovirus assembly process. An additional possible function for the adenovirus VPg is indicated by its ability to protect the DNA from attack by exonucleases (Dunsworth-Browne et al., 1980). Treatment of the DNA with proteinase altered its susceptibility to exonuclease (Sharp et al., 1976).

Upon introduction into a susceptible cell, a virus (+)RNA must serve as a template not only for (−)RNA synthesis but also for protein synthesis. Poliovirus messenger RNA lacks a VPg (Nomoto et al., 1977; Dorner et al., 1981), and proteolytic removal of most of the VPg polypeptide chain from the virion RNA does not seem to alter the in vitro translation products directed by it (Golini et al., 1978; Perez-Bercoff and Gander, 1978; Sangar et al., 1980). The translation of comovirus (Stanley et al., 1978) and nepovirus (Chu et al., 1981) RNAs similarly was unchanged by proteinase treatment. Nor was translation of a nepovirus satellite RNA influenced by prior treatment with proteinase (Koenig and Fritsch, 1982). The RNA of certain picornaviruses, with or without an attached VPg, can be an effective and competitive messenger RNA in systems that can recognize and use uncapped RNAs, whereas neither VPg-bearing RNA nor proteinase-treated (peptide-) RNA nor RNAs with free 5′ ends were translated in a cap-requiring system from sea urchin eggs (Sonenberg and Trachsel, 1982; Perez-Bercoff and Kaempfer, 1982; Winkler et al., 1983; references cited therein). The two or more regions of the virion RNA of Mengovirus (a picornavirus) that are protected by translation initiation factor eIF-2 or by ribosomes did not include the VPg-linked 5′-end or 5′-adjacent sequences (Perez-Bercoff and Kaempfer, 1982). Thus present evidence indicates no apparent role for VPgs in translation.

C. Implications of VPg Structure and Function

Linkage to a VPg is not limited to virus genomic nucleic acids. The satellite RNA of tomato blackring virus has the same VPg that is characteristic of its supporting virus (Mayo et al., 1982a). The satellite RNA does not replicate to a detectable level when it is inoculated alone, and a VPg has not been detected among its translation products. A supply of VPg is an element in the apparent reliance of the satellite RNA upon tomato blackring virus for replication functions. Since the VPg presumably acts as a primer, other enzymatic machinery of the supporting virus also would be expected to be involved to produce an RNA replication system that is able to sue the VPg.

Ghosh *et al.* (1981) discovered that the subgenomic coat protein messenger RNA of Southern bean mosaic virus particles is linked to a VPg that is the same as or similar to the VPg of the genomic RNA. The subgenomic RNA corresponds in nucleotide sequence to the 3′ approximately one-fourth of the genomic RNA. Several mechanisms for the formation of such a subgenomic RNA have been proposed, including (1) processing from the full-length (+)RNA, (2) transcription from template RNA that would represent the 5′ approximately one-fourth of full length (−)RNA, and (3) internal initiation of transcription on a full length (−)RNA. Clearly the first of these alternatives is not favored if the VPg serves as a primer for RNA synthesis in this system.

As has been indicated above, the poliovirus putative VPg precursor P3-9 appears to be present in the infected cell in significant excess over the amounts needed to meet the requirements of virus nucleic acid synthesis, indicating that the free VPg and/or its precursors may have as yet undiscovered functions. The groups of viruses known to have VPgs are diverse, and VPg-like proteins are not limited to virus systems. Kemble and Thompson (1982) have shown that the linear mitochondrial DNAs of some male sterile maize lines have terminally attached proteins. Hirohiko and Sakaguchi (1982) described a linear plasmid from *Streptomyces* species that has terminal proteins. Covalently 5′-linked proteins have been postulated to block the telomeric ends of eukaryotic chromosomes (Cavalier-Smith, 1983).

It is possible that VPgs or VPg-like molecules play a transient role in the synthesis of linear single- and double-stranded RNAs and DNAs not known to be VPg-linked, including the RNAs of uninfected plants that apparently are transcribed from RNAs (Dorssers *et al.*, 1982; Fraenkel-Conrat, 1983). An activity in uninfected cells can remove a VPg from picornavirus RNA (Ambros and Baltimore, 1978; Sangar *et al.*, 1981) possibly even before chain completion, leaving a 5′ phosphoryl group. The successful search for a transient VPg–nucleic acid complex might begin with the detection of a previously unknown, free VPg in an uninfected or infected cell. Such detection is likely to be difficult because of the small amounts of VPg that may be present and the known propensity of VPgs to bind to glass and plastic and to remain soluble under usual conditions for protein precipitation. Uninfected and infected cells are not the only potential sources of new VPgs. Some VPgs of well studied viruses may remain to be discovered.

The routine use of proteinases to increase the yield of virion nucleic acid may have prevented VPg detection in these systems (e.g., Persson and MacDonald, 1982).

References

Adler, C., Elzinga, M., and Wimmer, E. (1983). *J. Gen. Virol.* **64**, 349.
Ambros, V., and Baltimore, D. (1978). *J. Biol. Chem.* **253**, 5263.
Ambros, V., Pettersson, R., and Baltimore, D. (1978). *Cell* **15**, 1439.
Bamford, D., McGraw, T., MacKenzie, G., and Mindich, L. (1983). *J. Virol.* **47**, 311.
Baron, M. H., and Baltimore, D. (1982a). *Cell* **28**, 395.
Baron, M. H., and Baltimore, D. (1982b). *Cell* **30**, 745.
Baroudy, B. M., Venkatesan, S., and Moss, B. (1982). *Cell* **28**, 315.
Berkner, K. L., and Sharp, P. A. (1983). *Nucleic Acids Res.* **11**, 6003.
Bjornsti, M. A., Reilly, B. E., and Anderson, D. L. (1982). *J. Virol.* **41**, 508.
Blanco, L., García, J. A., Peñalva, M. A., and Salas, M. (1983). *Nucleic Acids. Res.* **11**, 1309.
Bolton, A. E., and Hunter, W. M. (1973). *Biochem. J.* **133**, 529.
Bruening, G. (1969). *Virology* **37**, 577.
Burroughs, J. N., and Brown, F. (1978). *J. Gen. Virol.* **41**, 443.
Cavalier-Smith, T. (1983). *Nature (London)* **301**, 112.
Challberg, M. D., and Kelly, J. (1981). *J. Virol.* **38**, 272.
Challberg, M. D., and Kelly, J. (1982). *Annu. Rev. Biochem.* **51**, 901.
Chase, C. D., and Benzinger, R. H. (1982). *J. Virol.* **42**, 176.
Chirgwin, J. M., Przybyla, A. E., MacDonald, R. J., and Rutter, W. J. (1979). *Biochemistry* **18**, 5294.
Chu, P., Boccardo, G., and Francki, R. I. B. (1981). *Virology* **109**, 428.
Colpan, M., Schumacher, J., Bruggemann, W., Sanger, H. L., and Riesner, D. (1983). *Anal. Biochem.* **131**, 257.
Crawford, N. M., and Baltimore, D. (1983). *Proc. Natl. Acad. Sci. U.S.A.* **80**, 7452.
Daubert, S. D., and Bruening, G. (1979). *Virology* **98**, 246.
Daubert, S. D., Bruening, G., and Najarian, R. C. (1978). *Eur. J. Biochem.* **92**, 45.
Desiderio, S. V., and Kelly, T. J. (1981). *J. Mol. Biol.* **145**, 319.
Dobos, P., Hill, B. J., Hallett, R., Kells, D. T. C., Becht, H., and Teninges, D. (1979). *J. Virol.* **32**, 593.
Donis-Keller, H., Maxam, A. M,. and Gilbert, W. (1977). *Nucleic Acids Res.* **4**, 2527.
Dorner, A. J., Rothberg, P. G., and Wimmer, E. (1981). *FEBS Lett.* **132**, 219.
Dorssers, L., Zabel, P., van der Meer, J., and van Kammen, A. (1982). *Virology* **116**, 236.
Dunsworth-Browne, M., Schell, R. E., and Berk, A. J. (1980). *Nucleic Acids Res.* **8**, 543.
Fahnstock, S. R., and Nomura, M. (1972). *Proc. Natl. Acad. Sci. U.S.A.* **69**, 363.
Feeney, R.E., Yamasaki, R. B., and Geoghegan, K. F. (1982). *In* "Modification of Proteins" (R. E. Feeney, J. Whitaker, eds.), p. 3. *Adv. Chem. Ser.* No. 160.
Flanegan, J. B., Pettersson, R. F., Ambros, V., Hewlett, M. J., and Baltimore, D. (1977). *Proc. Natl. Acad. Sci. U.S.A.* **74**, 961.
Forss, S., and Schaller, H. (1982). *Nucleic Acids Res.* **10**, 6441.
Fraenkel-Conrat, H. (1983). *Proc. Natl. Acad. Sci. U.S.A.* **80**, 422.
Fraker, P. J., and Speck, J. C. (1978). *Biochem. Biophys. Res. Commun.* **80**, 849.
García, E., Gomez, A., Ronda, C., Escarmis, C., and Lopez, R. (1983a). *Virology* **128**, 92.
García, J. A., Pastrana, R., Prieto, I., and Salas, M. (1983b). *Gene* **21**, 65.
Gerlich, W. E., and Robinson, W. S. (1980). *Cell* **21**, 801.

Ghosh, A., Dasgupta, R., Salerno-Rife, T., Rutgers, T., and Kaesberg, P. (1979). *Nucleic Acids Res.* **7**, 2137.

Ghosh, A., Rutgers, T., Ke-quiang, M., and Kaesberg, P. (1981). *J. Virol.* **39**, 87.

Goding, C. R., and Russell, W. C. (1983). *EMBO J.* **2**, 339.

Goldbach, R., and Rezelman, G. (1983). *J. Virol.* **46**, 614.

Goldbach, R., Rezelman, G., and van Kammen, A. (1980). *Nature (London)* **286**, 297.

Goldbach, R., Rezelman, G., Zabel, P., and van Kammen, A. (1982). *J. Virol.* **42**, 630

Golini, F., Nomoto, A., and Wimmer, E. (1978). *Virology* **89**, 112.

Hanecak, R., Semler, B. L., Anderson, C. W., and Wimmer, E. (1982). *Proc. Natl. Acad. Sci. U.S.A.* **79**, 3973.

Hari, V. (1981). *Virology* **112**, 391.

Harris, T. J. R. (1979). *Nucleic Acids Res.* **7**, 1765.

Hermoco, J. M., and Salas, M. (1980). *J. Mol. Biol.* **145**, 319.

Hewlett, M. J., Rose, J. K., and Baltimore, D. (1976). *Proc. Natl. Acad. Sci. U.S.A.* **73**, 327.

Hirohiko, H., and Sakaguchi, K. (1982). *Plasmid* **7**, 59.

Hruby, D. E., and Roberts, W. K. (1978). *J. Virol.* **25**, 413.

Hunter, W. M., and Greenwood, F. C. (1962). *Nature (London)* **74**, 59.

Jones, A. T., Mayo, M. A., and Duncan, G. H. (1983). *J. Gen. Virol.* **64**, 1167.

Kemble, R. J., and Thompson, R. D. (1982). *Nucleic Acids Res.* **10**, 8181.

Kiefer, M. C., Daubert, S. D., Schneider, I. R., and Bruening, G. (1982). *Virology* **121**, 262.

King, A. M. Q., Sangar, D. V., Harris, T. J. R., and Brown, F. (1980). *J. Virol.* **34**, 627.

Kitamura, N., Adler, C. J., Rothberg, P. G., Martinko, J., Nathenson, S. G., and Wimmer, E. (1980). *Cell* **21**, 295.

Kitamura, N., Semler, B. L., Rothberg, P., Larsen, G., Adler, S., Dorner, A., Emini, E., Hanecak, R., Lee, J., van der Werf, S., Anderson, C., and Wimmer, E. (1981). *Nature (London)* **291**, 547.

Klein, I., and Klootwijk, J. (1976). *Anal. Biochem.* **74**, 263.

Klootwijk, J., Klein, I., Zabel, P., and van Kammen, A. (1977). *Cell* **11**, 75.

Koenig, I., and Fritsch, C. (1982). *J. Gen. Virol.* **60**, 343.

Kornberg, A. (1980). "DNA Replication." Freeman, San Francisco.

Kummer, U., Thiel, E., Doxiadis, I., Eulitz, M., Sladoljev, S., and Thierfelder, S. (1981). *J. Immunol. Methods* **42**, 367.

Laemmli, U. K. (1970). *Nature (London)* **227**, 680.

Lee, Y. F., Nomoto, A., Detjen, B. M., and Wimmer, E. (1977). *Proc. Natl. Acad. Sci. U.S.A.* **74**, 59.

Lichy, J. H., Field, J., Horwitz, M. S., and Horwitz, J. (1982). *Proc. Natl. Acad. Sci. U.S.A.* **79**, 5225.

Lomonossoff, G. P., and Shanks, M. (1983). *EMBO J.* **2**, 2253.

Loomis, D. W., Lile, J. D., Sandstrom, R. D., and Burbott, A. J. (1979). *Phytochemistry* **18**, 1049.

McCollum, K., and Tabrosky, G. (1983). *Anal. Biochem.* **130**, 311.

Marion, P. L., and Robinson, W. S. (1983). *Curr. Top. Microbiol. Immunol.* **105**, 99.

Matthews, R. E. F. (1982). *Intervirology* **17**, 1.

Mayo, M. A., Barker, H., and Harrison, B. D. (1979). *J. Gen. Virol.* **43**, 735.

Mayo, M. A., Barker, H., and Harrison, B. D. (1982a). *J. Gen. Virol.* **59**, 149.

Mayo, M. A., Barker, H., Robinson, D. J., Tamada, T., and Harrison, B. D. (1982b). *J. Gen. Virol.* **59**, 163.

Mayo, M. A., Barker, H., and Robsinon, D. (1982c). *J. Gen. Virol.* **62**, 417.

Mellado, R., Peñalva, M., Inciarte, M., and Salas, M. (1980). *Virology* **104**, 84.

Merril, C. R., Goldman, D., Sedman, S. A., and Ebert, M. H. (1981). *Science* **211**, 1437.

Molnar-Kimber, K., Summers, J., Taylor, J. M., and Mason, W. S. (1983). *J. Virol.* **45**, 165.

Nagata, K., Guggenheimer, R., Enomoto, T., Lichy, J., and Hurwitz, J. (1982). *Proc. Natl. Acad. Sci. U.S.A.* **79**, 6438.

Nagata, K., Guggenheimer, R., and Hurwitz, J. (1983). *Proc. Natl. Acad. Sci. U.S.A.* **80**, 4266.

Najarian, R. C., and Bruening, G. (1980). *Virology* **106**, 301.

Nishimura, S. (1972). *Prog. Nucleic Acids Res. Mol. Biol.* **12**, 49.

Nole, D., Nikaido, K., and Ames, G. F. (1979). *Biochemistry* **18**, 4159.

Nomoto, A., Lee, Y. F., and Wimmer, E. (1976). *Proc. Natl. Acad. Sci. U.S.A.* **73**, 375.

Nomoto, A., Detjen, B., Pozzatti, R., and Wimmer, E. (1977). *Nature (London)* **268**, 208.

Nossal, N. G. (1983). *Annu. Rev. Biochem.* **53**, 581.

Ostrove, J. M., Rosenfeld, P., Williams, J., and Kelly, T. J. (1983). *Proc. Natl. Acad. Sci. U.S.A.* **80**, 935.

Pelham, H. (1979). *Virology* **96**, 463.

Peñalva, M., and Salas, M. (1982). *Proc. Natl. Acad. Sci. U.S.A.* **79**, 5522.

Perez-Bercoff, R., and Gander, M. (1978). *FEBS Lett.* **96**, 306.

Perez-Bercoff, R., and Kaempfer, R. (1982). *J. Virol.* **41**, 30.

Persson, R. H., and MacDonald, R. D. (1982). *J. Virol.* **44**, 437.

Prusiner, S. B., Bolton, D. C., Groth, D. F., Bowman, K. A., Cochran, S. P., and McKinley, M. P. (1982). *Biochemistry* **21**, 6942.

Racarmillo, V., and Baltimore, D. (1981). *Science* **214**, 916.

Reisman, D., and de Zoeten, G. (1982). *J. Gen. Virol.* **62**, 187.

Rekosh, D. M. K., Russell, W. C., Bellet, A. J. D., and Robinson, A. J. (1977). *Cell* **11**, 283.

Rezelman, G., Goldbach, R., and van Kammen, A. (1980). *J. Virol.* **36**, 366.

Rice, R. H., and Means, G. E. (1971). *J. Biol. Chem.* **246**, 831.

Richards, O. C., Ehrenfeld, E., and Manning, J. (1979). *Proc. Natl. Acad. Sci. U.S.A.* **76**, 676.

Robinson, D. L., Barker, H., Harrison, B. D., and Mayo, M. A. (1980). *J. Gen. Virol.* **51**, 317.

Roblin, R. (1968). *J. Mol. Biol.* **31**, 51.

Rothberg, P. G., Harris, T. J. R., Nomoto, A., and Wimmer, E. (1978). *Proc. Natl. Acad. Sci. U.S.A.* **75**, 4868.

Ruben, M., Bacchetti, S., and Graham, F. (1983). *Nature (London)* **301**, 172.

Salacinski, P. R. (1981). *Anal. Biochem.* **117**, 136.

Salas, M., and Viñuela, E. (1980). *Trends Biochem. Sci.* **5**, 191.

Salas, M., Mellado, R. P., Viñuela, E., and Sogo, J. M. (1978). *J. Mol. Biol.* **119**, 269.

Salas, M., García, J. A., Peñalva, M. A., Blanco, L., Prieto, I., Mellado, R. P., Lázaro, J. M., Pastrana, R., Escarmis, C., and Hermoso, J. M. (1983). *In* "Methods of DNA Replication and Recombination" (UCLA Symposium on Molecular and Cellular Biology, New Series) (N. R. Cozzarelli, ed.), Vol. 10, p. 203. Liss, New York.

Sangar, D. V., Rowlands, D. J., Harris, T. J., and Brown, F. (1977). *Nature (London)* **268**, 648.

Sangar, D. V., Black, D. N., Rowlands, D. J., Harris, T. J. R., and Brown, F. (1980). *J. Virol.* **33**, 59.

Sangar, D. V., Bryant, J., Harris, T. J. R., Brown, F., and Rowlands, D. (1981). *J. Virol.* **39**, 67.

Schaffer, F. L., Ehresmann, D. W., Fretz, M. K., and Soergel, M. E. (1980). *J. Gen. Virol.* **47**, 215.

Schneider, I. R., Hull, R., and Markham, R. (1972). *Virology* **47**, 320.

Sedat, J. W., Kelly, R. B., and Sinsheimer, R. L. (1967). *J. Mol. Biol.* **26**, 537.

Semler, B. L., Anderson, C., Hanecak, R., Dorner, L., and Wimmer, E. (1982). *Cell* **28**, 405.

Semler, B. L., Hanecak, R., Dorner, L. F., Anderson, C. W., and Wimmer, E. (1983). *Virology* **126**, 624.

Shabarova, Z. (1970). *Prog. Nucleic Acids Res. Mol. Biol.* **10**, 145.

Sharp, P. A., Moore, C., and Haverty, J. (1976). *Virology* **75**, 442.

Shatkin, A. (1976). *Cell* **9**, 645.

Shih, D. S., Kaesberg, P., and Hall, T. C. (1974). *Nature (London)* **249**, 353.

Shih, M., Watabe, K., and Ito, J. (1982). *Biochem. Biophys. Res. Commun.* **105**, 1031.

Silberklang, M., Gillum, A. M., and RajBhandary, U. L. (1977). *Nucleic Acids Res.* **4**, 4091.

Sogo, J. M., García, J. A., Peñalva, M. A., and Salas, M. (1982). *Virology* **116**, 1.

Sonenberg, N., and Trachsel, H. (1982). *Curr. Top. Cell. Regul.* **21**, 65.

Stanley, J., and van Kammen, A. (1979). *Eur. J. Biochem.* **101**, 45.

Stanley, J., Rottier, P., Davies, J. W., Zabel, P., and van Kammen, A. (1978). *Nucleic Acids Res.* **5**, 4505.

Stanley, J., Goldbach, R., and van Kammen, A. (1980). *Virology* **106**, 180.

Steiner, A., Helander, E., Fujitaka, J., Smith, L., and Smith, R. (1980). *J. Chromatogr.* **202**, 263.

Stillman, B. W. (1983). *Cell* **35**, 7.

Strauss, E. G., and Strauss, J. H. (1983). *Curr. Top. Microbiol. Immunol.* **105**, 1.

Tabrosky, G. (1974). *Adv. Protein Chem.* **28**, 1.

Takegami, T., Semler, B. L., Anderson, C. W., and Wimmer, E. (1983a). *Virology* **128**, 33.

Takegami, T., Kuhn, R. J., Anderson, C. W., and Wimmer, E. (1983b). *Proc. Natl. Acad. Sci. U.S.A.* **80**, 7447.

Tamanoi, F., and Stillman, B. W. (1982). *Proc. Natl. Acad. Sci. U.S.A.* **79**, 2221.

Tamanoi, F., and Stillman, B. W. (1983). *Proc. Natl. Acad. Sci. U.S.A.* **80**, 6446.

Thornton, G. B., Robberson, D. L., and Arlinghaus, R. B. (1981). *J. Virol.* **39**, 229.

van Bergen, B. G. M., van der Lay, P. A., van Driel, W., van Mansfeld, A. D. M., and van der Vliet, P. C. (1983). *Nucleic Acids Res.* **11**, 1975.

Van Dyke, T., and Flanegan, J. (1980). *J. Virol.* **35**, 732.

Vartapetian, A. B., Drygin, Y. F., Chumakov, K. M., and Bogdanov, A. A. (1980). *Nucleic Acids Res.* **8**, 3729.

Veerisetty, V., and Sehgal, O. P. (1980). *Phytopathology* **70**, 282.

Watabe, K., Shih, M., Sugnio, A., and Ito, J. (1982). *Proc. Natl. Acad. Sci. U.S.A.* **79**, 5245.

Watabe, K., Shih, M., and Ito, J. (1983). *Proc. Natl. Acad. Sci. U.S.A.* **80**, 4248.

Watson, J. D. (1972). *Nature (New Biol.)* **239**, 197.

Wengler, G., Wengler, G., and Gross, H. J. (1982). *Virology* **123**, 273.

Wimmer, E. (1982). *Cell* **28**, 199.

Wimmer, E., and Reichmann, M. E. (1968). *Science* **160**, 1452.

Winkler, M. M., Bruening, G., and Hershey, J. W. B. (1983). *Eur. J. Biochem.* **137**, 227.

Wu, M. N., Davidson, N., and Wimmer, E. (1978). *Nucleic Acids Res.* **5**, 4711.

Yoshikawa, H., and Ito, J. (1981). *Proc. Natl. Acad. Sci. U.S.A.* **78**, 2596.

Zabel, P., Moerman, M., Van Straatun, F., Goldbach, R., and van Kammen, A. (1982). *J. Virol.* **41**, 1083.

Zabel, P., Moerman, M., Lomonosoff, G., Shanks, M., and Beyreuther, K. (1984). *EMBO J.* (in press).

Index

A

Acetone, prions and PrP precipitation by, 319

Acid green 3-CI 42085, Luxol brilliant green BL preparation from, 229

Acid phosphatase, staining procedure for, 172

Acrolein, for plant tissue fixation, 148, 152

Acrylamide
  ultrapure, for oligonucleotide fingerprinting, 55, 58, 60–61
  stock solution, 56

Acrylic resins, as embedding media, 157

Actin filaments, fixation of, 149

Actinomycin D, poliovirus strain resistant to radiolabeling of RNA of, 51

Acute hemorrhagic conjunctivitis virus, fingerprint analysis of, 74

Acyclovir, effect on Epstein–Barr virus genome replication, 13

Adenovirus(es)
  cores of, 373
  DNA, 351
    exonuclease susceptibilities of, 350
  VPgs of, 356, 369
    characterization, 362–363
    possible functions, 374

Aedes albopictus mosquitoes
  photograph of, 283
  virus propagation in, 289

Aerosols, containment of, in scrapie agent studies, 340

Agaricus bisporus, viruslike particles from, immunosorbent electron microscopy studies of, 98

Agarose, low-gelling-temperature type, 66

Alcohols, prions and PrP precipitation by, 319

Aldehydes, for plant-tissue fixation, 146, 155

Alfalfa mosaic virus
  enzyme identification in, 173

immunosorbent electron microscopy of, 93
thin-section diagnosis of, 195

Alkaline phosphatase, in 5' end labeling of viral RNA, 51–52

Alphaviruses
  fingerprint analyses of, 76
  mosquito use to detect and propagate, 282, 289

Amino acids in inclusion bodies, radioactive material incorparation into, 175

Ammonium molybdate as stain for immunosorbent electron microscopy, 93

Ammonium persulfate for gel electrophoresis in oligonucleotide fingerprinting, 56, 57

Amorphous cytoplasmic inclusions (AI) of potyviruses
  purification procedure for, 244–246
  serological analyses of, 254–277

Amyloid, prions and, 295

Andean potato laten virus (APLV), thin-section diagnosis of, 186

Animal viruses
  genome-linked proteins of, 347–379
  mass–MW determinations on, 124–126

Anisometric viruses thin-section diagnosis of, 200–212

Aphid(s)
  -borne potyviruses, identification of, 207
  immunosorbent electron microscopy of extracts of, 91

Aphtho viruses, oligonucleotide fingerprinting of, 63

Arabis mosaic virus, VPg of, 355

Araldite as epoxy plant tissue embedding medium, 158–159

Araujia mosaic virus, cylindrical inclusions from, 238, 239

Arenaviruses, fingerprint analysis of, 77

Arthropod-borne viruses, replication in mosquitoes, 288

Artichoke yellow ringspot virus, thin-section diagnosis of, 190

N

O